CONTENTS

Credits and Sources:

Chapter 3
p. 37: Art by Raychel Ciemma
p. 38: Art by Raychel Ciemma and American
    Composition and Graphics

Chapter 4
p. 56: Art by Raychel Ciemma
p. 59: Art by L. Calver
p. 61: Art by Robert Demarest

Chapter 5
p. 69: Micrograph Ed Reschke. Art by Joel Ito
p. 70: Art by Robert Demarest
p. 72: Art by Raychel Ciemma

Chapter 6
p. 81: (b) Micrograph John D. Cunningham;
    (c) Micrograph D. W. Fawcett, *The Cell*, Philadelphia:
    W. B. Saunders Co., 1966; Art by Robert Demarest
p. 84: Art by Kevin Somerville

Chapter 7
p. 89: Art by Kevin Somerville

Chapter 9
p. 114: Art by Kevin Somerville
p. 116: Art by Raychel Ciemma
p. 121: Art by Robert Demarest, based on *Basic Human
    Anatomy*, by A. Spence, Benjamin-Cummings, 1982
p. 124: Art by Raychel Ciemma

Chapter 11
p. 143: Art by Kevin Somerville

Chapter 12
p. 157: Art by Robert Demarest

Chapter 13
p. 167: Art by Raychel Ciemma
p. 171: Art by Robert Demarest
p. 172: Art by Robert Demarest
p. 173: Art by Robert Demarest
p. 176: (b) Micrograph Manfred Kage/Peter Arnold, Inc.;
    Art by Robert Demarest
p. 179: C. Yokochi and J. Rohen, *Photographic Anatomy of
    the Human Body*, Second Edition, Igaku-Shoin Ltd.,
    1979

Chapter 14
p. 191: Art by Robert Demarest
p. 194: Art by Robert Demarest

Chapter 15
p. 201: Art by Kevin Somerville

Chapter 16
p. 215: Art by Raychel Ciemma, Micrograph Ed Reschke
p. 218: Art by Raychel Ciemma

Chapter 17
p. 231: Art by Raychel Ciemma
p. 232: Art by Raychel Ciemma
p. 236: Art by Raychel Ciemma
p. 236: Photographs from Lennart Nilsson, *A Child is
    Born*, © 1966, 1977 Dell Publishing Company, Inc.

Chapter 19
p. 256: Art by Raychel Ciemma
p. 261: Art by Raychel Ciemma

Chapter 22
p. 295: Art by Lisa Starr

Chapter 24
p. 314: Art after *Evolving*, by F. Ayala and J. Valentine,
    Benjamin-Cummings, 1979

STUDY GUIDE AND WORKBOOK: AN INTERACTIVE APPROACH

for Starr and McMillan's

HUMAN BIOLOGY

FIFTH EDITION

SHELLEY PENROD

North Harris College

ANN MAXWELL

Angelo State University

JANE B. TAYLOR

North Virginia Community College

JOHN D. JACKSON

North Hennepin Community College

THOMSON

BROOKS/COLE

Australia • Canada • Mexico • Singapore • Spain • United Kingdom • United States

Biology Publisher: Jack Carey
Signing Representative: Ragu Raghavan
Assistant Editor: Suzannah Alexander
Editorial Assistant: Jana Davis
Marketing Manager: Ann Caven
Marketing Assistant: Sandra Perin
Advertising Project Manager: Linda Yip

Print Buyer: Karen Hunt
Production Project Manager: Belinda Krohmer
Permissions Editor: Bob Kauser
Cover Image: eStock Photo, IT International Ltd.
Typesetting: G&S Typesetters, Inc.
Printing and Binding: Globus Printing Co.

Printed in the United States of America

1   2   3   4   5   6   7   06   05   04   03   02

ISBN 0-534-40203-8

For more information about our products,
contact us at:
Thomson Learning Academic Resource Center
1-800-423-0563

For permission to use material from this text,
contact us by:
Phone: 1-800-730-2214
Fax: 1-800-731-2215
Web: www.thomsonrights.com

Asia
Thomson Learning
5 Shenton Way, #01-01
UIC Building
Singapore 068808

Australia
Nelson Thomson Learning
102 Dodds Street
Southbank
Victoria 3006
Australia

Canada
Nelson Thomson Learning
1120 Birchmount Road
Toronto, Ontario M1K 5G4
Canada

Europe/Middle East/South Africa
Thomson Learning
High Holborn House
50/51 Bedford Row
London WC1R 4LR
United Kingdom

Latin America
Thomson Learning
Seneca, 53
Colonia Polanco
11560 Mexico D.F.
Mexico

Spain
Paraninfo Thomson Learning
Calle/Magallanes, 25
28015 Madrid, Spain

PREFACE

Tell me and I will forget, show me and I might remember, involve me and I will understand.
— Chinese Proverb

The proverb outlines three levels of learning, each successively more effective than the method preceding it. The writer of the proverb understood that humans learn most efficiently when they *involve* themselves in the material to be learned. This study guide is like a tutor; when properly used it increases the efficiency of your study periods. The interactive exercises actively involve you in the most important terms and central ideas of your text. Specific tasks ask you to recall key concepts and terms and apply them to life; they test your understanding of the facts and indicate items to reexamine or clarify. Your performance on these tasks provides an estimate of your next test score based on specific material. Most important, though, this biology study guide and text together help you make informed decisions about matters that affect your own well-being and that of your environment. In the years to come, human survival on planet Earth will demand administrative and managerial decisions based on an informed biological background.

HOW TO USE THIS STUDY GUIDE

Following this preface, you will find an outline that will show you how the study guide is organized and will help you use it efficiently. Each chapter begins with a title and an outline list of the 1- and 2-level headings in that chapter. The Interactive Exercises follow, wherein each chapter is divided into sections of one or more of the main (1-level) headings that are labeled 1.1, 1.2, and so on. *For easy reference to an answer or definition, each question and term in this unique study guide is accompanied by the appropriate text page(s), which appears in the form:* [p.352]. The Interactive Exercises begin with a list of Selected Words (other than boldfaced terms)

chosen by the authors as those that are most likely to enhance understanding. In the text chapters, the selected words appear in italics, quotation marks, or roman type. The list of Selected Words is followed by a list of Boldfaced, Page-Referenced Terms that appear in the text. These terms are essential to understanding each study guide section of a particular chapter. Space is provided by each term for you to formulate a definition in your own words. Next is a series of different types of exercises that may include completion, short answer, true/false, fill-in-the-blanks, matching, choice, dichotomous choice, label and match, problems, labeling, sequence, multiple choice, and completion of tables.

A Self-Quiz immediately follows the Interactive Exercises. This quiz is composed primarily of multiple-choice questions, although sometimes we present another examination device or some combination of devices. Any wrong answers in the Self-Quiz indicate areas of the text you need to reexamine. A series of Chapter Objectives/Review Questions follows each Self-Quiz. These are tasks that you should be able to accomplish if you have understood the assigned reading in the text. Some objectives require you to compose a short answer or long essay while others may require a sketch or may require you to supply the correct words.

The final part of each chapter is named Integrating and Applying Key Concepts. This section invites you to try your hand at applying major concepts to situations in which there is not necessarily a single pat answer — and so none is provided in the chapter answer section. Your text generally will provide enough clues to get you started on an answer, but this part is intended to stimulate your thought and to provoke group discussions.

A person's mind, once stretched by a new idea, can never return to its original dimension.
— Oliver Wendell Holmes

STRUCTURE OF THIS STUDY GUIDE
The outline below shows how each chapter in this study guide is organized.

Chapter Number ⟶

Chapter Title ⟶ CELLS

Chapter Outline ⟶ **CHAPTER INTRODUCTION**

3.1 CELLS: ORGANIZED FOR LIFE
All cells are alike in some ways
There are two basic kinds of cells
Why are cells small?
The structure of a cell's membranes reflects their function

3.2 THE PARTS OF A EUKARYOTIC CELL

3.3 THE CYTOSKELETON: SUPPORT AND MOVEMENT

3.4 THE PLASMA MEMBRANE: A LIPID BILAYER
The plasma membrane is a mix of lipids and proteins
Membrane proteins carry out most membrane functions

3.5 THE CYTOMEMBRANE SYSTEM
ER: A protein and lipid assembly line
Golgi bodies: Packing and shipping
A variety of vesicles

3.6 MOVING SUBSTANCES ACROSS MEMBRANES BY DIFFUSION AND OSMOSIS
Diffusion: A solute moves down a gradient
Osmosis: How water crosses membranes

3.7 OTHER WAYS SUBSTANCES CROSS CELL MEMBRANES
Small solutes cross membranes through transport proteins
Vesicles transport large solutes

3.8 *Focus on Our Environment: REVENGE OF EL TOR*

3.9 THE NUCLEUS
A nuclear envelope encloses the nucleus
Proteins and RNA are built in the nucleolus
DNA is organized in chromosomes

3.10 MITOCHONDRIA: THE CELL'S ENERGY FACTORIES
Mitochondria make ATP
ATP forms in an inner compartment of the mitochondrion

3.11 METABOLISM: DOING CELLULAR WORK
ATP — The cell's energy currency
There are two main types of metabolic pathways
A closer look at enzymes

3.12 HOW CELLS MAKE ATP
In glycolysis, glucose is broken down to pyruvate molecules
The Krebs cycle produces energy-rich transport molecules
Electron transport produces a large harvest of ATP

3.13 SUMMARY OF AEROBIC RESPIRATION

3.14 ALTERNATIVE ENERGY SOURCES IN THE BODY
Carbohydrate breakdown in perspective
Energy from fats
Energy from proteins

SUMMARY

Interactive Exercises ⟶ The interactive exercises are divided into numbered sections by titles of main headings and page references. Each section begins with a list of author-selected words that appear in the text chapter in italics, quotation marks, or roman type. This is followed by a list of important boldfaced, page-referenced terms from each section of the chapter. Each section ends with interactive exercises that vary in type and require constant interaction with the important chapter information.

Self-Quiz ⟶ This is usually a set of multiple-choice questions that sample important blocks of text information.

Chapter Objectives/ ⟶ This section gives combinations of relative objectives to be met and
Review Questions questions to be answered.

Integrating and ⟶ This section requires applications of text material to questions for which
Applying Key Concepts there may be more than one correct answer.

Answers to Interactive ⟶ Answers for all interactive exercises can be found at the end of this
Exercises and Self-Quiz study guide, organized by chapter and title and the main headings with their page references. Answers for the Self-Quiz follow.

1

LEARNING ABOUT HUMAN BIOLOGY

Interactive Exercises

Note: In the answer section of this book, a specific molecule is most often indicated by its abbreviation. For example, adenosine triphosphate is ATP.

CHAPTER INTRODUCTION

1.1. THE CHARACTERISTICS OF LIFE [p.2]

Selected Words: deoxyribonucleic acid [p.2], *homeostasis* [p.2]

In addition to the boldfaced terms, the text features other important terms essential to understanding the assigned material. "Selected Words" is a list of these terms, which appear in the text in italics, in quotation marks, and occasionally in roman type. Latin binomials found in this section are underlined and in roman type to distinguish them from other italicized words.

Boldfaced, Page-Referenced Terms

These terms are important; they were in boldface type in the chapter. Write a definition for each term in your own words without looking at the text. Next, compare your definition with that given in the chapter or in the text glossary. If your definition seems accurate, allow some time to pass and repeat this procedure until you can define each term rather quickly (how fast you answer is a gauge of your learning efficiency).

[p.2] DNA _____

[p.2] cell _____

_____

[p.2] homeostasis _____

_____

Choice

For examples 1–14, choose from the following characteristics of life:

a. DNA   b. taking in and using energy and materials   c. sensing and responding
d. reproduction and growth   e. cell   f. homeostasis

1. _____ An animal eating food or a plant capturing solar energy [p.2]

2. _____ Short for *deoxyribonucleic acid* [p.2]

3. _____ A molecule that living things share and nonliving things do not [p.2]

4. _____ A driver stops when a traffic light turns red [p. 2]

5. _____ The smallest entity that is alive [p.2]

6. _____ One of the small units of which all living things are made [p.2]

7. _____ Infancy, childhood, adolescence, adulthood [p.2]

8. _____ Energy and molecules from protein in a meal are used to build muscle tissue in the human eating the meal [p.2]

9. _____ Living things increase in size, mature, and produce more of their kind [p.2]

10. _____ Maintains the internal environment within life-supporting ranges [p.2]

11. _____ Sweating, shivering, and other human body mechanisms maintain body temperature at approximately 98.6°C in spite of heat or cold in the environment [p.2]

12. _____ Contains the instructions for using nonliving substances such as carbon to build and operate the organism [p.2]

13. _____ A dog comes when its master whistles [p.2]

14. _____ Means "staying the same" [p.2]

1.2. HUMANS IN THE LARGER WORLD [p.3]

Boldfaced, Page-Referenced Term

[p.3] evolution _____

_____

Fill-in-the-Blanks

Despite the unity that life forms share in their basic characteristics, evolution has produced incredible

(1) _____ [p.3] among living things. Biologists classify living things into five (2) _____

[p.3] according to characteristics that reflect their (3) _____ [p.3] heritage. The kingdom that includes the (4) _____ [p.3] is sometimes split into two kingdoms (Eubacteria and Archae-bacteria). The other four kingdoms are the (5) _____ [p.3], (6) _____ [p.3], (7) _____ [p.3], and (8) _____ [p.3]. Humans are in the kingdom consisting of all (9) _____ [p.3]. More specifically, humans and apes are in a group of closely related animals called (10) _____ [p.3]. Primates are one group of (11) _____ [p.3] that are in turn one type of "animals with backbones" called (12) _____ [p.3]. Although humans share many characteristics with other primates, we also have distinctive traits such as (13) _____ [p.3] dexterity, a highly evolved (14) _____ [p.3] allowing sophisticated thought processes, and complex (15) _____ [p.3] that have been the main theater of human evolution during the past 40,000 years.

1.3. LIFE'S ORGANIZATION [pp.4–5]

Boldfaced, Page-Referenced Term

[p.4] biosphere _____

_____

Matching

Choose the most appropriate description for each term.

1. _D_ organ system [p.4]
2. _C_ cell [p.4]
3. _E_ community [p.4]
4. _F_ nonliving components [p.4]
5. _G_ ecosystem [p.4]
6. _I_ population [p.4]
7. _J_ tissue [p.4]
8. _B_ biosphere [p.4]
9. _H_ multicellular organism [p.4]
10. _A_ organ [p.4]

A. Different tissues combined together to carry out a function in an organism
B. All parts of Earth's waters, crust, and atmosphere in which organisms live
C. The smallest living unit
D. Different organs working together in a coordinated manner
E. The populations of all species occupying the same area
F. Atoms and molecules
G. A community and its surrounding environment
H. An individual composed of cells arrayed in tissues, organs, and organ systems
I. A group of individuals of the same kind, such as all of Earth's humans
J. A group of cells organized to carry out a specific function

Sequence

Arrange the following levels of organization in nature in the correct hierarchic order. Find the letter of the least inclusive level and write it next to the number 11; write the letter of the most inclusive level next to the number 20; and so forth.

11. _____   A. Organ system [p.4]

12. _____   B. Biosphere [p.4]

13. _____   C. Organ [p.4]

14. _____   D. Ecosystem [p.4]

15. _____   E. Nonliving components [p.4]

16. _____   F. Multicellular organism [p.4]

17. _____   G. Population [p.4]

18. _____   H. Community [p.4]

19. _____   I. Cell [p.4]

20. _____   J. Tissue [p.4]

Fill-in-the-Blanks

Energy that enters the biosphere from the (21) _____ [p.4] is captured by organisms such as plants in a process called (22) _____ [p.4]. Because such organisms use solar energy to combine materials from the air, soil, and water into stored energy, they are the world's basic food (23) _____ [p.4]. To obtain necessary energy and materials, animals must eat plant parts or feed on other animals that have done so. Therefore, animals are called (24) _____ [p.4]. In addition to producers and consumers, ecosystems must have (25) _____ [p.5] such as bacteria and fungi. These organisms obtain the materials and energy they need by breaking down the remains of organisms, thus (26) _____ [p.5] substances back to producers. Because of these interconnections, ecosystems can be thought of as (27) _____ [p.5] of life in which events in one part impact the entire ecosystem.

1.4. WHAT IS SCIENCE? [pp.6–7]

1.5. SCIENCE IN ACTION: CANCER, BROCCOLI, AND MIGHTY MICE [p.8]

Selected Words: "science" [p.6], "if–then" process [p.6], *inductive reasoning* [p.6], *deductive reasoning* [p.6]

Boldfaced, Page-Referenced Terms

[p.6] scientific method _____

_____

[p.6] hypothesis _____

_____

[p.6] prediction _____

_____

[p.7] experiment _____

_____

[p.7] controlled experiment _____

_____

[p.7] control group _____

_____

[p.7] variable _____

_____

Complete the Table

1. Complete the following table of concepts important to understanding the scientific method of problem solving. Choose from *experiment, variable, prediction, control group,* and *hypothesis.*

Concept	Definition
a. [p.6]	An educated guess about what the answer to a question or the solution to a problem might be
b. [p.6]	A statement of what one should be able to observe about a problem if a hypothesis is valid; the "if–then" process
c. [p.7]	A test designed to allow the observer to predict what will happen if his or her hypothesis is not wrong or what won't happen if it is wrong
d. [p.7]	The control group is identical to the experimental group except for this key factor under study
e. [p.7]	Used in scientific experiments as a standard to which the experimental group is compared

Sequence

Arrange the following steps of the scientific method in the correct sequence. Find the letter of the first process and write it next to the number 2; write the letter of the final process next to the number 7; and so forth.

2. _____ A. Develop a hypothesis about what the answer to a question or the solution to a problem might be. [p.6]

3. _____ B. Devise ways to test the accuracy of predictions drawn from the hypothesis (use of observations, models, and experiments). [p.6]

4. _____ C. Repeat tests in order to assure accuracy through a large sample size, or devise new tests of the hypothesis. [p.6]

5. _____ D. Make a prediction by using the "if–then" process. [p.6]

6. _____ E. Objectively analyze and report the test results and conclusions. [p.6]

7. _____ F. Observe some aspect of the natural world and identify a question or problem to be explored. [p.6]

After arranging the six steps above in the proper order of the scientific method, now arrange the following six events from actual research in the correct order. Find the letter of the first step and write it next to the number 8; write the letter of the final step next to the number 13; and so forth. Then, to the right of each letter, identify the step. Choose from: *question, hypothesis, prediction, test, repeat testing, conclusion*.

8. _____ _____
A. Data showing that the experimental group of mice developed fewer tumors than the control group support the prediction made about the effects of sulforaphane. [pp.6,8]

9. _____ _____
B. If sulforaphane helps the body fight off cancer, then mice in the experimental group should develop significantly fewer tumors than mice in the control group. [pp.6,8]

10. _____ _____
C. The experiment is repeated, and a further experiment is added to test the effects of sulforaphane in mice with a mutation that makes them especially vulnerable to cancer. [pp.6,8]

11. _____ _____
D. It is expected that the compound sulforaphane, found in the cabbage family, boosts the body's defenses against cancer. [pp.6,8]

12. _____ _____
E. Observations indicate that diets rich in cabbage and other members of this vegetable family are associated with lower incidence of cancer. Researchers ask the question, "Does a substance in these vegetables called sulforaphane help prevent cancer?" [pp.6,8]

13. _____ _____
F. A control group of mice is fed a normal diet. The variable by which the experimental group of mice differs from the control group is the addition of large amounts of sulforaphane to its diet. [pp.6,8]

Completion

A young girl is stung by a bee that has yellow and black stripes. Several days later, a yellow and black striped wasp stings the child. It only takes a few more stings for the girl to conclude that all black and yellow striped insects give painful stings. This is an example of (14) _____ [p.6] reasoning. Later, the girl is playing outside with a younger friend when she sees a yellow and black striped flying insect. She pulls her friend away, explaining that "*if* you bother a yellow and black striped insect, *then* you will probably get a painful sting." This is an example of (15) _____ [p.6] reasoning.

1.6. SCIENCE IN PERSPECTIVE [p.9]

1.7. CRITICAL THINKING IN SCIENCE AND LIFE [p.10]

1.8. *Choices: Biology and Society:* HOW DO YOU DEFINE DEATH? [p.11]

Selected Words: fact [p.10], opinion [p.10], anencephaly [p.11]

Boldfaced, Page-Referenced Terms

[p.9] theory _____

_____

[p.10] critical thinking _____

_____

True/False

In the blank beside each statement below, write a "T" (true) or an "F" (false).

1. _____ A theory in science is simply a person's idea about something. [p.9]

2. _____ Once a theory is accepted by scientists, its accuracy is never challenged again. [p.9]

3. _____ A theory explains a broad range of related phenomena in the natural world. [p.9]

4. _____ With time and better technology, science will be able to investigate and answer any question that humans ask. [p.9]

5. _____ Science requires strict objectivity. [p.9]

6. _____ Science does not make value judgments. [p.9]

7. _____ Because science provides knowledge through discovery, it is fair to expect the scientific community to guide the use of this knowledge. [p.9]

Completion

A study shows that, as a group, women who jog regularly develop breast cancer less often than women who never jog. This data reveals a(n) (8) _____ [p.10] between jogging and lower incidence of breast cancer. However, the study did not control other variables besides exercise in the women's lifestyle (diet, stress, etc.), and does not provide enough information to conclude that jogging is a(n) (9) _____ [p.10] of lower cancer rates. The manufacturer of a home jogging machine publishes the results of the study on breast cancer in order to encourage women to buy its product. Checking the accuracy of the commercial and questioning the source are important aspects of (10) _____ [p.10] thinking. The manufacturer also claims that using the jogging machine will result in a happier lifestyle. Because this information cannot be verified, it is a(n) (11) _____ [p.10], not a (12) _____ [p.10].

Self-Quiz

Multiple Choice

_____ 1. The scientific study of life is called _____ . [p.1]
   a. scientific method
   b. biology
   c. anatomy
   d. ecology

_____ 2. About 12 to 24 hours after a meal, a person's blood sugar level normally varies from about 60 to 90 mg per 100 ml of blood, although it may attain 130 mg/100 ml after meals high in carbohydrates. The maintenance of blood sugar level within a fairly narrow range despite uneven intake of sugar is due to the body's ability to maintain a state of _____ . [p.2]
   a. adaptation
   b. diversity
   c. metabolism
   d. homeostasis

_____ 3. _____ is a change in details of the body plan and function of organisms through successive generations. [p.3]
   a. Homeostasis
   b. Reproduction
   c. Metabolism
   d. Evolution

_____ 4. Webs of life including producers, consumers, and decomposers that function in the physical environment are called _____ . [p.5]
   a. populations
   b. biomes
   c. ecosystems
   d. food chains

5. The information-gathering system used by scientists to gain knowledge about the natural world is _____ . [p.6]
   a. the scientific method
   b. inductive reasoning
   c. deductive reasoning
   d. the "if–then" process

6. The experimental group and the control group are identical except for _____ . [p.7]
   a. the number of variables studied
   b. the variable under study
   c. the number of test subjects
   d. the number of experiments performed on each group

7. A hypothesis should *not* be accepted as valid if _____ . [p.7]
   a. the sample studied is determined to be representative of the entire group
   b. a variety of different tools and experimental designs yield similar observations and results
   c. other investigators can obtain similar results when they conduct the same experiment under similar conditions
   d. more than one variable is identified in the control and experimental groups

8. Scientific theories are based on all of these except _____ . [p.9]
   a. systematic observations
   b. hypotheses
   c. "if–then" predictions
   d. repeated controlled tests
   e. correlations and opinions

Choice

For questions 9–10, choose from the following:

   a. theory   b. hypothesis   c. prediction   d. conclusion

9. _____ A _____ is a statement about what one can expect to observe in nature if a theory or hypothesis is correct. [p.6]

10. _____ A _____ is a statement about whether a hypothesis (or theory) should be accepted, rejected, or modified, based on tests of the predictions derived from it. [p.6]

Chapter Objectives/Review Questions

This section lists general and detailed chapter objectives that can be used as review questions. You can make maximum use of these items by writing answers on a separate sheet of paper. Fill in answers where blanks are provided. To check for accuracy, compare your answers with information in the chapter or glossary.

1. Having a special molecule called _____ sets living things apart from the nonliving world. [p.2]
2. In addition to having DNA, what are the characteristics that all living things share? [p.2]
3. To sustain life, body systems must maintain a dynamic state of internal equilibrium called _____ . [p.2]
4. A(n) _____ is an organized living unit that can survive and reproduce itself, given DNA instructions and the necessary energy and materials. [p.2]
5. List the five kingdoms of living things. Which is sometimes split into two kingdoms? To which kingdom do humans belong? [p.3]
6. List three characteristics that separate humans from their primate relatives. [p.3]
7. Name and arrange the levels of organization in nature, from the least inclusive to the most inclusive (nonliving cellular components to biosphere). [p.4]
8. Explain how the actions of producers, consumers, and decomposers create interdependency among organisms. [pp.4–5]

9. All organisms are part of webs of organization in nature, in that they depend directly or indirectly on one another for _____ and _____ . [p.5]
10. List and explain the six steps that describe what scientists generally do when they proceed with an investigation. [p.6]
11. Correlating specific events to form a general conclusion is _____ reasoning, while drawing specific inferences from a working hypothesis is _____ reasoning. [p.6]
12. What is the function of the control group in an experiment? How does the control group differ from the experimental group? What is a variable? [p.7]
13. How does a scientific theory differ from a scientific hypothesis? [p.9]
14. Explain why no theory is an "absolute truth" in science. [p.9]
15. Describe how the methods of science differ from methods using subjective thinking and systems of belief. [p.9]
16. Define "critical thinking" and list ways in which a piece of information is evaluated critically. [p.10]
17. After reading the essay "How Do You Define Death?" describe what course you would follow, given a choice, if you had a child born with anencephaly. [p.11]

Integrating and Applying Key Concepts

1. Humans have the ability to maintain body temperature very close to 37°C.
   a. What conditions tend to make body temperature drop?
   b. What measures do you think your body takes to raise body temperature when it drops?
   c. What conditions cause body temperature to rise?
   d. What measures do you think your body takes to lower body temperature when it rises?
2. What sorts of topics do scientists usually regard as untestable by the methods they generally use?

2

THE CHEMISTRY OF LIFE

Interactive Exercises

CHAPTER INTRODUCTION [p.15]

2.1. THE ATOM [p.16]

2.2. *Science Comes to Life:* **USING RADIOISOTOPES TO TRACK CHEMICALS AND SAVE LIVES** [p.17]

Selected Words: "atomic weight" [p.16], "radioactivity" [p.16]

Boldfaced, Page-Referenced Terms

[p.14] element _____

_____

[p.14] trace element _____

_____

[p.15] atom _____

_____

[p. 15] molecules _____

_____

[p. 15] compound _____

_____

[p.15] organic compounds _____

_____

[p.16] protons _____

_____

[p.16] electrons _____

_____

[p.16] neutrons _____

_____

[p.16] atomic number _____

_____

[p.16] mass number _____

_____

[p.16] isotope _____

_____

[p.16] radioisotope _____

_____

[p.17] half-life _____

_____

[p.17] tracer _____

_____

[p.17] radiation therapy _____

_____

Matching

1. _____ atom [p.15]
2. _____ atomic number [p.16]
3. _____ electrons [p.16]
4. _____ element [p.15]
5. _____ isotopes [p.16]
6. _____ mass number [p.16]
7. _____ neutrons [p.16]
8. _____ tracer [p.17]
9. _____ trace element [p.15]
10. _____ organic compounds [p.15]
11. _____ half-life [p.17]
12. _____ radioactivity [p.17]
13. _____ protons [p.16]
14. _____ radioisotopes [p.17]
15. _____ radiation therapy [p.17]

A. The smallest unit that has the properties of a given element
B. The number of protons in the nucleus of one atom of an element
C. Subatomic particles in the nucleus of an atom that have no charge
D. An element important to the body that represents less than 0.01 percent of body weight
E. Energy emitted from unstable isotopes
F. Positively charged subatomic particles in the nucleus of an atom
G. Molecules built of two or more elements on a framework of carbon atoms
H. Unstable isotopes that tend to spontaneously emit energy or subatomic particles and form stable isotopes; iodine 123 is an example
I. Combined number of protons and neutrons in the nucleus of an atom
J. The time it takes for half of a quantity of a radioisotope to decay into a more stable isotope
K. Negatively charged subatomic particles that move rapidly around the nucleus of an atom
L. Use of radioisotopes, such as radium 226 or cobalt 60, to treat some cancers
M. Fundamental form of matter; has mass and takes up space
N. Forms of an element, the atoms of which contain a different number of neutrons than other forms of the same element
O. A substance with a radioisotope attached so that its movement can be tracked

Complete the Table

16. Complete the following table by entering the name of the element and its symbol in the appropriate spaces. [p.16]

Element	Symbol	Atomic Number	Mass Number
a.		1	1
b.		6	12
c.		7	14
d.		8	16
e.		11	23
f.		17	35

Dichotomous Choice

17. For each element in the chart above, the (atomic number/mass number) is always the same. [p.16]
18. Atoms of an element with different (atomic numbers/mass numbers) are called isotopes. [p.16]
19. The atomic number is the number of (protons/neutrons). [p.16]
20. A neutral atom will have the same number of protons as it does (neutrons/electrons). [p.16]

2.3. WHAT IS A CHEMICAL BOND? [pp.18–19]

2.4. IMPORTANT BONDS IN BIOLOGICAL MOLECULES [pp.20–21]

2.5. *Focus on Your Health:* ANTIOXIDANTS — FIGHTING FREE RADICALS [p.22]

Selected Words: "shells" [p.18], "orbitals" [p. 18], *lowest* energy level [p.18], *higher* energy levels [p.18], *octet rule* [p.18], *inert* [p.18], "biological molecules" [p.20], *share* electrons [p.20], *single* covalent bond [p.20], *double* covalent bond [p.20], *triple* covalent bond [p.20], *nonpolar* covalent bond [p.21], *polar* covalent bond [p.21], "electronegative" [p.21], *net* charge [p.21], *oxidation* [p.22], *loses* electrons [p.22]

Boldfaced, Page-Referenced Terms

[p.18] chemical bonds _____

_____

[p.19] mixture _____

_____

[p.20] ion _____

_____

[p.20] ionic bond _____

_____

[p.20] covalent bond _____

_____

[p.21] hydrogen bond _____

[p.22] free radicals _____

[p.22] antioxidant _____

Short Answer

1. Oxygen gas consists of two atoms of oxygen bonded together. Water consists of two atoms of hydrogen and one atom of oxygen bonded together. Which is/are molecules? Which is/are compounds? Explain. [p.19] _____

_____

_____

Identification

2. Following the atomic model of helium gas (the first model shown below), where the number of protons and neutrons are shown in the nucleus and the electrons are arranged in shells, identify the electron configuration of the atoms illustrated by entering appropriate electrons in the form $2e^-$. [p.19]

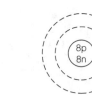

He

H

C

N

O

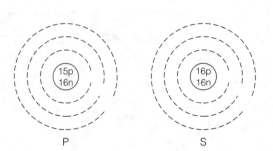

P   S

Short Answer

3. Of all the atoms shown above, which would not interact and form chemical bonds? Explain. [p.19] _____

_____

_____

_____

4. Of all the atoms shown above, which represent the four most abundant elements in the human body? Why will atoms of these elements form chemical bonds? [pp.15,19] _____

_____

_____

_____

Identification

5. Following the model of NaCl (sodium chloride), complete the sketch identifying the transfer of electron(s) (by arrows) to show how positive magnesium (Mg^+) and negative chloride (Cl^-) ions form ionic bonds to create a molecule of $MgCl_2$ (magnesium chloride). [p.20]

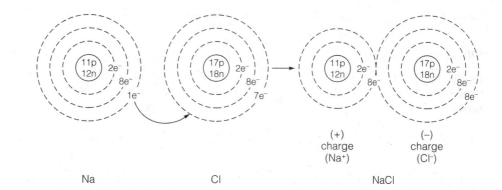

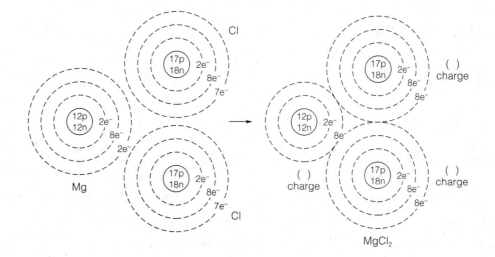

Short Answer

6. Distinguish between a nonpolar covalent bond and a polar covalent bond. Give one example of each. [pp.20–21] _____

_____

_____

_____

Identification

7. Following the model of hydrogen gas (next), complete the sketch by placing electrons (as dots) in the outer shells to identify the nonpolar covalent bonding that forms oxygen gas; similarly identify polar covalent bonds by completing electron structures to form a water molecule. [pp.20–21]

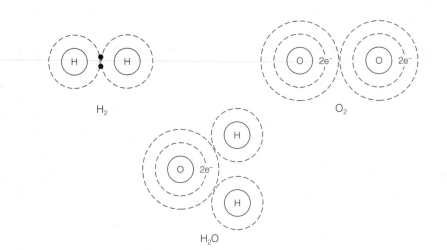

H_2

O_2

H_2O

Fill-in-the-Blanks

Whereas positively charged protons are found in the nucleus of an atom, negatively charged

(8) _____ [p.18] are found moving around the atomic nucleus in (9) "_____" [p.18].

Electrons within shells travel in (10) _____ [p.18]. Shells are equivalent to (11) _____

_____ [p.18]. Electrons in the shell nearest the nucleus are in the (12) _____ [p.18]

energy level, while those farther away are in (13) _____ [p.18] energy levels. The first shell

holds up to (14) _____ [p.18] electrons, while other shells can hold up to a maximum of (15)

_____ [p.18].

Atoms form a union called a(n) (16) _____ _____ [p.18] in order to fill up their

(17) _____ [p.18] shells. When atoms are joined together by chemical bonding, they form

(18) _____ [p.19], which may contain atoms of the same element. Molecules consisting of two or

more elements, always in the same proportions, are called (19) _____ [p.19].

An atom that has either lost or gained one or more electrons is a(n) (20) _____ [p.20]. An

association of two ions with opposing charges is called a(n) (21) _____ _____ [p.20].

In a(n) (22) _____ _____ [p.20], two atoms share a pair of electrons. H–H is a(n) (23) _____ [p.20] formula. The single line represents a(n) (24) _____ [p.20] covalent bond. Two lines indicate a double covalent bond, in which atoms share (25) _____ _____ [p.20] of electrons. In a(n) (26) _____ [p.20] covalent bond, three lines indicate three shared pairs of electrons.

In a polar covalent bond, the atom that exerts more pull and has a slight negative charge is said to be (27) _____ [p.21]. A(n) (28) _____ [p.21] bond is a weak attraction between an electronegative atom (often an (29) _____ [p.21] or (30) _____ [p.21] atom in a polar covalent bond) and a(n) (31) _____ [p.21] atom that is part of a second polar covalent bond. A(n) (32) _____ [p.21] line represents this weak attraction between hydrogen's (33) _____ [p.21] positive charge and the slight negative charge of the other atom. Hydrogen bonds may exist between two or more (34) _____ [p.21], or between different parts of the (35) _____ [p.21] molecule. There are many hydrogen bonds that hold together the two parallel strands of a(n) (36) _____ [p.21] molecule. The bonds are weak individually, but collectively they help (37) _____ [p.21] DNA's structure. Hydrogen bonds form between the molecules that make up (38) _____ [p.21] and contribute to its life-sustaining properties.

Short Answer

39. Explain how free radicals are produced and why they are a potential threat to human health. [p.22] ____

_____

_____

_____

40. What is an antioxidant? Name some antioxidants available in our diet. [p.22] _____

_____

_____

_____

2.6. LIFE DEPENDS ON WATER [p.23]

2.7. ACIDS, BASES, AND BUFFERS [pp.24–25]

Selected Words: heat capacity [p.23], solvent [p.23], acidic solutions [p.24], basic solutions [p.24], alkaline fluids [p.24], coma [p.25], tetany [p.25], acidosis [p.25], alkalosis [p.25]

Boldfaced, Page-Referenced Terms

[p.23] hydrophilic _____

_____

[p.23] hydrophobic _____

_____

[p.23] solute _____

_____

[p.24] hydrogen ions _____

_____

[p. 24] hydroxide ions _____

_____

[p.24] pH scale _____

_____

[p.24] acid _____

_____

[p.24] base _____

_____

[p.25] buffer system _____

_____

[p.25] salts _____

_____

Dichotomous Choice

Circle one of two possible answers given between parentheses in each statement.

1. By weight, your body is about (one-third/two-thirds) water. [p.23]
2. The polarity of water molecules allows them to form (hydrogen/covalent) bonds with one another and with other polar substances. [p.23]
3. All polar molecules are attracted to water and hence are said to be (hydrophobic/hydrophilic). [p.23]
4. The polarity of water repels oil and other nonpolar substances, which are (hydrophobic/hydrophilic). [p.23]
5. Hydrogen bonds between water molecules enable water to absorb a great deal of (cold/heat) energy before it significantly warms or evaporates. [p.23]

6. When the amount of heat energy is raised sufficiently, hydrogen bonds between water molecules break apart and (reform/do not reform); water then evaporates. [p.23]
7. Sweating followed by evaporation results in the body (gaining/losing) heat energy. [p.23]
8. Water is a superb (solute/solvent). [p.23]
9. Substances dissolved in water are (solutes/solvents). [p.23]
10. A substance is said to (precipitate/dissolve) as clusters of water molecules form around its individual ions or molecules. [p.23]

Fill-in-the-Blanks

Ionization of water molecules into hydrogen ions (H^+) and hydroxide ions (OH^-) is the basis for the

(11) _____ [p.24] scale. This scale is used to express the concentration of (12) _____

[p.24] ions in water, blood, and other fluids. On the pH scale, equal concentrations of H^+ and OH^- result in

a neutral pH given a value of (13) _____ [p.24]. Each change of pH by one unit corresponds to a

(14) _____ [p.24] change in H^+ concentration.

When dissolved in water, a(n) (15) _____ [p.24] donates protons (H^+) to other solutes or

water molecules. A solution with a pH (16) _____ [p.24] 7 is acidic. If a substance accepts protons

from solutes or water molecules, it is called a(n) (17) _____ [p.24]. A basic solution may also be

referred to as (18) _____ [p.24]. A solution with a pH (19) _____ [p.24] 7 is basic.

Slight shifts in pH can be controlled by a(n) (20) _____ [p.25] system. In one buffer system in

the body, carbonic acid (21) _____ [p.24] H^+ when the H^+ concentration in the blood decreases.

If the blood becomes too acidic, its partner base (22) _____ [p.24] binds the excess H^+.

Uncontrolled shifts in pH within body fluids can spell disaster. For example, an accumulation of

(23) _____ _____ [p.25] in the blood can lower pH severely as carbonic acid forms,

causing a condition called (24) _____ [p.25]. In contrast, an uncorrected increase in blood pH is

known as (25) _____ [p.25].

Compounds that release ions other than H^+ or OH^- are called (26) _____ [p.25]. The ions

released often have key functions in cells.

Short Answer

27. Explain how milk of magnesia acts as an antacid in the stomach. [p.24] _____

_____

_____

_____

Complete the Table

28. Fill in the missing information.

Fluid	pH Value	Acid/Base
a. blood [p.25]		
b. urine [p.24]		
c. gastric (stomach) fluid [p.24]		

2.8. ORGANIC COMPOUNDS: BUILDING ON CARBON ATOMS [pp.26–27]

Selected Words: "organic" compounds [p.26], "inorganic" [p.26], hydrocarbon [p.26], *dehydration synthesis* [p.27]

Boldfaced, Page-Referenced Terms

[p.26] functional groups _____

_____

[p.27] enzymes _____

_____

[p.27] condensation reaction _____

_____

[p.27] polymer _____

_____

[p.27] monomer _____

_____

[p.27] hydrolysis _____

_____

True/False

If a statement is true, write a "T" in the blank. If it is false, underline the incorrect word and write the correct word in the blank.

_____ 1. Molecules that have hydrogen and often other elements bonded to backbones of carbon are organic. [p.26]

_____ 2. The main types of organic molecules are carbohydrates, lipids, proteins, and water. [p.26]

_____ 3. Carbon forms two covalent bonds. [p.26]

_____ 4. Organic molecules can be chains or rings. [p.26]

_____ 5. A carbon backbone with only hydrogen atoms attached is a carbohydrate. [p.26]

_____ 6. Atoms or clusters of atoms that are covalently bonded to the carbon backbone and influence the chemical behavior of an organic compound are called enzymes. [p.26]

Labeling

Study the following structural formulas of organic compounds. Identify the circled functional groups (sometimes repeated) by entering the correct name in the blanks with matching numbers below the sketches; complete the exercise by circling the names of the compounds in the parentheses following each blank in which the functional group might appear.

7. _____ (fats — waxes — oils — sugars — amino acids — proteins — phosphate compounds, e.g., ATP) [p.26]

8. _____ (fats — waxes — oils — sugars — amino acids — proteins — phosphate compounds, e.g., ATP) [p.26]

9. _____ (fats — waxes — oils — sugars — amino acids — proteins — phosphate compounds, e.g., ATP) [p.26]

10. _____ (fats — waxes — oils — sugars — amino acids — proteins — phosphate compounds, e.g., ATP) [p.26]

11. _____ (fats — waxes — oils — sugars — amino acids — proteins — phosphate compounds, e.g., ATP) [p.26]

12. _____ (fats — waxes — oils — sugars — amino acids — proteins — phosphate compounds, e.g., ATP) [p.26]

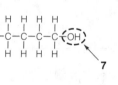

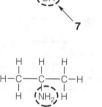

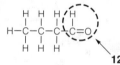

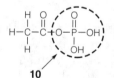

Short Answer

13. State the general role of enzymes as they relate to organic compounds. [p.27] _____

_____

_____

_____

Identification

14. The structural formulas for two adjacent amino acids are shown (next). By circling an H atom from one amino acid and an –OH group from the other, identify how enzyme action causes formation of a covalent bond and a water molecule (through a condensation reaction). Also circle the covalent bond that formed the peptide. [p.27]

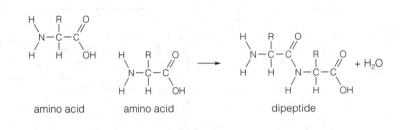

amino acid   amino acid   dipeptide

2.9. CARBOHYDRATES [pp.28–29]

Selected Words: "saccharide" [p.28], *mono*saccharide [p.28], *oligo*saccharide [p.28], *di*saccharide [p.28], "complex carbohydrates" [p.29], polysaccharides [p.29], "fiber" [p.29]

Boldfaced, Page-Referenced Terms

[p.28] carbohydrate _____

_____

[p.28] monosaccharides _____

_____

[p.28] oligosaccharides _____

_____

[p.28] polysaccharides _____

_____

Identification (application exercise)

1. In the following diagram, identify condensation reaction sites between the two glucose molecules by circling the components of the water removed that allow a covalent bond to form between the glucose molecules. Note that the reverse reaction is hydrolysis and that both condensation and hydrolysis reactions require enzymes in order to proceed. [p.28]

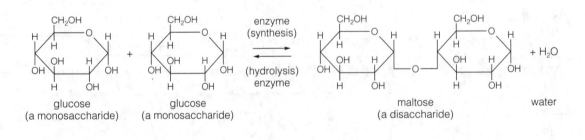

glucose   glucose   maltose   water
(a monosaccharide)   (a monosaccharide)   (a disaccharide)

Complete the Table

2. In the following table, enter the name of the carbohydrate described by its carbohydrate class and function(s). Select from *glucose, sucrose, lactose, glycogen, cellulose, starch,* and *deoxyribose.*

Carbohydrate	Carbohydrate Class	Function
a. [p.28]	Oligosaccharide (disaccharide)	Most plentiful sugar in nature; table sugar
b. [p.28]	Monosaccharide	Main energy source for body cells; precursors and building blocks of many organic compounds
c. [p.29]	Polysaccharide	Tough, insoluble structural material in plant cell walls
d. [p.28]	Monosaccharide	Five-carbon sugar; occurs in DNA
e. [p.28]	Oligosaccharide (disaccharide)	Sugar present in milk
f. [p.29]	Branched polysaccharide	Storage form of sugar in animals, including humans
g. [p.29]	Polysaccharide	Storage form of glucose in plants

2.10. LIPIDS [pp.30–31]

Selected Words: *unsaturated* tail [p.30], *saturated* tail [p.30], "vegetable oils" [p.30], *trans fatty acids* [p.30], "hydrogenated" [p.30]

Boldfaced, Page-Referenced Terms

[p.30] lipid _____

_____

[p.30] fats _____

_____

[p.30] fatty acids _____

_____

[p.30] triglycerides _____

_____

[p.31] phospholipid _____

_____

[p.31] sterols _____

_____

Identification

1. Combine glycerol with three fatty acids (see following illustration) to form a triglyceride by circling the participating atoms that will identify three covalent bonds; also circle the covalent bonds in the triglyceride. [p.30]

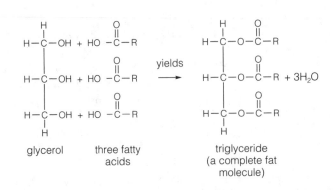

glycerol   three fatty acids   triglyceride (a complete fat molecule)

Labeling

2. In the appropriate blanks, label the molecule shown as either *saturated* or *unsaturated*. [p.30]

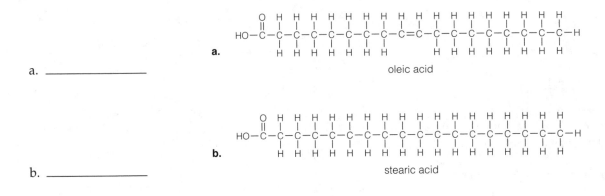

a.   oleic acid

a. _____

b.   stearic acid

b. _____

Short Answer

3. Define *phospholipid*; describe the structure and biological functions of phospholipids. [p.31] _____

_____

_____

_____

4. Describe how sterols are structurally different from fats. [p.31]

_____

_____

_____

Matching

5. _____ fatty acids [p.30]

6. _____ triglycerides [p.30]

7. _____ phospholipids [p.31]

8. _____ trans fatty acids [p.31]

9. _____ sterols [p.31]

A. Partially saturated (hydrogenated)
B. Up to 36 carbons long with unsaturated or saturated tails
C. Vitamin D, bile salts, steroid hormones
D. Main components of cell membranes
E. Body's richest source of energy; e.g., butter, lard, and oils

2.11. PROTEINS: BIOLOGICAL MOLECULES WITH MANY ROLES [pp.32–33]

2.12. A PROTEIN'S FUNCTION DEPENDS ON ITS SHAPE [pp.34–35]

Selected Words: peptide bonds [p.32], *primary* structure [p.32], *secondary* structure [p.34], *tertiary* structure [p.34], *quaternary* structure [p.35], *denatured* [p.35]

Boldfaced, Page-Referenced Terms

[p.32] proteins _____

_____

[p.32] amino acid _____

_____

[p.32] polypeptide chain _____

_____

[p.35] lipoproteins _____

_____

[p.35] glycoproteins _____

_____

[p.35] denaturation _____

_____

Matching

For questions 1–3, match the major parts of every amino acid by entering the letter of the part in the blank corresponding to the number on the molecule. [p.32]

1. _____

2. _____

3. _____

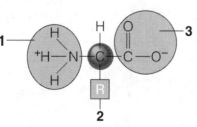

A. R group (a symbol for a characteristic group of atoms that differ in number and arrangement from one amino acid to another)
B. Carboxyl group (ionized)
C. Amino group (ionized)

Identification

4. In the following illustration of four amino acids in cellular solution (ionized state), circle the atoms and ions that form water to allow identification of covalent (peptide) bonds between adjacent amino acids to form a polypeptide. On the completed polypeptide, circle the newly formed peptide bond. [p.33]

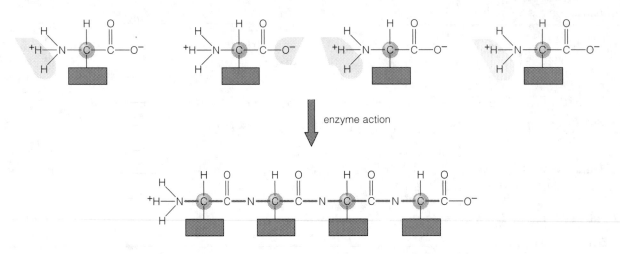

Matching

5. _____ primary protein structure [p.32]

6. _____ secondary protein structure [p.34]

7. _____ tertiary protein structure [p.34]

8. _____ quaternary protein structure [p.35]

A. A coiled or extended pattern, based on regular hydrogen-bonding interactions
B. Incorporates two or more polypeptide chains, yielding a protein that is globular, fibrous, or both
C. Unique sequence of amino acids in the polypeptide chain of a specific protein
D. Bending and looping of the polypeptide chain, caused by the interaction of R groups

Fill-in-the-Blanks

(9) _____ [p.35] form when cholesterol, triglycerides, and phospholipids in the blood combine with proteins. (10) _____ [p.35] are proteins with oligosaccharides bonded to them. This group includes proteins at the surface of your (11) _____ [p.35], protein hormones, and blood proteins.

Short Answer

12. Describe protein denaturation. [p.35] _____

_____

_____

_____

2.13. NUCLEOTIDES AND NUCLEIC ACIDS [p.36]

2.14. *Focus on Our Environment:* FOOD PRODUCTION AND A CHEMICAL ARMS RACE [p.37]

Selected Words: release energy [p.36], *require* energy [p.36], *herbicides, insecticides, fungicides* [p.37]

Boldfaced, Page-Referenced Terms

[p.36] nucleotides _____

_____

[p.36] ATP _____

_____

[p.36] coenzyme _____

_____

[p.36] nucleic acids _____

_____

[p.36] DNA _____

_____

[p.36] RNA _____

_____

Matching

For 1–3, match the following answers to the parts of a nucleotide shown in the accompanying diagram. [p.36]

1. _____

2. _____

3. _____

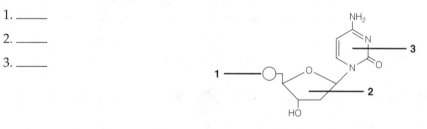

A. A five-carbon sugar (ribose or deoxyribose)
B. Phosphate group
C. A nitrogen-containing base that has either a single-ring or a double-ring structure

Identification

4. In the accompanying diagram of a single-stranded nucleic acid molecule, encircle as many *complete* nucleotides as possible. How many complete nucleotides are present? _____ [p.36]

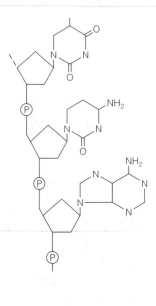

Matching

5. _____ adenosine triphosphate (ATP) [p.36]

6. _____ RNA [p.36]

7. _____ DNA [p.36]

8. _____ NAD⁺ and FAD [p.36]

A. Single nucleotide strand; plays a key role in processes by which genetic instructions are used to build the body's proteins
B. Nucleotide that provides energy for cellular reactions by the transfer of a phosphate group
C. Coenzymes that have nucleotides as subunits; transport hydrogen ions and their associated electrons from one cell reaction site to another
D. Helical double-nucleotide strand; encodes genetic information with base sequences

Short Answer

9. What is the benefit of chemical agents used to protect crops? What are the drawbacks? [p.37] _____

_____

_____

_____

Self-Quiz

Complete the Table

1. Complete the following table by entering the correct name of the major cellular organic compounds suggested in the "types" column (choose from *carbohydrates, proteins, nucleic acids,* and *lipids*).

Cellular Organic Compounds	Types
a. [p.31]	Phospholipids
b. [p.33]	Keratin found in hair
c. [p.32]	Enzymes, defense, transport, and regulatory molecules
d. [p.36]	Genetic instructions
e. [pp.28–29]	Glycogen, starch, and cellulose
f. [p.30]	Triglycerides
g. [p.35]	Glycoproteins
h. [p.30]	Glycerol plus fatty acids
i. [p.36]	Nucleotides including ATP
j. [pp.30–31]	Cholesterol and steroid hormones
k. [p.35]	Lipoproteins
l. [pp.28–29]	Glucose and sucrose

Multiple Choice

_____ 2. A molecule is _____ ; a compound is _____ . [p.19]
   a. a combination of two or more atoms; a molecule consisting of two or more elements in proportions that never vary
   b. the smallest unit of matter peculiar to a particular element; less stable than its constituent atoms
   c. a combination of two or more atoms; a very large molecule
   d. a carrier of one or more extra neutrons; a substance in which atoms of the same element are present in different proportions

_____ 3. If lithium has an atomic number of 3 and an atomic mass of 7, it has _____ neutrons in its nucleus; if a chlorine atom has an atomic number of 17, it will have _____ shells containing electrons. [pp.16,18]
   a. three; two
   b. four; two
   c. three; three
   d. four; three
   e. seven; three

_____ 4. A hydrogen bond is _____ ;
a covalent bond is _____ .
[pp.20,21]
   a. a sharing of a pair of electrons between a hydrogen nucleus and an
   oxygen nucleus; a bond in which electrons are gained or lost
   b. a sharing of a pair of electrons between a hydrogen nucleus and either
   an oxygen or a nitrogen nucleus; a
   bond that forms ions
   c. a weak attraction between an electronegative atom of a molecule and a
   neighboring hydrogen atom that is
   also part of a polar covalent bond; one
   in which electrons are shared
   d. none of the above

_____ 5. The pH scale is a measure of the concentration of _____ in a liquid; a
substance that releases H^+ in water is
a(n) _____ . [p.24]
   a. ionized water; acid
   b. OH^-; base
   c. hydrogen atoms; carbon compound
   d. OH^-; base
   e. H^+; acid

_____ 6. Any molecule that combines with or releases hydrogen ions, or both, and helps
to counter slight shifts in pH is known as
a(n) _____ ; a blood pH of 7.3
is _____ . [pp.24,25]
   a. neutral molecule; acid
   b. salt; basic
   c. base; neutral
   d. acid; acid
   e. buffer; basic

_____ 7. Carbon is part of so many different substances because _____ ; ^{12}C
and ^{14}C atoms differ in their numbers of
_____ . [pp.16,26]
   a. carbon forms two covalent bonds with
   a variety of other atoms; protons
   b. carbon forms four covalent bonds
   with a variety of atoms; neutrons
   c. carbon ionizes easily; electrons
   d. carbon is a polar compound; neutrons

_____ 8. Amino, carboxyl, phosphate, and hydroxyl are examples of _____ ;
_____ form the structural
elements of bones, muscles, and hair.
[pp.26,32]
   a. functional groups; carbohydrates
   b. sugar units; nucleic acids
   c. functional groups; proteins
   d. coenzymes; lipids

_____ 9. _____ are compounds used by
cells as transportable packets of quick energy, storage forms of energy, and structural materials; fats, oils, and steroids are
_____ . [pp.28–29, 30–31]
   a. Lipids; carbohydrates
   b. Nucleic acids; proteins
   c. Carbohydrates; lipids
   d. Proteins; nucleic acids

_____ 10. Hydrolysis could be correctly described
as the _____ ; glycogen, plant
starch, and cellulose are _____ .
[pp.27,29]
   a. heating of a compound in order to
   drive off its excess water and concen-
   trate its volume; proteins
   b. breaking of a polymer into its subunits by adding water molecule
   components to separated monomers;
   carbohydrates
   c. linking of two or more molecules by
   the removal of one or more water molecules; lipids
   d. constant removal of hydrogen atoms
   from the surface of a carbohydrate;
   nucleic acids

_____ 11. Genetic information is encoded in the sequence of the bases of _____ ;
molecules of _____ function
in processes using genetic instructions to
build the body's proteins. [p.36]
   a. DNA; DNA
   b. DNA; RNA
   c. RNA; DNA
   d. RNA; RNA

Chapter Objectives/Review Questions

This section lists general and detailed chapter objectives that can be used as review questions. You can make maximum use of these items by writing answers on a separate sheet of paper. Fill in answers where blanks are provided. To check for accuracy, compare your answers with information in the chapter or glossary.

1. A(n) _____ is a fundamental form of matter that has mass and takes up space. [p.15]
2. Describe how matter is organized in terms of elements, atoms, molecules, and compounds. [p.15]
3. Draw the structure of the carbon atom, with the correct spatial arrangement of protons, neutrons, and electrons. [p.19]
4. How does atomic number differ from mass number? [p.16]
5. ^{12}C and ^{14}C represent _____ of carbon. [p.16]
6. _____ are unstable and tend to spontaneously emit subatomic particles or energy in order to achieve stability. [p.16]
7. _____ - _____ _____ (PET) provides information about abnormalities in the metabolic functions of specific tissues. [p.17]
8. Explain the relationship between orbitals and shells of an atom. [p.18]
9. Atoms with an unfilled outer shell tend to form chemical _____ with other atoms and so fill their outer shell. [p.18]
10. What is the octet rule? [p.18]
11. A(n) _____ bond is an attraction between two oppositely charged ions. [p.20]
12. "A pair of electrons shared between two atoms" defines a(n) _____ bond. [p.20]
13. Describe how hydrogen bonds are formed; cite one example of a large molecule in which the hydrogen bond is important to you. [p.21]
14. Define and list examples of *antioxidants*; what is a free radical? [p.22]
15. Polar molecules are attracted to water and are said to be _____ . [p.23]
16. Nonpolar molecules are repelled by water and are known as _____ . [p.23]
17. Explain why water is such a good solvent. [p.23]
18. A substance that releases H^+ when it dissolves in water is a(n) _____ ; any substance that accepts H^+ when it dissolves in water is a(n) _____ . [p.24]
19. Na^+ and K^+ are components of _____ , substances that dissociate into ions other than H^+ and OH^- in solution. [p.25]
20. Acids often combine with bases to form _____ and water. [p.25]
21. Describe the structure and use of the pH scale. [p.24]
22. Define *buffer system*; describe how the carbonic acid/bicarbonate buffer system works in the blood. [p.25]
23. A(n) _____ is a carbon backbone with only hydrogen atoms attached. [p.26]
24. $-OH$, $-COOH$, and $-NH_3$ are examples of _____ groups. [p.26]
25. Define the terms *condensation* and *hydrolysis,* and *polymer* and *monomer*, and explain the first two in terms of their relationship with the second two. [p.27]
26. List the four main types of organic compounds that represent the molecules characteristic of life. [p.26]
27. List the ways that cells use carbohydrates. [p.28]
28. List the three classes of carbohydrates; how is each distinguished from the other two? [pp.28–29]
29. State the major functions of lipids in most organisms. [p.30]
30. Distinguish an unsaturated fatty acid from a saturated fatty acid. [p.30]
31. Butter, lard, and oils are all examples of _____ . [p.30]
32. Describe trans fatty acids, what foods they are found in, and a possible concern about their use. [p.31]
33. The main materials of cell membranes are _____ . [p.31]
34. List the major functions of proteins; all proteins are constructed from about 20 different kinds of _____ acids. [p.32]
35. One group of proteins, the _____ , make metabolic reactions proceed much faster than they otherwise would. [p.32]
36. Define *peptide bond* and *polypeptide chain*. [p.32]

37. Distinguish between the primary, secondary, tertiary, and quaternary structure of a protein molecule. [p.32]
38. The loss of a protein's three-dimensional shape following disruption of bonds is called _____ . [p.35]
39. How are nucleotides related to nucleic acids? Cite the names and functions of two major nucleic acids important to life. [p.36]

Integrating and Applying Key Concepts

1. Explain what would happen if water were a nonpolar molecule instead of a polar molecule. Would water be a good solvent for the same kinds of substances? Would the nonpolar molecule's heat capacity likely be higher or lower than that of water? Would its ability to form hydrogen bonds change?
2. Humans can obtain energy from many different food sources. Do you think this ability is an advantage or a disadvantage in terms of long-term survival? Why?
3. Most DNA information tells the cell how to make proteins, while none of this information represents instructions for other types of organic molecules. What type of protein allows reactions to occur that make or break down other organic molecules? Explain why damage to DNA instructions for making a specific enzyme may result in a genetic disease.

3

CELLS

Interactive Exercises

Selected Words: compound light microscope [p.40], *transmission electron microscope* [p.41], *scanning electron microscope* [p.41], "motor proteins" [p.46], "crosslinking proteins" [p.46]

Boldfaced, Page-Referenced Terms

[p.41] cell theory _____

_____

[p.42] plasma membrane _____

_____

[p.42] DNA-containing region _____

_____

[p.42] cytoplasm _____

_____

[p.42] cytosol _____

_____

[p.42] prokaryotic cell _____

_____

[p.42] eukaryotic cell _____

_____

[p.42] organelles _____

_____

[p.43] surface-to-volume ratio _____

_____

[p.43] lipid bilayer _____

_____

[p.46] cytoskeleton _____

_____

[p.46] microtubules _____

_____

[p.46] microfilaments _____

_____

[p.46] intermediate filaments _____

_____

[p.46] flagellum (plural: flagella) _____

_____

[p.46] cilium (plural: cilia) _____

_____

[p.46] centrioles _____

_____

[p.46] basal body _____

_____

Fill-in-the-Blanks

By studying magnified cells through (1) _____ [p.41], biologists developed a set of three

generalizations that constitute the (2) _____ _____ [p.41]. First, all organisms,

including humans, consist of one or more (3) _____ [p.41]. Second, the cell is the smallest unit

having the properties of (4) _____ [p.41]. Third, the continuity of life depends on the concept

that cells come from (5) _____ [p.41] cells.

Matching

6. _____ transmission electron microscope [p.41]

7. _____ compound light microscope [p.41]

8. _____ scanning electron microscope [p.41]

A. Lenses enlarge image formed by light rays passing through a thin specimen
B. Uses a magnetic field to channel a stream of electrons
C. Produces images with extraordinary depth

Complete the Table

9. To complete the following table, identify the three basic characteristics of any cell. [p.42]

Cell Part	Function
a.	Inherited genetic instructions
b.	Everything between the plasma membrane and the region of DNA; consisting of the cytosol and other components
c.	Thin, outermost covering; allows life-sustaining activities to proceed inside the cell apart from the environment

Short Answer

10. Distinguish prokaryotic cells from eukaryotic cells. [p.42] _____

_____

_____

_____

11. Explain why most cells are so small. How do cells that are not small compensate for their larger size?
   [pp.42] _____

_____

_____

_____

12. Describe the "heads-out, tails-in" structure of a cell membrane, including the terms "hydrophobic" and
   "hydrophilic." [p.43] _____

_____

_____

_____

Fill-in-the-Blanks

In eukaryotic cells, the problem of having incompatible reactions occurring at the same time within a single cell has been solved by the presence of (13) _____ [p.44]. The barriers that physically separate incompatible reactions are the (14) _____ [p.44] that surround each organelle.

The system of interconnected fibers, threads, and lattices in the cytosol is called the (15) _____ [p.46]. Two major elements of this system are the (16) _____ [p.46] and (17) _____ [p.46], on which nearly all cell movements depend. Microtubules consist of protein subunits called (18) _____ [p.46], whereas microfilaments are made up of two chains of the protein (19) _____ [p.46]. Some cells contain (20) _____ _____ [p.46], which strengthen cells and tissues.

"Motor proteins" have roles in cell (21) _____ [p.46]. (22) "_____ _____" [p.46] create a meshwork that plays a role in support and movement.

The "9 + 2 array" is a system involved in the bending of (23) _____ [p.46] and (24) _____ [p.46]. The microtubules that comprise these structures arise from (25) _____ [p.46], which remain at the base of the structure as a (26) _____ _____ [p.46].

3.4. THE PLASMA MEMBRANE: A LIPID BILAYER [p.47]
3.5. THE CYTOMEMBRANE SYSTEM [pp.48–49]

Selected Words: *rough* ER [p.48], *smooth* ER [p.48]

Boldfaced, Page-Referenced Terms

[p.48] cytomembrane system _____

_____

[p.48] endoplasmic reticulum (ER) _____

_____

[p.48] ribosomes _____

_____

[p.48] Golgi bodies _____

_____

[p.49] vesicles _____

_____

[p.49] lysosome _____

_____

[p.49] peroxisomes _____

_____

Short Answer

1. Why is it necessary for the plasma membrane to be "fluid" rather than rigid? Why is the membrane described as a "mosaic" structure? [p.47] _____

_____

_____

_____

Complete the Table

2. Study the following illustration and complete the table by entering the name of each membrane structure involved with the function described. Choose from: *lipid bilayer, transport proteins, receptor proteins, recognition proteins,* and *adhesion proteins.* [p.47]

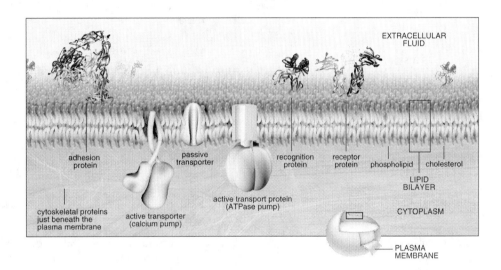

Membrane Structure	Function
a.	Diverse types at the cell surface are like molecular fingerprints
b.	The phospholipid "sandwich" that makes a cell membrane somewhat fluid
c.	Proteins that span the bilayer and allow water-soluble substances to move through their interior
d.	Help cells of the same type stick together and stay positioned within the proper tissue
e.	Bind hormones and other extracellular substances that trigger changes in cell activities; there are different combinations on different cells

Matching

Study the following illustration and match each component of the cytomembrane system with its description below.

3. _____ smooth ER [p.48]

4. _____ rough ER [p.48]

5. _____ Golgi body [p.48]

6. _____ vesicle [p.49]

7. _____ lysosome [p.49]

8. _____ peroxisome [p.49]

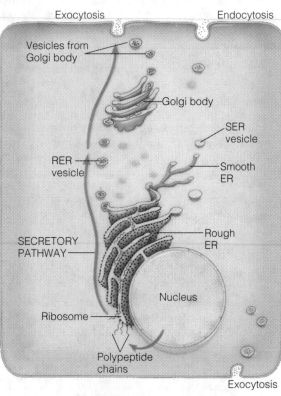

A. Type of vesicle containing enzymes that break down fatty acids, amino acids, and alcohol
B. Organelle in which enzymes finish proteins and lipids, sort them, and package them within vesicles
C. Contain enzymes that digest cells, cell parts, and bacteria
D. Tiny membranous sac that moves through the cytoplasm
E. Pipe-like structure in which many cells assemble lipids or inactivate drugs and toxins
F. Abundant in cells that secrete finished proteins

Sequence

9. Assume that the cell diagram above is of a pancreas cell that secretes digestive enzymes. In the first blank below, list the first cellular structure involved in the production, processing, and secretion of these proteins, followed by the remainder of cell structures or events involved in the correct order. Choose from: *Golgi body, rough ER, exocytosis into small intestine, rough ER vesicle, Golgi body vesicle, ribosome.*

a. _____   d. _____

b. _____   e. _____

c. _____   f. _____

Dichotomous Choice

10. The "fluid" property of a cell membrane is created by the (phospholipids/proteins). [p.47]
11. Most cell functions are carried out by (phospholipids/proteins). [p.47]
12. The (endoplasmic reticulum/Golgi body) connects with the membrane surrounding the nucleus. [p.48]
13. The vesicles that break away from Golgi bodies carry substances throughout the (body/cytoplasm). [p.48]

3.6. MOVING SUBSTANCES ACROSS MEMBRANES BY DIFFUSION AND OSMOSIS [pp.50–51]

3.7. OTHER WAYS SUBSTANCES CROSS CELL MEMBRANES [pp.52–53]

3.8. *Focus on Our Environment:* REVENGE OF EL TOR [p.53]

Selected Words: "concentration" [p.50], "gradient" [p.50], "dynamic equilibrium" [p.50], *electric gradient* [p.50], *pressure gradient* [p.50], *tonicity* [p.51], *isotonic* [p.51], *hypotonic* [p.51], *hypertonic* [p.51], *osmotic pressure* [p.51], *hydrostatic pressure* [p.51], "facilitated diffusion" [p.52], "membrane pumps" [p.52], "passive" [p.52], *Vibrio cholerae* [p.53], *oral rehydration therapy (ORT)* [p.53]

Boldfaced, Page-Referenced Terms

[p.50] selective permeability _____

_____

[p.50] concentration gradient _____

_____

[p.50] diffusion _____

_____

[p.51] osmosis _____

_____

[p.52] passive transport _____

_____

[p.52] active transport _____

_____

[p.53] exocytosis _____

_____

[p.53] endocytosis _____

_____

[p.53] phagocytosis _____

_____

Matching

1. _____ osmosis [p.51]
2. _____ pressure gradient [p.50]
3. _____ endocytosis [p.53]
4. _____ selective permeability [p.50]
5. _____ active transport [p.52]
6. _____ diffusion [p.50]
7. _____ electric gradient [p.50]
8. _____ exocytosis [p.53]
9. _____ tonicity [p.51]
10. _____ passive transport [p.52]
11. _____ phagocytosis [p.53]
12. _____ concentration gradient [p.50]
13. _____ dynamic equilibrium [p.50]

A. Difference in pressure between two adjoining regions
B. Property of allowing some substances but not others to cross a cell membrane
C. Net movement of like molecules or ions from an area of higher concentration to an area of lower concentration
D. Difference in electric charge across a cell membrane
E. Diffusion of water across a selectively permeable membrane in response to solute concentration gradients
F. Refers to the relative solute concentrations in two fluids
G. Solutes are made to cross membranes against their own concentration gradients
H. A small section of plasma membrane indents, surrounds some organic matter outside of the cell, balloons inward, and pinches off, producing a vesicle
I. A vesicle moves to the cell membrane and fuses with it, releasing its contents to the outside of the cell
J. Difference in the number of molecules or ions of a given substance in two neighboring regions
K. Literally means "coming inside the cell"
L. Solutes diffuse across a cell membrane through a transport protein down their own concentration gradients
M. refers to the nearly uniform net distribution of molecules throughout two adjoining regions

True/False

If the statement is true, write a "T" in the blank. If the statement is false, underline the incorrect word(s) and write the correct word in the answer blank.

_____ 14. Diffusion occurs faster when the concentration gradient is steep. [p.50]

_____ 15. In a solution with more than one solute, each solute diffuses separately from others, according to its own concentration gradient. [p.50]

_____ 16. In a dynamic equilibrium, the number of molecules moving one way is greater than the number moving the other way. [p.50]

_____ 17. Osmotic pressure refers to water's tendency to pass from a hypotonic solution to a hypertonic one. [p.51]

_____ 18. Osmosis occurs in response to a concentration gradient of solute that results in equal concentrations of water molecules. [p.51]

_____ 19. If an animal cell were placed in a hypertonic solution, it would swell and perhaps burst. [p.51]

_____ 20. Physiological saline is 0.9 percent NaCl; red blood cells placed in such a solution will not gain or lose water; therefore, one could state that the fluid in red blood cells is hypertonic. [p.51]

_____ 21. Cell membranes display selective permeability. [p.50]

_____ 22. Red blood cells shrivel and shrink when placed in a hypotonic solution. [p.51]

_____ 23. *Vibrio cholerae* is a fungus that causes disease when it enters human drinking water supplies. [p.53]

_____ 24. Cellular products are released to the outside of a cell by phagocytosis. [p.53]

_____ 25. Exocytosis occurs when a cell takes in a substance by engulfing it into a vesicle derived from the plasma membrane. [p.53]

Complete the Table

Complete the table by filling in the empty boxes.

Substance	Process	Where Substance Crosses	Active or Passive
small nonpolar molecules like CO_2 or O_2	(26)	(27)	passive [p.50]
(28)	osmosis	(29)	passive [p.51]
ions, solutes with any amount of charge	(30)	transport proteins	passive [p.52]
ions, solutes with any amount of charge	(31)	(32)	active [p.52]

3.9. THE NUCLEUS [p.54]

3.10. MITOCHONDRIA: THE CELL'S ENERGY FACTORIES [p.55]

Selected Words: nucleoplasm [p.54], crista (plural: cristae) [p.55]

Boldfaced, Page-Referenced Terms

[p.54] nucleus _____

_____

[p.54] nuclear envelope _____

_____

[p.54] nucleolus _____

_____

[p.54] chromatin _____

_____

[p.54] chromosome _____

_____

[p.55] mitochondrion (plural: mitochondria) _____

_____

Matching

Match each of the following terms with the appropriate phrase or description.

1. _____ chromosome [p.54]
2. _____ mitochondrion [p.55]
3. _____ nucleolus [p.54]
4. _____ chromatin [p.54]
5. _____ crista [p.55]
6. _____ nucleoplasm [p.54]
7. _____ nucleus [p.54]
8. _____ pore [p.54]

A. Organelle with a double membrane; separates DNA from cell cytoplasm
B. A cell's collection of DNA and its associated proteins
C. Fold of the inner member of a mitochondrion
D. Dense mass within the nucleus where subunits of ribosomes are built
E. Passageway across the nuclear envelope for the movement of ions and small molecules
F. Fluid portion of the nucleus
G. A single DNA molecule with its associated proteins
H. Organelle in which ATP is produced

Fill-in-the-Blanks

In mitochondria, many (9) _____ [p.55] molecules form when (10) _____ [p.55] compounds are broken down to (11) _____ _____ [p.55] and (12) _____ [p.55]. These kinds of ATP-forming reactions cannot be completed without plenty of (13) _____ [p.55] taken in by breathing. Only (14) _____ [p.55] cells contain mitochondria.

Short Answer

15. Distinguish between chromatin and chromosomes. [p.54] _____

_____

_____

_____

16. List three reasons why many biologists believe that mitochondria evolved from ancient bacteria consumed by a larger cell. [p.55] _____

_____

_____

_____

3.11. METABOLISM: DOING CELLULAR WORK [pp.56–57]

Selected Words: *"phosphorylation"* [p.56], *intermediate* [p.56], *end products* [p.56], *energy carriers* [p.56], *transport proteins* [p.56], *coupled* reactions [p.56]

Boldfaced, Page-Referenced Terms

[p.56] metabolism _____

_____

[p.56] ATP/ADP cycle _____

_____

[p.56] metabolic pathways _____

_____

[p.56] anabolism _____

_____

[p.56] catabolism _____

_____

[p.56] substrate _____

_____

[p.56] enzymes _____

_____

[p.57] active site _____

_____

[p.57] coenzymes _____

_____

[p.57] NAD$^+$ _____

_____

[p.57] FAD _____

Matching

1. _____ active site [p.57]
2. _____ catabolism [p.56]
3. _____ energy carrier [p.56]
4. _____ metabolic pathway [p.56]
5. _____ anabolism [p.56]
6. _____ substrate [p.56]
7. _____ intermediate [p.56]
8. _____ phosphorylation [p.56]
9. _____ coupled reaction [p.56]
10. _____ enzyme [pp.56–57]
11. _____ end product [p.56]

A. Speeds the rate of a specific chemical reaction
B. Any transfer of a phosphate group to a molecule
C. A surface crevice on an enzyme where it interacts with a substrate
D. Pathway where large molecules are broken down into products that have less energy
E. ATP or a few other compounds that activate other substances
F. Any substance that forms between the start and end of a pathway
G. A series of reactions occurring in orderly steps that are catalyzed by enzymes
H. Steps in a pathway that are linked by a transfer of electrons from one substance to another, such as energy-releasing reactions in mitochondria
I. Substance that remains at the end of a reaction or pathway
J. Any substance that enters a reaction; also called a reactant or a precursor
K. Pathway in which small molecules are put together into more complex, higher-energy molecules

Short Answer

12. List the three features that all enzymes share. [p.57] _____

_____

_____

Matching

Match the items on the following sketch with the correct description. [p.57]

13. _____

14. _____

15. _____

16. _____

A. Substrate molecules
B. Active site of the enzyme
C. Enzyme, a protein with catalytic power
D. Product molecule

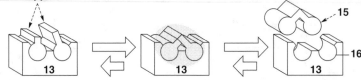

Dichotomous Choice

Circle one of two possible answers given between parentheses in each statement.

17. The reaction diagrammed above is (anabolic/catabolic). [p.56]
18. In the human body, enzymes function best within a pH range of (7.35–7.4/6.35–8.4). [p.57]
19. The pathways that build complex carbohydrates, proteins, and other large molecules are all (anabolic/catabolic). [p.56]
20. Nearly all enzymes are (carbohydrates/proteins). [pp.56–57]
21. NAD^+ and FAD are examples of (transport proteins/coenzymes). [p.57]

Fill-in-the-Blanks

Living cells require (22) _____ [p.56] to accomplish cellular work, called (23) _____

[p.56]. Our cells' mitochondria convert the "raw" energy from organic compounds in (24) _____

[p.56] into a chemical form of energy, called (25) _____ [p.56]. ATP consists of a five-carbon

sugar, the base adenine, and (26) _____ [p.56] phosphate groups. Enzymes break the (27)

_____ _____ [p.56] bond between the outer two (28) _____ [p.56]

groups of an ATP molecule. The group is then attached to another substance. This transfer is called (29)

_____ [p.56]. When ATP transfers a phosphate group, the remaining molecule is (30)

_____ [p.56]. When a phosphate group is attached to adenosine diphosphate, the result is

(31) _____ [p.56]. This is the (32) _____ [p.56] cycle.

In a(n) (33) _____ _____ [p.56], enzymes catalyze an orderly series of chemical

reactions. Those pathways that build more complex molecules from small ones are collectively called

(34) _____ [p.56]. Those that break large molecules into smaller products are called (35) _____ [p.56]. A (36) _____ [p.56] is any substance that enters a reaction. It is also called a (37) _____ or _____ [p.56]. Substances formed between the beginning and end of a pathway are known as (38) _____ [p.56], whereas those that remain at the end of a pathway are (39) _____ _____ [p.56]. ATP represents a(n) (40) _____ _____ [p.56], which donates energy to chemical reactions of metabolism. Reactions that are linked by a transfer of electrons from one substance to another are known as (41) _____ [p.56] reactions.

(42) _____ [pp.56–57] speed up chemical reactions. The specific substance that an enzyme modifies is its (43) _____ [p.57]. The interaction between this substance and the enzyme occurs at the enzyme's (44) _____ _____ [p.57].

(45) _____ [p.57] are organic molecules that help enzymes with reactions. Two examples are (46) _____ [p.57] and (47) _____ [p.57], both of which are derived from vitamins.

The body must have ways to boost or slow the activity of enzymes, or to control their synthesis. This is necessary to maintain (48) _____ [p.58].

3.12. HOW CELLS MAKE ATP [pp.58–59]
3.13. SUMMARY OF AEROBIC RESPIRATION [pp.60]
3.14. ALTERNATIVE ENERGY SOURCES IN THE BODY [pp.61]

Selected Words: "aerobic" [p.58], *anaerobic* [p.58], "phosphorylations" [p.58], *adipose* tissue [p.62]

Boldfaced, Page-Referenced Terms

[p.58] aerobic respiration _____

_____

[p.58] fermentation pathways _____

_____

[p.58] glycolysis _____

_____

[p.58] pyruvate _____

_____

[p.58] PGAL _____

_____

[p.58] Krebs cycle _____

_____

[p.59] electron transport systems _____

_____

Dichotomous Choice

1. Cells of the human body typically form ATP by way of (anaerobic/aerobic) respiration, which requires oxygen. [p.58]
2. Cells make ATP by breaking (ionic/covalent) bonds in carbohydrates especially, but also proteins and lipids. [p.58]
3. During breakdown reactions, energy associated with (calories/electrons) drives ATP formation. [p.58]
4. Aerobic pathways start with a set of reactions known as (electron transport/glycolysis). [p.58]
5. Glycolysis occurs in the (mitochondria/cytoplasm). [p.58]
6. In glycolysis, glucose is partially broken down into two molecules of (pyruvate/ATP). [p.58]
7. For every glucose molecule entering glycolysis, (one/two) three-carbon pyruvate molecule(s) is/are produced. [p.58]
8. The first steps of glycolysis are (energy-releasing/energy-requiring). [p.58]
9. The first steps of glycolysis proceed only when two ATP molecules each transfer a phosphate group to (pyruvate/glucose), raising its energy content. [p.58]
10. The glucose molecule is then split into two molecules of (PGAL/pyruvate), which is then converted to an unstable intermediate. [p.58]
11. During glycolysis, intermediates donate phosphate groups to ADP, resulting in (four/36) ATP molecules. [p.58]
12. The *net* yield of ATP from glycolysis is (two/four) ATP molecules. [p.58]
13. NAD^+ picks up electrons and hydrogen ions released from (end products/intermediates) and becomes NADH, which can give up H^+ and electrons to become NAD^+ again. [p.58]
14. The oxygen needed to make ATP is used in the (cytoplasm/mitochondria). [p.58]
15. Following glycolysis, the two pyruvates enter the (inner/outer) compartments of a mitochondrion. [p.58]
16. In the preparatory steps, a carbon is removed from each pyruvate, leaving a fragment that is combined with (coenzyme A/carbon dioxide), resulting in acetyl-CoA. [p.58]
17. The two-carbon fragment is transferred to (pyruvate/oxaloacetate), which starts the Krebs cycle. [p.58]
18. The carbons in the two pyruvates are lost in the Krebs cycle as (NADH/CO_2). [p.58]
19. Reactions during the Krebs cycle produce (two/36) molecules of ATP. [p.58]
20. Reactions prior to and during the Krebs cycle produce a large number of (phosphate/coenzyme) molecules such as NADH and $FADH_2$. [p.58]
21. $FADH_2$ and NADH molecules from the first and second stages of aerobic respiration are used in the (Krebs cycle/electron transport system). [p.58]
22. The ATP-producing mechanisms of the final stage of aerobic respiration involve (electron transport systems/osmosis). [p.59]
23. Electron transport systems and neighboring transport proteins are embedded in the (outer/inner) membrane that divides the mitochondrion into two compartments. [p.59]
24. The transfer of electrons in the electron transport systems results in the movement of (water/H^+) from the inner to the outer compartment of a mitochondrion. [p.59]
25. An H^+ gradient forms in the outer compartment of the mitochondrion after the pumping of H^+ from the inner compartment; H^+ then flows back into the inner compartment, where (ADP/ATP) is formed. [p.59]
26. At the end of the electron transport systems, oxygen withdraws electrons and then combines with water, resulting in the formation of (carbon dioxide/water). [p.59]
27. For each molecule of glucose entering the aerobic pathway including glycolysis, (32/36) molecules of ATP are formed. [p.60]

Fill-in-the-Blanks

The reactions of aerobic respiration take place in three stages. The first stage, (28) _____ [p.58], occurs in the (29) _____ [p.58] outside of the mitochondrion. Stage 2, the (30) _____ [p.58] cycle, occurs in the inner compartment of a (31) _____ [p.58]. These first two stages produce only four net (32) _____ [p.58], but hydrogens and electrons are stripped from intermediate compounds; coenzymes deliver these to a(n) (33) _____ _____ [pp.58–59] system. This is the third and final stage, which occurs on the (34) _____ [p.59] mitochondrial membrane (cristae); a high ATP yield is produced. (35) _____ [p.59] accepts the "spent" (36) _____ [p.59] from the transport system, forming (37) _____ [p.59]. Including the yield from glycolysis, the Krebs cycle, and the electron transport systems, the total number of ATPs produced by aerobic respiration is (38) _____ [p.60].

Short Answer

39. Why is the brain considered a glucose "hog"? [p.61] _____

_____

_____

_____

40. Discuss glycogen. (What is it? When is it produced? Where is most stored? What percentage of the human body's total energy store does it represent?) [p.61] _____

_____

_____

_____

41. What is the function of glucagons? [p.61] _____

_____

_____

_____

42. What percentage of the human body's total energy is stored as fat? In what form does the body store most fat? [p.61] _____

_____

43. When are triglycerides used as an energy source? [p.61] _____

_____

_____

_____

44. Why does a fatty acid yield more ATP than a glucose molecule? [p.61] _____

_____

_____

_____

45. A toxic by-product from the breakdown of excess protein is _____ . [p.61]

Self-Quiz

_____ 1. _____ are organelles of digestion within a cell; they bud from _____ membranes. [p.49]
a. Peroxisomes; ER
b. Mitochondria; cytoskeleton
c. Lysosomes; Golgi
d. Centrioles; basal body

_____ 2. The _____ is free of ribosomes, curves through the cytoplasm, and is the main site of lipid synthesis. [p.48]
a. lysosome
b. Golgi body
c. smooth ER
d. rough ER

_____ 3. Which of the following is _not_ one of the three fundamental features of cells? [p.42]
a. cell wall
b. plasma membrane
c. cytoplasm
d. DNA

_____ 4. Mitochondria convert energy obtained from _____ to forms that the cell can use, principally ATP. [p.55]
a. water
b. glucose and other organic molecules
c. NADPH$_2$
d. carbon dioxide

_____ 5. In a lipid bilayer, phospholipid tails point inward and form a(n) _____ layer that excludes water. [p.43]
a. acidic
b. hydrophilic
c. basic
d. hydrophobic

_____ 6. A protistan adapted to life in a freshwater pond is collected in a bottle and transferred to a saltwater bay. Which of the following is likely to happen? [pp.50–51]
a. The cell will burst.
b. Salts will flow out of the protistan cell.
c. The cell will shrink.
d. Enzymes will flow out of the protistan cell.

_____ 7. Calcium pumps move calcium ions against their concentration gradient, keeping a much lower concentration inside muscle cells than outside. This is an example of _____ . [p.52]
a. passive transport
b. endocytosis
c. exocytosis
d. active transport

_____ 8. O$_2$ and CO$_2$ and other small, nonpolar molecules move across the cell membrane by _____ . [p.50]
a. facilitated diffusion
b. endocytosis
c. diffusion
d. active transport

_____ 9. Ions and small, water-soluble molecules cross the plasma membrane down their concentration gradients by _____ , also called "facilitated diffusion." [p.50]
a. passive transport
b. endocytosis
c. simple diffusion
d. active transport

_____ 10. Phagocytosis is a special type of _____ used by immune system cells to engulf harmful viruses and bacteria. [p.53]
a. exocytosis
b. passive transport
c. endocytosis
d. hydrolysis

_____ 11. Glycolysis would quickly halt if the process ran out of _____ , which serves as the hydrogen and electron acceptor. [p.58]
a. coenzyme A
b. oxygen
c. NAD$^+$
d. H$_2$O

_____ 12. The final electron acceptor at the end of aerobic respiration is _____ . [p.59]
a. NADH
b. carbon dioxide (CO$_2$)
c. ATP
d. oxygen

_____ 13. When glucose is used as an energy source, the largest amount of ATP is generated by the _____ portion of the respiratory process. [p.60]
a. glycolysis
b. acetyl-CoA
c. Krebs cycle
d. electron transport systems

_____ 14. Which of these is *not* necessary for aerobic respiration to occur? [pp.58–59]
a. glucose
b. oxygen
c. mitochondrion
d. carbon dioxide

Chapter Objectives/Review Questions

This section lists general and detailed chapter objectives that can be used as review questions. You can make maximum use of these items by writing answers on a separate sheet of paper. Fill in answers where blanks are provided. To check for accuracy, compare your answers with information in the chapter or glossary.

1. List the three generalizations that together constitute the cell theory. [p.40]
2. Name and describe the three basic parts of cells. [p.42]
3. Describe and contrast the distinguishing features of prokaryotic and eukaryotic cells. [p.42]
4. Phospholipid molecules have _____ heads and _____ tails. [p.43]
5. Describe the structure of a cell membrane. [p.47]
6. Describe the functions of transport proteins, adhesion proteins, recognition proteins, and receptor proteins. [p.47]
7. When molecules move down a concentration gradient, they move from a region where they are _____ concentrated to a region where they are _____ concentrated. [p.51]
8. Compare and contrast the processes of osmosis and diffusion. [pp.50–51]
9. Define *electric gradient* and *pressure gradient*. [p.50]
10. Two fluids are said to be _____ when solute concentrations are equal on both sides of a cell membrane; when solute concentrations are unequal, one fluid is _____ (fewer solutes) and the other is _____ (more solutes). [p.51]
11. Describe the mechanisms involved in active and passive transport (facilitated diffusion) across the cell membrane. [p.52]
12. Explain the role of vesicles during exocytosis and endocytosis. [p.52]
13. _____ physically separate incompatible cellular chemical reactions in time and cytoplasmic space. [p.44]
14. Give the functions of the following basic eukaryotic organelles and structures: nucleus, nucleolus, nuclear envelope, chromosomes, chromatin, endoplasmic reticulum (rough and smooth), Golgi bodies, lysosomes, peroxisomes, vesicles, mitochondria, ribosomes, cytoskeleton, microtubules, microfilaments, intermediate filaments, flagellum, cilium, centrioles, and basal body. [pp.44–55]
15. The outermost part of the nucleus, the nuclear _____, has two lipid bilayers. [p.54]
16. Assembly of protein and RNA subunits of ribosomes occurs in the _____ . [p.54]
17. A DNA molecule and its associated proteins is a(n) _____ . [p.42]
18. Explain how the ER, Golgi bodies, and certain vesicles function as the cytomembrane skeleton. [p.48]
19. _____ are enzyme sacs that break down fatty acids and amino acids. [p.49]
20. _____ are the ATP-producing powerhouses of eukaryotic cells. [p.55]
21. The _____ gives cells their shape and internal organization. [pp.44,46]
22. Dynamic cellular chemical activity is called _____ . [p.56]
23. Distinguish between biosynthetic metabolic pathways (anabolism) and degradative metabolic pathways (catabolism). [p.56]
24. Define and state the role of each of the following participants in a metabolic pathway: *substrate, intermediate, enzyme, cofactor, coenzyme, energy carrier,* and *end products.* [pp.56–57]
25. _____ are molecules on which enzymes act. [p.56]
26. Substrates interact with enzymes on the _____ sites of enzymes. [p.56]
27. The main energy carrier in cells is _____ . [pp.41,56]

28. The mechanism whereby four ATP molecules are formed in glycolysis is _____ _____ phosphorylation. [p.58]
29. The actual ATP-making mechanism of the electron transport system is known as _____ . [p.60]
30. Describe how ATP is produced with a pumping mechanism across the inner membrane of the mitochondrion. [p.60]
31. List (in order) and cite the highlights of the three major stages of aerobic respiration. [pp.58–61]
32. Describe how alternative energy sources such as fats and proteins may enter the energy-releasing pathways. [pp.62–63]

Integrating and Applying Key Concepts

1. If there were no such thing as active transport, how would the lives of organisms be affected?
2. Exactly how does being deprived of oxygen, such as being suffocated with a pillow, kill a person?

4

INTRODUCTION TO TISSUES, ORGAN SYSTEMS, AND HOMEOSTASIS

Interactive Exercises

CHAPTER INTRODUCTION [p.65]

4.1. EPITHELIUM: THE BODY'S COVERING AND LININGS [pp.66–67]

4.2. CONNECTIVE TISSUE: BINDING, SUPPORT, AND OTHER ROLES [pp.72–73]

4.3. MUSCLE TISSUE: MOVEMENT [p.74]

4.4. NERVOUS TISSUE: COMMUNICATION [p.75]

4.5. *Science Comes to Life:* **THE HUMAN BODY SHOP: HOPE AND CONTROVERSY**
[pp.72–73]

Selected Words: simple epithelium [p.66], stratified epithelium [p.66], *pseudostratified* epithelium [p.66], *squamous epithelium* [p.66], *cuboidal epithelium* [p.66], *columnar epithelium* [p.66], "ground substance" [p.68], *hyaline cartilage* [p.69], *elastic cartilage* [p.69], *fibrocartilage* [p.69], *striated* [p.70], "involuntary" [p.70], "bionic" [p.72], "bioartificial" [p.72], "seeding" [p.72], "bioreactor" [p.73], "organ donor" critters [p.73]

Boldfaced, Page-Referenced Terms

[p.65] tissue _____

_____

[p.65] organ _____

_____

[p.65] organ system _____

_____

[p.65] internal environment _____

_____

[p.66] epithelium _____

_____

[p.66] microvilli (singular: microvillus) _____

_____

[p.66] basement membrane _____

_____

[p.66] gland _____

_____

[p.66] exocrine glands _____

_____

[p.67] endocrine glands _____

_____

[p.68] connective tissue _____

_____

[p.68] loose connective tissue _____

_____

[p.68] dense, irregular connective tissue _____

_____

[p.68] dense, regular connective tissue _____

_____

[p.68] cartilage _____

_____

[p.69] bone tissue _____

_____

[p.69] adipose tissue _____

_____

[p.69] blood _____

_____

[p.70] muscle tissue ——

——

[p.70] skeletal muscle ——

——

[p.70] smooth muscle ——

——

[p.70] cardiac muscle ——

——

[p.71] nervous tissue ——

——

[p.71] neurons ——

——

[p.71] nerve ——

——

[p.71] neuroglia ——

——

Fill-in-the-Blanks

In living things, cells working together produce a (1) ———————— [p.65]. Different tissues function together in a(n) (2) ———————— [p.65], and organs work together as an (3) ———————— ———————— [p.65]. A group of cells and intercellular substances interacting to perform one or more functions is called a(n) (4) ———————— [p.65]. The four basic types include (5) ———————— [p.65], (6) ———————— [p.65], (7) ———————— [p.65], and (8) ———————— [p.65]. Any combination of these tissue types in specific and predictable proportions and patterns makes up a(n) (9) ———————— [p.65]. Two or more organs interacting to perform a common task comprise a(n) (10) ———————— [p.65] system. The blood and fluids that bathe body cells are collectively referred to as the body's (11) ———————— ———————— [p.65]. Organ systems work together to help maintain (12) ———————— [p.65] — stable operating conditions in the internal environment.

(13) ———————— [p.66] tissue has a free surface, which faces either a body cavity or the outside environment. The other surface adheres to a(n) (14) ———————— [p.66] membrane, a noncellular layer packed with proteins and polysaccharides. The most abundant and widely distributed of all body tissues are (15) ———————— [p.68] tissues, comprised of two categories: (16) ———————— ———————— [p.68] tissue and (17) ———————— ———————— [p.68] tissue. (18) ———————— [p.70] tissues are involved in moving the body through the environment; they maintain and change positions of body parts. The three types of muscle tissue include (19) ———————— [p.70], (20) ———————— [p.70], and (21) ———————— [p.70]. The basic tissue type that has the most control over the body's ability to respond to changing conditions is (22) ———————— [p.71] tissue, which is made up of tens of thousands of (23) ———————— [p.71].

Complete the Table

24. Complete the following table to associate the major types of epithelium (use terms for layer number and shapes, such as stratified cuboidal) with sites where they may be found in the human body. [p.66]

Epithelium Type	Typical Location in the Human Body
a.	Male urethra, ducts of salivary glands
b.	Glands and their ducts, surface of ovaries, pigmented epithelium of eye
c.	Ducts of sweat glands
d.	Throat, nasal passages, sinuses, trachea, male genital ducts
e.	Linings of blood vessels, lung alveoli (sites of gas exchange)
f.	Skin (keratinized), mouth, throat, esophagus, vagina (nonkeratinized)
g.	Stomach, intestines, uterus

Fill-in-the-Blanks

(25) ——————— [p.66] glands secrete substances onto an epithelial surface through ducts or tubes. Examples of exocrine gland secretions include (26) ——————— [p.66], (27) ——————— [p.66], (28) ——————— [p.66], (29) ——————— [p.66], (30) ——————— [p.66], and (31) ——————— [p.66]. (32) ——————— [p.67] glands do not secrete substances through tubes or ducts. Their products, (33) ——————— [p.67], are secreted directly into the extracellular fluid bathing the glands. Examples include (34) ——————— [p.67], (35) ——————— [p.67], and (36) ——————— [p.67] glands.

Choice

For questions 37–42, choose from the following:

    a. dense, regular connective tissue   b. loose connective tissue   c. dense, irregular connective tissue

37. _____ Often supports epithelia [p.68]
38. _____ Consists of fibers oriented every which way, mostly collagen, and fibroblasts [p.68]
39. _____ Parallel collagen fibers in tendons and ligaments that resist tearing and allow joint movement [p.68]
40. _____ Includes loosely arranged fibers, fibroblasts, and infection-fighting white blood cells [p.68]
41. _____ Connective tissue that composes elastic ligaments that attach bone to bone, and collagen-containing tendons, which attach muscle to bone [p.68]
42. _____ Connective tissue that forms protective capsules around organs [p.68]

Choice

For questions 43–49, choose from the following:

a. hyaline cartilage   b. elastic cartilage   c. fibrocartilage

43. _____ The most common type of cartilage [p.69]

44. _____ Forms cartilage "cushions" in joints such as the knee and in the disks that separate vertebrae of the spinal column [p.69]

45. _____ Serves as the forerunner of bone in the developing skeleton of an embryo [p.69]

46. _____ Provides a friction-reducing cover at the ends of freely movable mature bones where they articulate in joints [p.69]

47. _____ Occurs in places where a flexible yet rigid structure is required; has collagen and elastic fibers [p.69]

48. _____ Packed with collagen fibers arranged in thick bundles to withstand tremendous pressure [p.69]

49. _____ Type of cartilage found in the ear and the epiglottis [p.69]

Choice

For questions 50–56, choose from the following:

a. skeletal   b. smooth   c. cardiac

50. _____ Found in the walls of blood vessels, the stomach, and the intestines [p.70]

51. _____ Found only in the heart [p.70]

52. _____ Found in muscles attached to bones [p.70]

53. _____ Bundled groups of fascicles [p.70]

54. _____ Contractile cells tapered at both ends [p.70]

55. _____ Communication junctions allow contraction as a unit [p.70]

56. _____ Typically striated [p.70]

Choice

For questions 57–60, choose from the following:

a. neurons   b. neuroglia   c. nerve

57. _____ Cells that are organized as communication lines that extend through the body [p.71]

58. _____ A cluster of processes from several neurons that conduct messages between the central nervous system (CNS) and muscles and glands [p.71]

59. _____ Accessory cells located around neurons; provide physical support and insulation [p.71]

60. _____ Possess cell bodies and two types of processes called dendrites and axons [p.71]

Short Answer

61. Distinguish between "bionic" and "bioartificial" organs. [p.72] _____

_____

_____

_____

62. Explain how cartilage and bone tissues are grown in a desired shape. [p.72] _____

_____

_____

63. Why are stem cells from early embryos valued more in research than are adult stem cells? [p.73] _____

_____

_____

_____

4.6. CELL JUNCTIONS: HOLDING TISSUES TOGETHER [p.74]

4.7. MEMBRANES: THIN, SHEETLIKE COVERS [p.75]

Selected Words: epithelial membranes [p.75], *connective tissue membranes* [p.75], *mucous membranes* [p.75], *serous membranes* [p.75], *cutaneous membranes* [p.75], *synovial membranes* [p.75]

Boldfaced, Page-Referenced Terms

[p.74] tight junctions _____

_____

[p.74] adhering junctions _____

_____

[p.74] gap junctions _____

_____

Labeling-Matching

Most epithelial cells and cells of other tissues adhere strongly to one another by means of specialized attachment sites. Identify each example of cell junctions by entering the correct name in the blank below each sketch. Complete the exercise by matching and entering the letter of the correct description in the parentheses following each label.

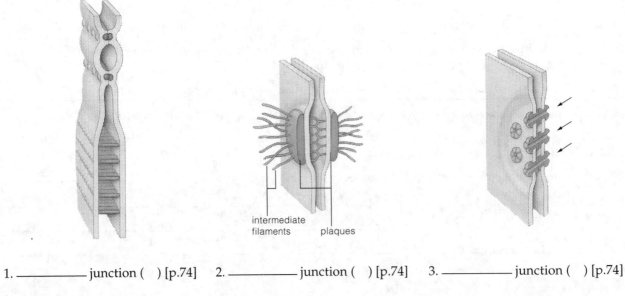

intermediate
filaments   plaques

1. _____ junction (   ) [p.74]   2. _____ junction (   ) [p.74]   3. _____ junction (   ) [p.74]

A. Spot-weld junctions at the plasma membranes of two adjacent cells. They are anchored to the cytoskeleton in each cell and help hold cells together in tissues subject to stretching, such as epithelium of the skin and stomach.
B. Strands of protein forming gasketlike seals that help stop substances from leaking across a tissue. This allows epithelial cells to control what enters the body.
C. Helps cells communicate by promoting the rapid transfer of ions and small molecules through channels directly linking the cytoplasm of adjacent cells.

Complete the Table

4. Complete the following table by filling in the type of membrane to which each phrase refers. [p.75]

Membrane Type	Description
a.	Occur in paired sheets separated by a thin film of fluid; help anchor organs and provide a lubricated surface to prevent abrasion of organs
b.	Line tendon sheaths and cavities around some joints; no epithelial cells present
c.	Hardy, dry membrane; part of the integumentary system
d.	Pink, moist membranes; line the body's tubes and cavities; most have glands that secrete substances such as mucus

Matching

Match the types of membranes with the locations in which they may be found.

5. _____ lining the sheaths of tendons [p.75]

6. _____ found where absorption and glandular secretion occur, such as lining the stomach [p.75]

7. _____ lining capsulelike cavities such as around the knee joint [p.75]

8. _____ covering the outside of the body [p.75]

9. _____ lining the digestive, respiratory, urinary, and reproductive tracts [p.75]

10. _____ enclosing organs such as the heart and lungs [p.75]

A. serous
B. cutaneous
C. synovial
D. mucous

4.8. ORGAN SYSTEMS [p.76]

4.9. THE SKIN: EXAMPLE OF AN ORGAN SYSTEM [pp.78–79]

4.10. *Focus on our Environment:* SUN, SKIN, AND THE OZONE LAYER [p.79]

Selected Words: ectoderm [p.77], mesoderm [p.77], endoderm [p.77], *Langerhans cells* [p.78], *Granstein cells* [p.79], *sebaceous glands* [p.79], *acne* [p.79], *herpes simplex* [p.79]

Boldfaced, Page-Referenced Terms

[p.77] cranial cavity _____

_____

[p.77] spinal cavity _____

_____

[p.77] thoracic cavity _____

_____

[p.77] abdominal cavity _____

_____

[p.77] pelvic cavity _____

_____

[p.78] integument _____

_____

[p.78] epidermis _____

_____

[p.78] dermis _____

_____

[p.78] keratinocytes _____

_____

[p.78] melanocytes _____

_____

[p.79] hair _____

_____

Matching

Match the most appropriate function with the letter for each system shown on the accompanying illustrations.

1. _____ Female: production of eggs; provision of a protected nutritive environment for development. Male: production and transfer of sperm to the female. Both systems have hormonal influences on other organs. [p.77]

2. _____ Ingestion of food and water; breakdown and absorption of food molecules; elimination of food residues from the body. [p.77]

3. _____ Movement of body and its internal parts; maintenance of posture; generation of heat. [p.76]

4. _____ Detection of external and internal stimuli; control and coordination of responses to stimuli; integration of activities of all organ systems. [p.76]

5. _____ Protection from injury and dehydration; some defense against pathogens; body temperature control; excretion of some wastes; reception of external stimuli. [p.76]

6. _____ Delivery of oxygen to tissue fluids; removal of carbon dioxide wastes produced by cells; regulation of pH of body fluids. [p.77]

7. _____ Support and protect body parts; sites for muscle attachment, red blood cell production, and calcium and phosphorus storage. [p.76]

8. _____ Hormonal control of body functioning; works with nervous system in integrating short-term and long-term activities. [p.76]

9. _____ Maintenance of volume and composition of fluids in the internal environment; excretion of excess fluid and blood-borne wastes. [p.77]

10. _____ Rapid internal transport of many materials to and from cells; helps stabilize internal temperature and pH. [p.76]

11. _____ Return of some tissue fluid to blood; roles in immunity (defense against infection and tissue damage). [p.77]

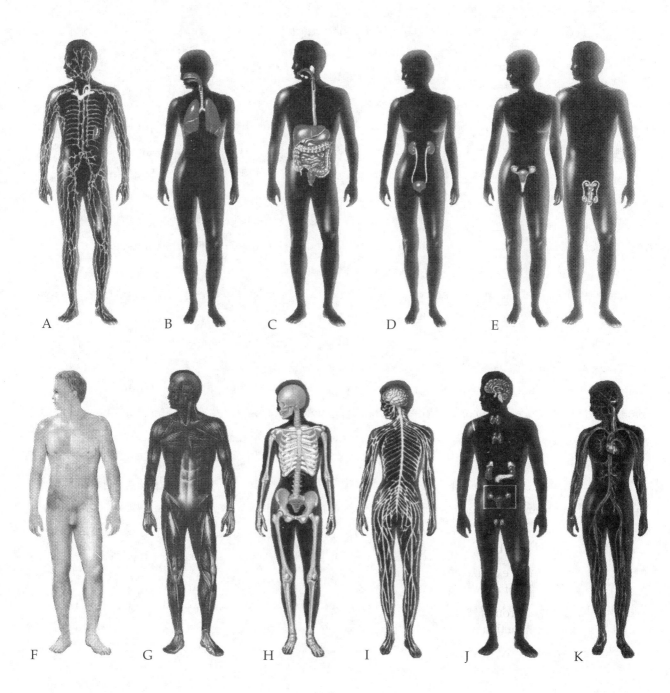

A   B   C   D   E

F   G   H   I   J   K

Fill-in-the-Blanks

The (12) _____ [p.77] cavity and the (13) _____ [p.77] cavity house the central nervous system — the brain and spinal cord. The heart and lungs are located in the (14) _____ [p.77] cavity. A muscular diaphragm separates the thoracic cavity from the (15) _____ [p.77] cavity, which holds the stomach, liver, most of the intestine, and other organs. Reproductive organs, bladder, and rectum are located within the (16) _____ [p.77] cavity. The (17) _____ [p.77] plane divides the body into right and left halves. The (18) _____ [p.77] plane divides the body into (19) _____ [p.81] (front) and posterior (back) parts. The (20) _____ [p.77] plane divides it into superior (upper) and (21) _____ [p.77] (lower) parts.

Skin is a covering that maintains its shape, blocks harmful rays from the (22) _____ [p.78], kills many (23) _____ [p.78] on contact, holds in (24) _____ [p.78], fixes small cuts and burns, and lasts as long as its owner. It also helps control the internal body (25) _____ [p.78]. Signals from the skin's (26) _____ [p.78] receptors help the brain assess what's going on in the outside world. Skin produces vitamin (27) _____ [p.82] or cholecalciferol required for calcium absorption in the intestine. Skin is part of the body's (28) _____ [p.78] and is the body's largest (29) _____ [p.78]. Skin has two distinct regions — an outer (30) _____ [p.78] and an underlying (31) _____ [p.78] made of connective tissue. Beneath that is a subcutaneous layer, the (32) _____ [p.78], a loose connective tissue that anchors the skin and yet allows it some freedom of movement. (33) _____ [p.78] stored in the hypodermis insulates the body and cushions some body parts. Epidermis consists of layers; it is a(n) (34) _____ _____ _____ [p.78], consisting of cells knitted together by cell junctions. Most cells of the epidermis are keratinocytes that make (35) _____ [p.78], a tough, water-insoluble protein. The outermost layer of skin is the tough, waterproof (36) _____ _____ [p.78], which consists of dead keratinocytes that eventually abrade off or flake away. In the deepest epidermal layer, cells called (37) _____ [p.78] produce the brown-black pigment called melanin. This is transferred to keratinocytes, forming a shield against harmful (38) _____ [p.82] radiation. All humans have the same number of melanocytes, but (39) _____ _____ [p.78] varies due to differences in distribution and activity. The epidermis also contains two types of cells involved in immunity. They are the phagocytic (40) _____ [p.78] cells and the (41) _____ [p.79] cells. Dense connective tissue makes up most of the (42) _____ [p.79]. Sweat glands, oil glands, and the husklike structures called hair (43) _____ [p.79] reside mostly in the dermis, even though they are derived from epidermal tissue. Oil glands known as (44) _____ [p.79] glands soften and lubricate the hair and skin.

(45) _____ [p.79] results from increased melanin levels in the epidermis and provides some protection from UV radiation, but after many years, tanning causes (46) _____ [p.79] fibers in the dermis to clump together. The skin loses resiliency and begins to look leathery. Excessive exposure to UV

radiation also suppresses the (47) _____ [p.79] system. UV radiation from sunlight or from the lamps of tanning salons can activate (48) _____ _____ [p.79] in the skin. These bits of DNA can trigger (49) _____ [p.79]. With loss of the ozone layer in the stratosphere, the rate of skin cancers now is rapidly (50) _____ [p.79]. Experts estimate that even if all ozone-depleting substances were banned tomorrow, it would take about (51) _____ [p.79] years for the planet to recover.

Labeling

Label the numbered parts of the accompanying illustration. [p.82]

52. _____ _____
53. _____ _____
54. _____
55. _____ _____
56. _____ _____
57. _____ _____
58. _____ _____
59. _____ _____
60. _____ _____
61. _____
62. _____
63. _____

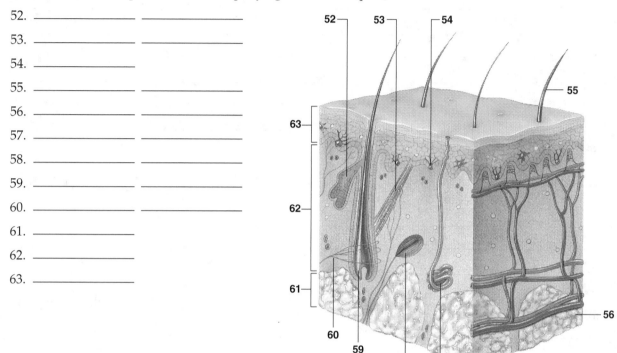

4.11. HOMEOSTASIS: THE BODY IN BALANCE [pp.80-81]

Selected Words: "internal environment" [p.80], *interstitial fluid* [p.80], "set points" [p.80]

Boldfaced, Page-Referenced Terms

[p.80] extracellular fluid _____

_____

[p.80] homeostasis _____

_____

[p.80] sensory receptors _____

_____

[p.80] stimulus _____

_____

[p.80] integrator _____

_____

[p.80] effectors _____

_____

[p.80] negative feedback _____

_____

[p.81] positive feedback _____

_____

Fill-in-the-Blanks

(1) _____ [p.80] fluid is the fluid outside of cells. Much of this fluid is (2) _____ [p.80], meaning it occupies spaces between cells and tissues. Substances move between the interstitial fluid and the (3) _____ [p.80] that it bathes as the cells draw nutrients from it and dump metabolic wastes into it. Drastic changes in its composition can have drastic effects on cell activities. The type and number of (4) _____ [p.80] are especially crucial. (5) _____ [p.80] means "staying the same." Homeostatic mechanisms operate to maintain (6) _____ [p.80] in the volume and composition of extracellular fluid.

Sensory (7) _____ [p.80] are cells or cell parts that can detect a(n) (8) _____ [p.80], a specific change in the environment. Your brain is a(n) (9) _____ [p.80], a control point where different bits of information are pulled together in selecting a response. It can send signals to muscles and glands (or both). Muscles and glands are (10) _____ [p.80]— they carry out the response. When information from receptors shows that conditions have deviated from (11) "_____ _____ " [p.80], the brain functions to bring conditions back within proper operating range.

(12) _____ [p.80] mechanisms are controls that help keep physical and chemical aspects of the body within tolerable ranges. In (13) _____ [p.80] feedback, an activity alters a condition in the internal environment, and this triggers a response that reverses the altered condition. An example is a furnace with a thermostat or the body temperature of humans being maintained within a normal range. In (14) _____ [p.80] feedback, a chain of events is set in motion that intensifies a change from an original condition; childbirth is an example.

Labeling-Matching

Identify the numbered items on the diagram of components necessary for negative feedback. Enter your choices in the numbered blanks. Choose from *effector, receptor, stimulus, integrator,* and *response.* Complete the exercise by matching the lettered examples to each component. [p.80]

15. _____ ( )

16. _____ ( )

17. _____ ( )

18. _____ ( )

19. _____ ( )

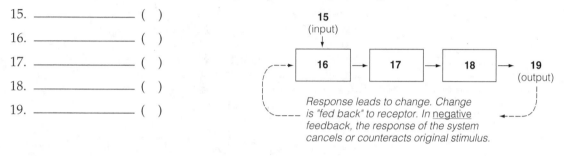

A. Brain
B. Excessively salty food
C. Muscles related to the function of the jaw, esophagus, and stomach
D. Regurgitate food
E. Taste buds

Choice

For questions 20–26, choose from the following:

a. negative feedback mechanisms   b. positive feedback mechanisms

20. _____ Set in motion a chain of events that intensify a change from an original condition [p.81]

21. _____ A furnace with a thermostat [p.80]

22. _____ After a limited time, the intensification reverses the change [p.81]

23. _____ Childbirth [p.81]

24. _____ Maintaining body temperature in a normal range [p.80]

25. _____ Associated with instability in a system [p.81]

26. _____ An activity alters a condition in the internal environment, and this triggers a response that reverses the altered condition [p.80]

Self-Quiz

Labeling-Matching

Identify each of the following illustrations by labeling it with one of the following: *connective, epithelial, muscle,* or *nervous*. Complete the exercise by matching and entering the capital letter from the first group of choices in the first set of parentheses and a matching lowercase letter from the second group of choices in the set of second parentheses following each label.

1. _____ ( ) ( ) [pp.68–69]
2. _____ ( ) ( ) [pp.66–67]
3. _____ ( ) ( ) [p.70]
4. _____ ( ) ( ) [p.70]
5. _____ ( ) ( ) [pp.68–69]
6. _____ ( ) ( ) [pp.68–69]
7. _____ ( ) ( ) [pp.68–69]
8. _____ ( ) ( ) [pp.66–67]
9. _____ ( ) ( ) [pp.68–69]
10. _____ ( ) ( ) [p.71]
11. _____ ( ) ( ) [p.70]
12. _____ ( ) ( ) [pp.66–67]
13. _____ ( ) ( ) [pp.68–69]

A. Adipose
B. Bone
C. Cardiac
D. Dense, regular
E. Loose
F. Simple columnar
G. Simple cuboidal
H. Simple squamous
I. Smooth
J. Skeletal
K. Blood
L. Cartilage
M. Neurons

a. Pumps circulatory blood; involuntary and striated
b. Single layer of cubelike cells that function in secretion and absorption
c. Communication by means of electrochemical signals
d. Fibers loosely arranged; cells include fibroblasts; functions in elasticity and diffusion
e. Contract to propel substances along internal pathways; involuntary
f. Transport of nutrients and waste products to and from cells
g. Attaches muscle to bone and bone to bone
h. Large, densely clustered cells specialized for fat storage; insulation and padding
i. Chondroblasts found inside solid matrix; hyaline, elastic, and fibrocartilage types
j. A single layer of flattened cells; found in blood vessel walls and lung air sacs; functions in diffusion
k. Provides support, protection, and attachment sites for skeletal muscles
l. Contraction for voluntary movements; striated and voluntary
m. Single layer of tall cells that may be ciliated; found in parts of the gut, uterus, and respiratory linings; function in secretion and absorption

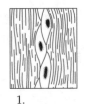

2.

1.

3.

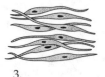

4.

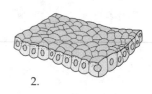

5.

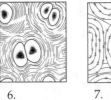

6.

7.

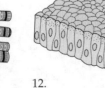

8.

lymphocyte
platelets
neutrophils —
erythrocytes

9.

dendrite

axon

10.

11.

12.

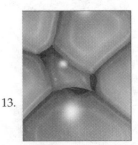

13.

Multiple Choice

_____ 14. Which of the following is not a connective tissue? [pp.68–69]
   a. bone
   b. adipose
   c. cartilage
   d. skeletal muscle

_____ 15. All connective tissues contain cells separated by _____ . [p.68]
   a. muscle and nerves
   b. fluid and proteins
   c. fibers and ground substance
   d. blood

_____ 16. Blood is considered to be a(n) _____ tissue. [p.69]
   a. epithelial
   b. muscular
   c. connective
   d. none of these

_____ 17. Which group is arranged correctly from smallest structure to largest? [p.70]
   a. Muscle cells, muscle bundle, muscle
   b. Muscle cells, muscle, muscle bundle
   c. Muscle bundle, muscle cells, muscle
   d. None of the above

_____ 18. Involuntary muscle consisting of tapered cells is _____ . [p.70]
   a. cardiac
   b. skeletal
   c. striated
   d. smooth

_____ 19. An organized group of cells and intercellular substances that take part in one or more tasks is a(n) _____ . [p.65]
   a. organ
   b. organ system
   c. tissue
   d. cuticle

_____ 20. A tissue whose cells are striated and bound by specialized junctions that allow many cells to contract as a single unit is _____ tissue. [p.70]
   a. smooth muscle
   b. dense fibrous connective
   c. supportive connective
   d. cardiac muscle

_____ 21. Most cells of the epidermis produce the protein _____ ; the dermis contains the fibrous proteins _____ . [p.78]
   a. melanin; keratin and collagen
   b. keratin; collagen and elastin
   c. melanin; keratin and hemoglobin
   d. keratin; hemoglobin and melanin

_____ 22. Which of the following is _not_ associated with a negative feedback mechanism? [pp.80–81]
   a. Maintains tolerable ranges of physical and chemical aspects of the body.
   b. It is similar to the function of a thermostat and furnace.
   c. A chain of events is set in motion that intensifies a change from an original condition.
   d. Mechanisms that keep the body from overheating.

Chapter Objectives/Review Questions

This section lists general and detailed chapter objectives that can be used as review questions. You can make maximum use of these items by writing answers on a separate sheet of paper. To check for accuracy, compare your answers with the information given in the chapter or glossary.

1. Cells are the basic units of life; in a multicellular animal, similar interacting cells and their intercellular substances are grouped into a(n) —————— . [p.65]
2. State the structural relationships among cells, tissues, organs, and organ systems. [p.65]
3. —————— always has a free surface, which faces a body cavity or the outside environment. [p.66]
4. —————— glands have no ducts; the thyroid gland is an example. [p.67]
5. Describe the general characteristics of connective tissue and explain how they enable connective tissue to carry out its various tasks. [p.68]
6. Describe the structural differences and functions of loose connective tissue and dense, irregular connective tissue. [p.68]
7. —————— —————— connective tissue is the type found in ligaments and tendons. [p.68]
8. The most common type of cartilage in the human body is —————— cartilage. [p.69]
9. Cartilage in the human ear is —————— cartilage. [p.69]
10. Cartilage that cushions the knee and other joints is —————— . [p.69]
11. The specialized function of cells in adipose tissue is for —————— storage. [p.69]
12. Distinguish among skeletal, cardiac, and smooth muscle in terms of location, structure, and function. [p.70]
13. Muscle tissues contain cells that —————— when they receive outside stimulation. [p.70]
14. Neurons can transmit nerve impulses and thus serve as lines of —————— . [p.71]
15. Neuron branch processes called —————— pick up incoming chemical messages; others called —————— conduct outgoing messages. [p.71]
16. A(n) —————— is a cluster of processes from several neurons. [p.71]
17. State the general functions of tight junctions, adhering junctions, omit, and gap junctions. [p.74]
18. Various types of thin, sheetlike —————— cover or line organs and may be of the mucous or serous type. [p.75]
19. List the cavities of the human body and generally name the organs found in each. [pp.76–77]
20. List each of the eleven principal organ systems in humans and match each to its main task. [pp.76–77]
21. Explain how keratin in the epidermis protects the rest of your body. [p.78]
22. Describe the two-layered structure of human skin and identify the structures located in each layer. [pp.78–79]
23. Describe the ways by which extracellular fluid helps cells survive. [p.80]
24. Describe the relationships among sensory receptors, integrators, and effectors in a negative feedback system. [pp.80–81]
25. Describe a positive feedback system and give an example. [p.81]

Integrating and Applying Key Concepts

The condition known as osteoporosis primarily affects postmenopausal women. It involves a reduction in the deposition of calcium into the body's bones. Explain how this affects the functions of bones.

5

THE SKELETAL SYSTEM

Interactive Exercises

CHAPTER INTRODUCTION [p.85]

5.1. BONE — MINERALIZED CONNECTIVE TISSUE [pp.86–87]

5.2. THE SKELETON: THE BODY'S BONY FRAMEWORK [pp.88–89]

Selected Words: osteoblasts [p.86], lacunae [p.86], *osteoclasts* [p.87], *osteoporosis* [p.87]

Boldfaced, Page-Referenced Terms

[p.85] skeletal system _____

_____

[p.86] osteocytes _____

_____

[p.86] compact bone _____

_____

[p.86] osteon _____

_____

[p.86] spongy bone _____

_____

[p.86] epiphyseal plate _____

_____

[p.87] bone remodeling _____

_____

[p.88] bone marrow _____

_____

[p.88] axial skeleton _____

_____

[p.88] appendicular skeleton _____

_____

[p.88] ligaments _____

_____

[p.88] tendons _____

_____

Fill-in-the-Blanks

Bone is a blend of (1) _____ [p.86] cells and (2) _____ [p.86] minerals. There are two kinds of bone tissue — (3) _____ [p.86] bone, which looks solid and smooth, and (4) _____ [p.86] bone, which looks lacy and delicate. During embryonic development, bone-forming cells called (5) _____ [p.86] secrete matrix that will eventually become mineralized into bone. In the process, these cells are trapped inside the matrix, where they become mature living bone cells called (6) _____ [p.86]. As a human grows, long bones elongate at cartilaginous regions at each end called (7) _____ _____ [p.86]. When growth stops, (8) _____ [p.86] replaces the cartilage plates. Over a lifetime, bones change as minerals, such as (9) _____ [p.87] and (10) _____ [p.87], are constantly being deposited and withdrawn, a process known as (11) _____ _____ [p.87]. During this process, cells called (12) _____ [p.87] deposit bone, while cells called (13) _____ [p.87] break it down. Extreme decreases in bone density result in (14) _____ [p.87], particularly among older women.

Cavities within some bones contain (15) _____ _____ [p.88], which is involved in the production of (16) _____ _____ [p.88]. The complete human skeleton consists of (17) _____ [p.88] bones. Bones are connected at joints by (18) _____ [p.88]. Stronger (19) _____ [p.88] attach muscles to bones. Bone tissue serves as a "pantry" for (20) _____ [p.88] and (21) _____, [p.88] which may be deposited or withdrawn to support metabolic activities involving these minerals.

Labeling

Identify each indicated part of the accompanying illustrations. [pp.86,88]

22. _____ _____

23. _____ _____

24. _____ _____

25. _____ _____

26. _____ _____ _____

27. _____

28. _____ _____

29. _____

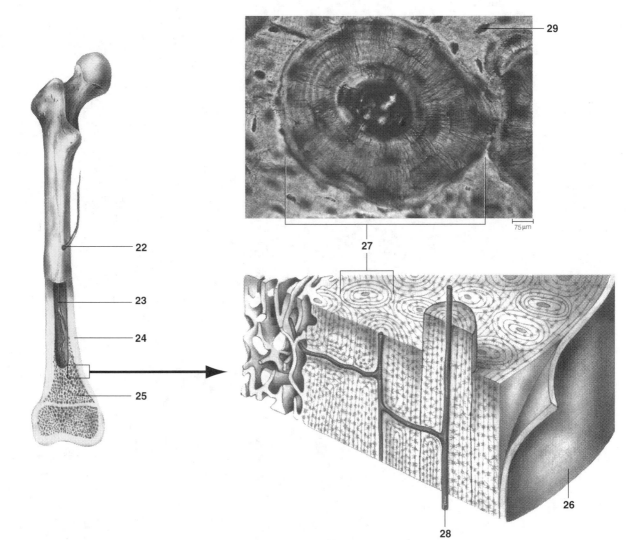

Identification

Write the scientific name of the numbered bones on this page in the blanks provided. [p.93]

30. _____ _____

31. _____

32. _____

33. _____

34. _____

35.. _____

36. _____

37. _____

38. _____

39. _____

5.3. THE AXIAL SKELETON [pp.90–91]

5.4. THE APPENDICULAR SKELETON [pp.92–93]

Selected Words: *frontal bone* [p.90], *sinusitis* [p.90], *temporal bones* [p.90], *sphenoid bone* [p.90], *ethmoid bone* [p.90], *parietal bones* [p.90], *occipital bone* [p.90], *foramen magnum* [p.90], *maxillary bones* [p.90], *zygomatic bones* [p.90], *lacrimal bone* [p.90], *palatine bones* [p.90], *vomer bone* [p.90], *cervical* vertebrae [p.91], *thoracic* vertebrae [p.91], *lumbar* vertebrae [p.91], sacrum [p.91], coccyx [p.91], *herniate* [p.91], *carpal* [p.92], *carpal tunnel syndrome* [p.92], *metacarpals* [p.92], *phalanges* [p.92], *coxal bones* [p.93], *pubic arch* [p.93], *iliac* [p.93], *pelvis* [p.93], *tibia* [p.93], *fibula* [p.93], *patella* [p.93], *tarsal* [p.93], *metatarsals* [p.93]

Boldfaced, Page-Referenced Terms

[p.90] brain case _____

_____

[p.90] sinuses _____

_____

[p.90] mandible _____

_____

[p.91] intervertebral disks _____

_____

[p.91] ribs _____

_____

[p.91] sternum _____

_____

[p.92] pectoral girdle _____

_____

[p.92] scapula _____

_____

[p.92] clavicle _____

_____

[p.92] humerus _____

_____

[p.92] radius _____

_____

[p.92] ulna _____

_____

[p.93] pelvic girdle _____

_____

[p.93] femur _____

_____

Identification

1. _____ bone [p.90]

2. _____ bone [p.90]

3. _____ bone [p.90]

4. _____ bone [p.90]

5. _____ [p.90]

6. _____ [p.90]

7. _____ bone [p.90]

8. _____ bone [p.90]

9. _____ _____ [p.90]

10. _____ vertebrae [p.91]

11. _____ vertebrae [p.91]

12. _____ vertebrae [p.91]

13. _____ (vertebrae) [p.91]

14. _____ (vertebrae) [p.91]

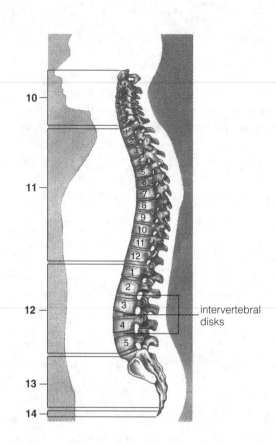

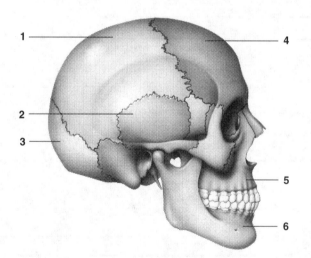

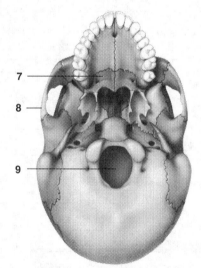

Fill-in-the-Blanks

The (15) _____ [p.90] skeleton includes the skull, vertebral column, ribs, and sternum. The frontal bone (of the skull) contains air spaces called (16) _____ [p.90] lined with mucous membrane. (17) _____ [p.90] bones surround the ear canals. The inner eye socket is formed by both the (18) _____ [p.90] bone and the (19) _____ [p.90] bone. The (20) _____ [p.90] bones form a large part of the skull, as does the (21) _____ [p.90] bone that makes up its back and base. The lower jaw is the (22) _____ [p.90], while the upper jaw is made of two (23) _____ [p.90] bones. The (24) _____ [p.90] bones form the "cheekbones" and outer eye sockets. Tear ducts pass between the (25) _____ [p.90] and (26) _____ [p.90] bones and drain into the nasal cavity. The (27) _____ [p.90] bones form the nasal cavity and the "roof" of the mouth. The nasal cavity is divided into two sections by the thin nasal septum, formed partially by the (28) _____ [p.90] bone.

Typically, humans have seven (29) _____ [p.91] vertebrae, twelve (30) _____ [p.91] vertebrae, five (31) _____ [p.91] vertebrae, a(n) (32) _____ [p.91], and a(n) (33) _____ [p.91]. About one-fourth of the length of the human vertebral column consists of tough, compressible (34) _____ _____ [p.91]. These shock absorbers can "slip out" or (35) _____ [p.91]. The "rib cage" consists of twelve pairs of (36) _____ [p.91] and the paddle-shaped (37) _____ [p.91]. It helps protect internal (38) _____ [p.91] and is important in (39) _____ [p.91].

The humerus is a part of the (40) _____ [p.92] girdle, which comprises a portion of the (41) _____ [p.92] skeleton. The humerus joins the radius and ulna, which join the wristbones known as (42) _____ [p.92]. The bones of the hand include (43) _____ [p.92] and (44) _____ [p.92]. "Hipbones" are the upper iliac regions of the (45) _____ [p.93] bones. The legs include the femur, (46) _____ [p.93], and (47) _____ [p.93]. The ankle is made up of (48) _____ [p.93]. The foot consists of five (49) _____ [p.93] and (50) _____ [p.93] in the toes.

5.5. JOINTS — CONNECTIONS BETWEEN BONES [pp.94–95]
5.6. SKELETAL DISEASES, DISORDERS, AND INJURIES [pp.96–97]
5.7. *Science Comes to Life:* REPLACING JOINTS [p.97]

Selected Words: joints [p.94], *synovial fluid* [p.94], *sutures* [p.94], strain [p.96], *prosthesis* [p.97]

Boldfaced, Page-Referenced Terms

[p.94] synovial joint _____

_____

[p.94] cartilaginous joint _____

_____

[p.94] fibrous joint _____

_____

[p.96] osteoarthritis _____

_____

[p.96] rheumatoid arthritis _____

_____

[p.96] tendinitis _____

_____

[p.96] carpal tunnel syndrome _____

_____

[p.96] sprain _____

_____

Identification

1. _____ [p.95]
2. _____ [p.95]
3. _____ [p.95]
4. _____ [p.95]
5. _____ [p.95]
6. _____ [p.95]
7. _____ [p.95]
8. _____ [p.95]

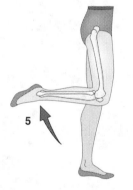
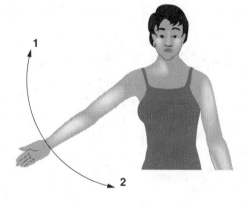
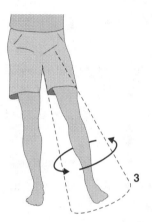
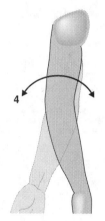

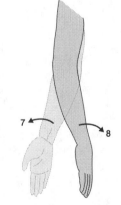

Fill-in-the-Blanks

(9) _____ [p.94] joints are freely movable and are lubricated by (10) _____ [p.94] fluid secreted into a capsule made of dense connective tissue that surrounds the bones of the joint. Synovial joints include (11) _____ [p.94] joints such as the knee and elbow, as well as (12) _____ [p.94] joints such as the hip joints. (13) _____ [p.94] joints such as those between vertebrae allow only slight movement. An adult's (14) _____ [p.94] joints, such as those holding teeth in their sockets, don't allow movement.

As a person ages, the (15) _____ [p.96] covering the ends of bones at freely movable joints may wear away, resulting in (16) _____ [p.96]. When a person's immune system attacks tissues in a joint, causing the synovial membrane to thicken and cartilage to erode away, the degenerative condition (17) _____ _____ [p.96] results. (18) _____ [p.96] develops when tendons and synovial membranes around joints become inflamed due to repeated movements. Tearing a ligament or tendon leads to a (19) _____ [p.96]. When a joint is (20) _____ [p.96], the two bones will no longer be in contact. A (21) _____ [p.96–97] fracture is just a crack in a bone. In a (22) _____ [p.97] fracture, the bone is completely separated into two pieces. A (23) _____ [p.97] fracture involves broken ends of bone puncturing the skin.

When a joint such as a hip or knee is seriously damaged, it can be replaced with an artificial joint called a(n) (24) _____ [p.97].

Self-Quiz

Choice

For questions 1–10, choose from the following answers:

      a. carpals   b. metatarsals   c. femur   d. humerus   e. ulna   f. tarsals   g. radius
           h. fibula   i. phalanges (used twice)   j. metacarpals   k. tibia

The upper arm bone (1. _____ [p.92]) is connected to the lower arm bones (2. _____ and _____ [p.92]). The lower arm bones are connected to the wristbones (3. _____ [p.92]). The wristbones are connected to the hand bones (4. _____ [p.92]). The hand bones are connected to the finger bones (5. _____ [p.92]).

The upper leg bone (6. _____ [p.93]) is connected to the lower leg bones (7. _____ and _____ [p.93]). The lower leg bones are connected to the anklebones (8. _____ [p.93]). The anklebones are connected to the foot bones (9. _____ [p.93]). The foot bones are connected to the toe bones (10. _____ [p.93]).

Choice

For questions 11–15, choose from the following answers:

a. ligaments   b. osteoblasts   c. osteoclasts   d. bone marrow   e. tendons   f. osteon

11. _____ Secrete bone-dissolving enzymes [p.87]

12. _____ Major site of blood cell formation [p.88]

13. _____ Secrete matrix that will become bone [p.86]

14. _____ Compact bone surrounding central canal [p.86]

15. _____ Attach bone to bone [p.88]

For questions 16–19, choose from the following answers:

a. synovial   b. cartilaginous   c. fibrous

16. _____ Nonmoving joints (sutures) between skull bones [p.94]

17. _____ Hips and shoulders [p.94]

18. _____ Knees and elbows [p.94]

19. _____ Connections of some ribs to the breastbone [p.94]

Chapter Objectives/Review Questions

This section lists general and detailed chapter objectives that can be used as review questions. You can make maximum use of these items by writing answers on a separate sheet of paper. To check for accuracy, compare your answers with information given in the chapter or glossary.

1. Explain the various roles of osteoblasts, osteoclasts, cartilage models, long bones, and epiphyseal plates in the development of human bones. [pp.86–87]
2. Identify the human bones by name and location, including the bones of the skull, the rib cage, the vertebral column, the pectoral girdle and upper appendages, and the pelvic girdle and lower appendages [pp.89–91]. Which of these are parts of the axial skeleton? Which are parts of the appendicular skeleton?
3. Give an example of each of these synovial joint movements: flexion, extension, circumduction, rotation, abduction, adduction, supination, pronation. [p.95]
4. Explain the difference between tendinitis and a sprain. [p.96]

Integrating and Applying Key Concepts

Explain why a fracture at an epiphyseal plate in a child might be more serious than a fracture in the center of a long bone.

6

THE MUSCULAR SYSTEM

CHAPTER INTRODUCTION

MUSCLES INTERACT WITH THE SKELETON
    Skeletal muscles pull on bones
    Many muscles are arranged as pairs or in groups

HOW MUSCLES CONTRACT
    A muscle contracts when its cells shorten
    Muscle cells shorten when actin filaments slide

ENERGY FOR MUSCLE CELLS

THE NERVOUS SYSTEM CONTROLS MUSCLE
  CONTRACTION
    Calcium ions: The key to contraction
    Neurons act on muscle cells at neuromuscular
      junctions

PROPERTIES OF WHOLE MUSCLES
    Muscle tension is the force skeletal muscles exert
      on bones
    Muscle cells are organized into motor units
    "Fast" and "slow" muscle

Focus on Your Health: MUSCLE MATTERS
    Exercise: Making the most of your muscles
    Bo-toxicating muscle
    Uses — and abuses — of muscle-building
      substances

Interactive Exercises

CHAPTER INTRODUCTION [p.101]

6.1. MUSCLES INTERACT WITH THE SKELETON [pp.102–103]

6.2. HOW MUSCLES CONTRACT [pp.104–105]

Selected Words: "musculoskeletal system" [p.101], "muscle fiber" [p.102], *reciprocal innervation* [p.102]

Boldfaced, Page-Referenced Terms

[p.101] muscular system _____

_____

[p.102] skeletal muscle _____

_____

[p.102] origin _____

_____

[p.102] insertion _____

_____

[p.102] tendon _____

_____

[p.104] sarcomere _____

_____

[p.104] myofibrils _____

_____

[p.104] actin _____

_____

[p.105] myosin _____

_____

[p.105] sliding-filament model _____

_____

[p.105] cross-bridge _____

_____

Fill-in-the-Blanks

Of the three muscle types, only (1) _____ [p.102] is included in the "muscular system." This type of muscle interacts with the (2) _____ [p.102] to move the body or parts of the body. Skeletal muscles also help stabilize (3) _____ [p.102] between bones.

The body has more than (4) _____ [p.102] skeletal muscles. The end of the muscle that is attached to the bone that remains fairly motionless during a movement is the (5) _____ [p.102]. The end that is attached to the bone that moves the most is the (6) _____ [p.102]. When a skeletal muscle contracts, it (7) _____ [p.102] on the bones to which it is attached.

Muscle cells, also called muscle (8) _____ [p.102], are bundled together into groups by (9) _____ [p.102] tissue. A(n) (10) _____ [p.102] is a cord of dense connective tissue that extends from a group of bundled muscle cells and attaches the muscle to a(n) (11) _____ [p.102].

Many muscles are arranged as (12) _____ [p.102] or groups. While some muscles work synergistically, others work in (13) _____ [p.102]. An example would be the muscles of the upper arm. When you bend your arm at the elbow, the biceps muscle (14) _____ [p.102] and the triceps muscle relaxes. When you straighten your arm, the biceps (15) _____ [p.102] and the triceps (16) _____ [p.102].

To accomplish this, one muscle group is stimulated by nerves, while no signals are sent to the opposing group, so it does not contract. This cooperative action results partly from (17) _____ [p.102] by nerves coming from the spinal cord.

Identification

Name each numbered muscle on the accompanying illustration. In the parentheses provided, write the letter of its specific movement. [p.101]

18. _____ ( )
19. _____ ( )
20. _____ _____ ( )
21. _____ _____ ( )
22. _____ ( )
23. _____ _____ ( )
24. _____ _____ ( )
25. _____ ( )
26. _____ _____ ( )
27. _____ _____ ( )

A. Bends lower leg at the knee when walking, extends the foot when jumping
B. Flexes the foot toward the shin
C. Raises the arm
D. Depresses the thoracic cavity, compresses the abdomen, bends the backbone
E. Bends the thigh at the hip, bends lower leg at the knee, rotates thigh in an outward direction
F. Draws the arm forward and in toward the body
G. Extends and rotates thigh outward when walking, running, and climbing
H. Lifts the shoulder blade, braces the shoulder, draws the head back
I. Flexes the thigh at the hip, extends the leg at the knee
J. Draws thigh backward, bends the knee

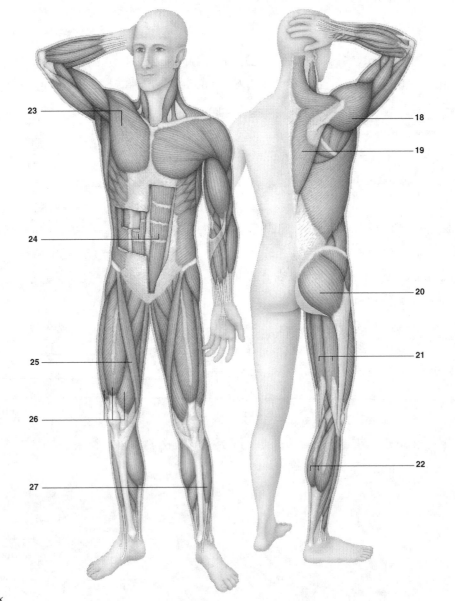

Fill-in-the-Blanks

The basic unit of muscle contraction in skeletal muscle cells is the (28) _____ [p.104]. A muscle cell contains many threadlike (29) _____ [p.104], each consisting of sarcomeres separated by (30) _____ [p.104] bands. Attached to each Z band of a sarcomere are thin filaments made of the protein (31) _____ [p.104]. Within a sarcomere, there are also thick filaments made of bundled fibers of another protein called (32) _____ [p.105]. Double (33) _____ [p.105] from the myosin molecules extend outward from the thick filaments.

The thick and thin filaments interact in the (34) _____ _____ [p.105] mechanism of contraction. Myosin heads form (35) _____ _____ [p.105] with binding sites on actin and pull the thin filaments of the sarcomere inward. This shortens the sarcomeres of each myofibril within a muscle cell. Coordinated shortening of the muscle cells within muscle bundles allows an entire muscle to (36) _____ [p.105].

Labeling

Label the numbered parts of the accompanying illustrations. [p.104]

37. _____ _____
38. _____ _____ _____
39. _____ _____
40. _____
41. _____
42. _____ _____

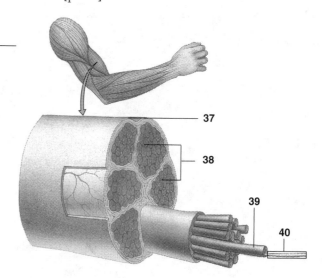

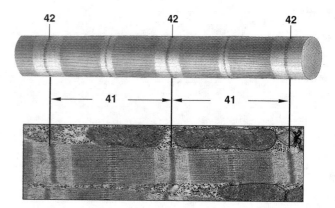

6.3. ENERGY FOR MUSCLE CELLS [p. 106]

6.4. THE NERVOUS SYSTEM CONTROLS MUSCLE CONTRACTION [pp.106–107]

6.5. PROPERTIES OF WHOLE MUSCLES [pp.108–109]

6.6. *Focus on Your Health:* MUSCLE MATTERS [p.110]

Selected Words: *axons* [p.107], *synapse* [p.107], *neurotransmitter* [p.107], *isometrically* [p.108], *isotonically* [p.108], *lengthening* contraction [p.108], *tetanus* [p.109], "slow" or "red" muscle [p.109], "fast" or "white" muscle [p.109], *atrophy* [p.110], *aerobic exercise* [p.110], *strength training* [p.110], "botox" [p.110], *Clostridium botulinum* [p.110], *botulism* [p.110], anabolic steroid [p.110], "andro" [p.111], creatine [p.111], *'roid rage* [p.111], *bodybuilder's psychosis* [p.111]

Boldfaced, Page-Referenced Terms

[p.106] creatine phosphate _____

_____

[p.106] muscle fatigue _____

[p.106] oxygen debt _____

[p.106] T tubules _____

[p.107] sarcoplasmic reticulum _____

[p.107] neuromuscular junctions _____

[p.108] muscle tension _____

[p.108] motor unit _____

[p.108] muscle twitch _____

[p.109] temporal summation _____

[p.109] tetany _____

[p.109] all-or-none principle _____

[p.109] muscle tone _____

_____

[p.110] exercise _____

_____

Short Answer

1. Explain the function of creatine phosphate in muscle cells. _____

_____

_____

_____

2. During prolonged, moderate exercise, what does the muscle cell use as a source of energy to make ATP by aerobic respiration? About how long does each source of energy last? _____

_____

_____

_____

3. Under what conditions is only glycolysis used to make ATP? _____

_____

_____

_____

4. During hard exercise, why would a muscle lose its ability to contract, accompanied by a person's breathing deeply and rapidly? _____

_____

_____

_____

Fill-in-the-Blanks

The nervous system delivers orders to muscles by way of (5) _____ [p.106] neurons. Neural

signals spread across the plasma membrane of a muscle cell, reaching tubelike (6) _____

_____ [p.106]. These are entrances into the (7) _____ _____ [p.107], which is

modified endoplasmic reticulum. An incoming nerve impulse triggers the release of (8) _____ [p.107]

ions from the SR. These ions bind to (9) _____ [p.107] on the surface of (10) _____ [p.107]

filaments and expose binding sites. This allows (11) _____ _____ _____ [p.107]

to attach to the exposed sites and cause a contraction. The muscle cell relaxes when the nerve impulse ends

and (12) _____ [p.107] is actively transported back into the SR. The actin binding site is covered

again so that (13) _____ [p.107] can no longer form cross-bridges.

A(n) (14) _____ _____ [p.107] is a place where motor nerve endings called (15)

_____ [p.107] abut a muscle cell membrane. The slight gap between the endings and the muscle

cell membrane is called a(n) (16) _____ [p.107]. The nervous message is sent across this gap by

way of a (17) _____ [p.107] called ACh (acetylcholine). When ACh binds to receptors on the

muscle cell membrane, it sets in motion the events that cause the muscle cell to (18) _____ [p.107].

Labeling

Identify the numbered parts of the accompanying illustration. [p.108]

19. _____ _____

20. _____ _____

21. _____ _____

22. _____ _____

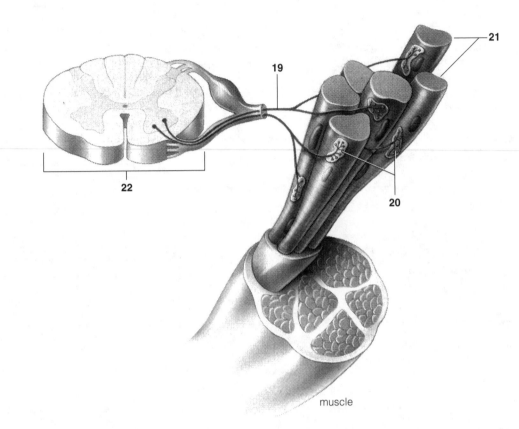

muscle

Fill-in-the-Blanks

A stimulated muscle will only shorten when (23) _____ _____ [p.108] exceeds the muscle's load. A(n) (24) _____ [p.108] contracting muscle develops tension but does not shorten. A(n) (25) _____ [p.108] contracting muscle shortens and moves its load.

The muscle cells controlled by a given motor unit form a(n) (26) _____ _____ [p.108]. A(n) (27) _____ _____ [p.108] is a response in which a muscle contracts briefly when reacting to a single, brief stimulus and then relaxes. The strength of the muscular contraction depends on how far the (28) _____ [p.108] response has proceeded by the time another signal arrives. This additive effect of continued nervous stimulation is called (29) _____ _____ [p.109]. (30) _____ [p.109] is the state of muscles operating near or at maximum temporal summation. To postpone muscle (31) _____ [p.109], motor units within a muscle "take turns" sustaining the contraction. Individual cells in a motor unit always contract according to a(n) (32) _____-_____-_____ [p.109] principle. Even when muscles are relaxed, some motor units remain contracted. This steady, low-level contracted state is called (33) _____ _____ [p.109].

(34) "_____" [p.109] or "red" muscle has cells packed with (35) _____ [p.109] and served by a large number of tiny (36) _____ [p.109]. These muscles contract fairly (37) _____ [p.109], but the contractions are sustained for a(n) (38) _____ [p.109] time. Muscles called (39) "_____" [p.109] or "white" have fewer capillaries, fewer mitochondria, and less myoglobin. These muscles can contract (40) _____ [p.109] and powerfully for (41) _____ [p.109] periods of time.

Short Answer

42. Name three effects that aerobic exercise has on muscles. _____

_____

_____

_____

43. Strength training produces muscles that are large and strong. What do muscles produced in this way lack? _____

44. Why does "botox" smooth wrinkles out? _____

_____

_____

_____

45. What does an anabolic steroid mimic? List negative side effects of anabolic steroids. _____

_____

_____

_____

Self-Quiz

For questions 1–5, choose from the following answers:

> a. isometric contraction(s)   b. cross-bridge formation   c. the sliding-filament model
> d. reciprocal innervation   e. isotonic contraction(s)

1. _____ Helps coordinate the contraction and relaxation of antagonistic pairs of muscles. [p.102]

2. _____ Muscle shortens and moves a load (e.g., walking downstairs). [p.108]

3. _____ Muscle develops tension; does not shorten during contraction (e.g., holding a glass of lemonade in one position). [p.108]

4. _____ Process assisted by calcium ions and ATP. [p.105]

5. _____ Explains how myosin filaments move to the center of a sarcomere and back. [p.105]

For questions 6–9, choose from the following answers:

> a. actin   b. myofibril   c. myosin   d. sarcomere   e. sarcoplasmic reticulum

6. _____ Each consists of many repetitive units of muscle contraction; many found in a muscle cell. [p.104]

7. _____ Basic unit of muscle contraction. [p.104]

8. _____ Thin filaments within a sarcomere, attached to Z bands. [p.104]

9. _____ Stores calcium ions and releases them in response to neural signals. [p.107]

Chapter Objectives/Review Questions

This section lists general and detailed chapter objectives that can be used as review questions. You can make maximum use of these items by writing answers on a separate sheet of paper. To check for accuracy, compare your answers with information given in the chapter or glossary.

10. Define *reciprocal innervation* and explain how it helps coordinate motor elements. [p.102]
11. Refer to Figure 6.4 of your main text and name (a) a muscle used in sit-ups, (b) another used in dorsally flexing the foot, and (c) another used in flexing the elbow joint. [p.103]
12. Describe the fine structure of a muscle cell, using the terms *myofibril, sarcomere, actin,* and *myosin.* [pp.104–105]
13. Explain what occurs in skeletal muscle cells when there is not enough oxygen to allow ATP production by aerobic respiration. When does muscle fatigue occur and why? [p.106]
14. List, in sequence, the biochemical and mechanical events that occur following an incoming neural signal to contract and explain how the cell relaxes. [pp.106–107]
15. Distinguish a twitch contraction from tetany, using the term *temporal summation.* [p.108–109]
16. Name two natural substances that are used by athletes to increase muscle mass and strength and explain how each works. [p.111]

Integrating and Applying Key Concepts

Explain how an individual's genetically determined amounts of "fast" versus "slow" muscle cells may influence success at sprinting versus long-distance athletic events.

7

DIGESTION AND NUTRITION

Interactive Exercises

Selected Words: diabetes mellitus [p.115], diverticula [p.115], "starvation" [p.115], lumen [p.116], mucosa [p.117], submucosa [p.117], smooth muscle [p.117], serosa [p.117], sphincters [p.117], caries [p.118], gingivitis [p.118], periodontal membrane [p.118], parotid gland [p.118], submandibular glands [p.118], sublingual glands [p.118], mucins [p.118], trachea [p.118], swallowing reflex [p.119], epiglottis [p.119], intrinsic factor [p.120], "heartburn" [p.120], gastrin [p.121], "gastric mucosal barrier" [p.120], Helicobacter pylori [p.120], peptic ulcer [p.120], rugae [p.120]

Boldfaced, Page-Referenced Terms

[p.115] nutrition _____

_____

[p.116] digestive system _____

_____

[p.116] gastrointestinal (GI) tract _____

_____

[p.117] mechanical processing _____

_____

[p.117] secretion _____

_____

[p.117] digestion _____

_____

[p.117] absorption _____

_____

[p.117] elimination _____

_____

[p.118] oral cavity _____

_____

[p.118] salivary glands _____

_____

[p.118] salivary amylase _____

_____

[p.118] bolus _____

_____

[p.118] palate _____

_____

[p.118] pharynx _____

_____

[p.118] esophagus _____

_____

[p.120] stomach _____

_____

[p.120] pepsins _____

_____

[p.120] gastric fluid _____

_____

[p.120] chyme _____

_____

[p.120] peristalsis _____

_____

Labeling

Identify the numbered parts of the accompanying illustration. [p.116]

1. _____ _____
2. _____ _____
3. _____
4. _____ _____
5. _____
6. _____ _____
7. _____
8. _____
9. _____
10. _____
11. _____

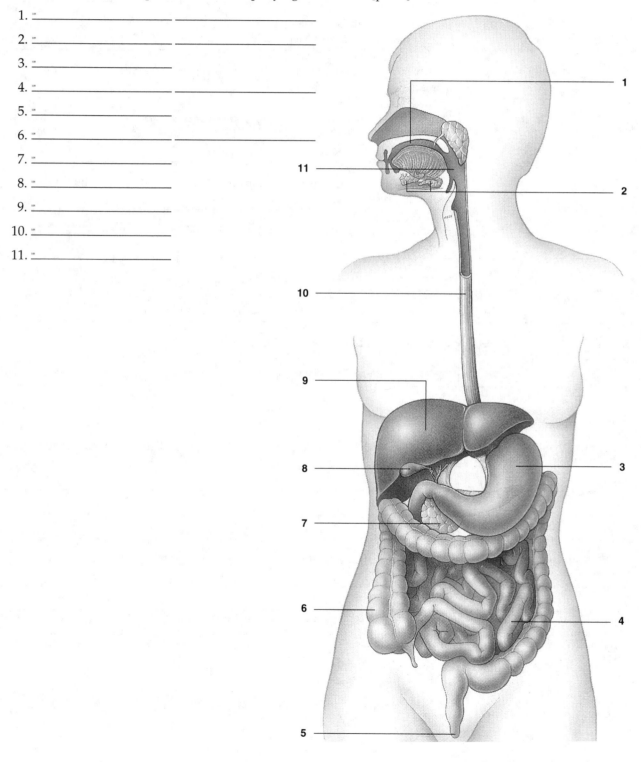

Complete the Table

12. Complete the following table by naming the organs described. [p.114]

Organ	Main Functions
a.	Food is moistened and chewed; polysaccharide digestion starts
b.	Secrete saliva containing digestive enzymes, buffers, and mucus
c.	Passageway to both the tubular part of the digestive system and the respiratory system; moves food forward by contracting sequentially
d.	Muscular tube, moistened by saliva, which moves food from the pharynx to the stomach
e.	Stores food; kills many microorganisms; starts protein digestion
f.	Digests and absorbs most nutrients
g.	Secretes enzymes that break down all major food molecules; produces buffers against hydrochloric acid from stomach
h.	Secretes bile for fat emulsification; roles in metabolism of carbohydrates, fats, and proteins
i.	Stores and concentrates bile produced by liver
j.	Concentrates and stores undigested matter by absorbing ions and water
k.	Distension stimulates expulsion of feces
l.	Terminal opening for expelling feces

Short Answer

13. List the five tasks that the various parts of the human digestive system accomplish by working together. [p.117] _____

_____

_____

14. Name each of the four basic layers of the digestive tube from the inside outward. [p.117] _____

_____

_____

Matching

15. _____ bolus [p.118]

16. _____ caries [p.118]

17. _____ esophagus [p.118]

18. _____ gingivitis [p.118]

19. _____ oral cavity [p.118]

20. _____ palate [p.118]

21. _____ periodontal disease [p.118]

22. _____ pharynx [p.118]

23. _____ salivary amylase [p.118]

24. _____ parotid, submandibular, and sublingual [p.118]

A. Destruction of the bone around the teeth that loosens the teeth
B. A muscular tube leading from the pharynx to the stomach
C. A softened, lubricated ball of food formed in the mouth
D. A term synonymous with *throat*
E. Tooth decay caused by bacteria living on food residues in the mouth
F. Three pairs of saliva-producing glands in the vicinity of the ears, lower jaw, and under the tongue
G. An enzyme that breaks down starch
H. A term synonymous with *mouth*
I. An inflammation of the gums caused by bacterial infection
J. Hardened roof of the mouth providing a hard surface against which the tongue can press food as it mixes with saliva

Dichotomous Choice

Circle one of two possible answers given between parentheses in each statement about the sequence of events called *swallowing*.

25. Swallowing begins when voluntary movements push a bolus into the (esophagus/pharynx), stimulating sensory receptors in its wall. [p.119]
26. The sensory receptors trigger (voluntary/involuntary) muscle contractions that prevent food from entering the nose and the trachea. [p.119]
27. The vocal cords stretch across the entrance to the larynx, and a flaplike valve, the (uvula/epiglottis), closes the opening to the respiratory tract while food moves into the esophagus. [p.119]
28. The (Heinrich/Heimlich) maneuver is an emergency procedure that can save a person choking on food. [p.119]

Fill-in-the-Blanks

The (29) _____ [p.120] is a muscular, expandable sac that stores food, helps break it down, and controls its passage into the (30) _____ _____ [p.120]. (31) _____ [p.120] fluid contains hydrochloric acid (HCl), mucus, and pepsinogens (precursors of digestive enzymes known as (32) _____ [p.120]). Gland cells in the stomach lining also secrete (33) _____ [p.120] factor, a protein necessary for vitamin B_{12} absorption in the small intestine. Combined with stomach contractions, the acidity in the stomach helps convert swallowed boluses into a thick mixture called (34) _____ [p.120]. (35) _____ [p.120] digestion begins in the stomach. The high acidity due to HCl secretion (36) _____ [p.120] proteins and exposes their peptide bonds. The acidity also converts inactive pepsinogens to active (37) _____ [p.120], which break the peptide bonds. The secretion of HCl and pepsinogen is stimulated by the hormone (38) _____ [p.120], which is secreted from gland cells. The (39) "_____ _____ _____" [p.120] of mucus and bicarbonate prevents the HCl and pepsin from breaking down the inner surface of the stomach lining. When the barrier is disrupted, a particular (40) _____ [p.120], *Helicobacter pylori,* may infect the stomach wall and cause the inner surface to break down, thus allowing hydrogen ions and pepsins to diffuse into the lining. The resulting open sore is a(n) (41) _____ _____ [p.120]. In the stomach, waves of contraction and relaxation called (42) _____ [p.120] mix the chyme and build up force as they approach the (43) _____ [p.120] sphincter. Strong contractions of the stomach wall (44) _____ [p.120] the sphincter. This squeezes most of the chyme back, so that only a small amount moves into the (45) _____ _____ [p.120] with each contraction. From two to six hours later, the stomach is empty and its walls crumple into folds called (46) _____ [p.120]. Only (47) _____ [p.120] and a few other substances begin to be absorbed across the stomach wall.

7.4. THE SMALL INTESTINE: SPECIALIZED FOR DIGESTION AND ABSORPTION [p.121]

7.5. HOW NUTRIENTS ENTER THE BODY [pp.122–123]

7.6. THE MULTIPURPOSE LIVER [p.124]

7.7. THE LARGE INTESTINE [p.125]

7.8. DIGESTION CONTROLS AND DISRUPTIONS [p.126]

Selected Words: duodenum [p.121], jejunum [p.121], ileum [p.121], "pancreatic juice" [p.121], *bile* [p.121], "brush border" [p.122], gallstones [p.124], Escherichia coli [p.125], "coliform bacteria" [p.125], *constipation* [p.125], hemorrhoids [p.125], *appendicitis* [p.125], *peritonitis* [p.125], *insoluble fiber* [p.125], *soluble* fiber [p.125], *diverticulitis* [p.125], *diarrhea* [p.125], *gastrin* [p.126], *somatostatin* [p.126], *secretin* [p.126], *cholecystokinin* [p.126], *glucose insulinotropic peptide* [p.126], lactose intolerance [p.126], *cystic fibrosis* [p.126], *Crohn's disease* [p.126], food allergies [p.126], *anaphylaxis* [p.126]

Boldfaced, Page-Referenced Terms

[p.121] small intestine _____

_____

[p.121] pancreas _____

_____

[p.121] liver _____

_____

[p.121] gallbladder _____

_____

[p.122] villus (plural: villi) _____

_____

[p.122] microvillus _____

_____

[p.123] segmentation _____

_____

[p.123] micelle _____

_____

[p.123] lacteals _____

_____

[p.124] hepatic portal vein _____

_____

[p.125] large intestine _____

_____

[p.125] cecum _____

_____

[p.125] colon _____

_____

[p.125] rectum _____

_____

[p.125] anal canal _____

_____

[p.125] anus _____

_____

[p.125] appendix _____

_____

[p.125] bulk _____

_____

[p.125] malabsorption disorder _____

_____

Fill-in-the-Blanks

The small intestine is about 1½ inches in diameter and 6 meters long, some (1) _____ [p.121] feet. It has three sections, the (2) _____ [p.121], the jejunum, and the ileum. Digestion in the small intestine depends on secretions from three accessory organs: the (3) _____ [p.121], the (4) _____ [p.121], and the (5) _____ [p.121]. A part of "pancreatic juice," the enzymes trypsin and chymotrypsin digest the polypeptide chains of proteins into (6) _____ [p.121] fragments. These fragments are further degraded to (7) _____ _____ [p.121] by the enzymes carboxypeptidase and aminopeptidase. The pancreas also secretes the buffer (8) _____ [p.121], which helps neutralize HCl arriving from the stomach. This maintains a favorable environment in which pancreatic (9) _____ [p.121] can function. Besides enzymes, fat digestion requires

(10) _____ [p.121] secreted by the liver and stored in the (11) _____ [p.121]. Bile salts speed up fat digestion by means of a process known as (12) _____ [p.121].

Digestion is completed, and the great majority of (13) _____ [p.122] are absorbed in the small intestine. The mucosa of the small intestine is densely folded, tremendously increasing the (14) _____ _____ [p.122] available for absorbing nutrients. The folds have smaller projections, about a millimeter long each, called (15) _____ [p.122]. The epithelial cells covering each villus usually have threadlike projections of the plasma membrane called (16) _____ [p.122].

(17) _____ [p.123] is the passage of nutrients, water, salts, and vitamins into the internal environment. The action of smooth muscle arranged in rings (referred to as (18) _____ [p.123]) facilitates absorption. (19) _____ [p.122] proteins in the plasma membrane of brush border cells actively move monosaccharides and amino acids across the intestinal lining. (20) _____ [p.123] salts assist the absorption of fatty acids and monoglycerides following lipid digestion by clumping together with cholesterol and other substances to form tiny droplets known as (21) _____ [p.123]. Nutrient molecules diffuse from the droplets into epithelial cells. There, fatty acids and monoglycerides recombine into (22) _____ [p.123]. Then triglycerides combine with proteins into particles, called (23) _____ [p.123], which leave cells by (24) _____ [p.123] and enter tissue fluid. Once glucose and amino acids are absorbed, they enter (25) _____ [p.123] vessels directly. The triglycerides enter lymph vessels called (26) _____ [p.123], which drain into blood vessels.

Dichotomous Choice

Circle one of two possible answers given between parentheses in each statement.

27. Nutrient-laden blood in intestinal villi flows to the (hepatic portal vein/hepatic vein). [p.124]
28. In the liver, excess glucose is taken up before a (hepatic portal vein/hepatic vein) returns the blood to the general circulation. [p.124]
29. The liver converts and stores much of the glucose to (glycogen/glucagon). [p.124]
30. The digestive role of the liver is to secrete (glycogen/bile). [p.124]
31. When there is no food moving through the GI tract, a sphincter closes off the main bile duct and bile backs up into the (small intestine/gallbladder) for temporary storage. [p.124]
32. Liver cells use (glycogen/cholesterol) to synthesize bile salts. [p.124]
33. Excess cholesterol that precipitates in the gallbladder forms (gallstones/urea). [p.124]
34. Toxic ammonia, NH_3, produced when cells break down amino acids, is carried by the circulatory system to the liver, where it is converted to less toxic (glucose/urea), which is excreted in the urine. [p.124]

Matching

Choose the most appropriate description for each term.

35. _____ anal canal [p.125]

36. _____ appendicitis [p.125]

37. _____ appendix [p.125]

38. _____ bulk [p.125]

39. _____ cecum [p.125]

40. _____ constipation [p.125]

41. _____ diarrhea [p.125]

42. _____ *Escherichia coli* [p.125]

43. _____ feces [p.125]

44. _____ insoluble fiber [p.125]

45. _____ colon (or large intestine) [p.125]

46. _____ peritonitis [p.125]

47. _____ rectum [p.125]

48. _____ soluble fiber [p.125]

A. Feces become dry and hard and defecation becomes difficult

B. The volume of fiber and other undigested food material that cannot be decreased by absorption in the colon

C. Connects with the sigmoid colon; feces here move toward the outside of the body through the anal canal

D. A serious internal infection resulting from an infected appendix that bursts, releasing bacteria into the abdominal cavity

E. Plant carbohydrates that swell or dissolve in water

F. Receives material not absorbed in the small intestine

G. An irritated mucosal lining that secretes more water and salts than the large intestine can absorb

H. A slender projection from the cecum with no known digestive function

I. Feces within this structure are moving directly toward the outside of the body

J. Normally inhabit intestines for nourishment but also furnish useful fatty acids and vitamins, such as vitamin K

K. Cellulose and other plant compounds that do not easily dissolve in water and cannot be digested by humans

L. A serious inflammation due to fecal obstruction of blood flow to the appendix

M. A mixture of undigested and unabsorbed food material, water, and bacteria

N. A blind pouch that is the beginning of the large intestine

Matching

Choose the most appropriate description for each term.

49. _____ cholecystokinin (CCK) [p.126]

50. _____ cystic fibrosis [p.126]

51. _____ food allergies [p.126]

52. _____ gastrin [p.126]

53. _____ glucose insulinotropic peptide [p.126]

54. _____ malabsorption disorder [p.126]

55. _____ nervous system, local nerve network, and endocrine system [p.126]

56. _____ secretin [p.126]

57. _____ somatostatin [p.126]

A. Hormone that stimulates the pancreas to secrete bicarbonate
B. Gastrointestinal hormone released in response to the presence of glucose and fat in the small intestine; stimulates insulin release
C. Gastrointestinal hormone that inhibits acid secretion
D. Sufferers do not produce pancreatic enzymes necessary for normal digestion and absorption of fats and other nutrients
E. Foods are interpreted by the body as "invaders"
F. The digestive controls that act before food is absorbed
G. Gastrointestinal hormone released in response to fat in the small intestine; enhances the actions of secretin and stimulates gallbladder contractions
H. Hormone that stimulates the secretion of acid into the stomach in the presence of peptides and amino acids
I. Anything that interferes with the uptake of nutrients across the lining of the small intestine, such as lactose intolerance

7.9. NUTRIENT PROCESSING AFTER AND BETWEEN MEALS [p.127]

7.10. THE BODY'S NUTRITIONAL REQUIREMENTS [pp.128–129]

7.11. VITAMINS AND MINERALS [pp.130–131]

7.12. CALORIES COUNT: FOOD ENERGY AND BODY WEIGHT [pp.132–133]

7.13. *Choices: Biology and Society:* MALNUTRITION AND UNDERNUTRITION [pp.134–135]

Selected Words: "fake fats" [p.129], *complete* proteins [p.129], *incomplete* proteins [p.129], *inorganic* [p.130], *body mass index* [p.132], *ob gene set point* [p.133], *anorexia nervosa* [p.133], *bulimia* [p.133], *undernutrition (starvation)* [p.134], *protein-energy malnutrition* [p.134], *kwashiorkor* [p.134], *marasmus* [p.135], *xerophthalmia* [p.135]

Boldfaced, Page-Referenced Terms

[p.128] kilocalories _____

_____

[p.128] food pyramids _____

_____

[p.129] essential fatty acids _____

_____

[p.129] essential amino acids _____

_____

[p.130] vitamins _____

_____

[p.130] minerals _____

_____

[p.132] obesity _____

_____

[p.132] basal metabolic rate (BMR) _____

_____

[p.134] malnutrition _____

_____

Dichotomous Choice

Circle one of two possible answers given between parentheses in each statement.

1. Living cells (store/break apart) most of their carbohydrates, lipids, and proteins and use the products as energy sources or building blocks. [p.127]
2. Carbohydrates that are not required by the cell are transformed into (glycogen/fats) that are stored in adipose tissue. [p.127]
3. Cells first use (protein/glucose) as an energy source before tapping into their stores of fat. [p.127]

Short Answer

Study the food pyramid on page 128 of your main text in order to answer the questions below.

4. What kind of foods should you have the most servings of each day? _____

_____

_____

_____

5. What size is one serving of poultry? _____

_____

_____

_____

6. How many servings of vegetables should an adult male have each day? _____

_____

_____

_____

7. Does an adult male or a teenager require more servings of milk, cheese, or yogurt? _____

_____

_____

8. What section of the food pyramid should be avoided? Why? _____

_____

_____

_____

Fill-in-the-Blanks

Starch and, to a lesser extent, glycogen should be the main (9) _____ [p.128] in the human diet.
Complex carbohydrates are preferable to simple sugars because they are high in (10) _____
[p.128] and contain (11) _____ [p.128] and minerals. Lipids, including fats, are essential in the
diet because they are a necessary component of cell (12) _____ [p.129], serve as (13)
_____ [p.129] reserves, (14) _____ [p.129] many organs, provide (15)
_____ [p.129] beneath the skin, and store fat-soluble (16) _____ [p.129]. Essential
fatty acids are those that cannot be produced by the (17) _____ [p.129] and therefore must come
from the diet. Butter and other (18) _____ [p.129] fats raise the level of cholesterol in the blood,
which can result in serious problems with circulation. Of the 20 common amino acids in proteins,
(19) _____ [p.129] are essential amino acids. (20) _____ [p.129] proteins have ratios of
amino acids that match human nutritional needs. Plant proteins are (21) _____ [p.129] because
they lack one or more of the essential amino acids. Protein deficiency is most damaging among the (22)
_____ [p.129].

(23) _____ [p.130] are organic substances essential for growth and survival. There are
(24) _____ [p.130] essential vitamins, each with specific metabolic functions. Minerals are
(25) _____ [p.130] substances that are essential in the diet.

Short Answer

26. About how many Americans die each year from weight-related conditions? Name three of these
    conditions. [p.132] _____

_____

_____

_____

Complete the Table

27. Complete the following table by determining how many kilocalories the people described should take in daily, given the stated exercise level, to maintain their weight. Consult page 132 of the text.

Height	Age	Sex	Level of Physical Activity	Present Weight	Number of Kilocalories
5'6"	25	Female	Moderately active	138	
5'10"	18	Male	Very active	145	
5'8"	53	Female	Not very active	143	

Fill-in-the-Blanks

Genes apparently play a role in an individual's (28) _____ _____ [p.133], the weight from which the body seems to counteract any deviation. Evidence for involvement of genes in weight gain comes from a gene isolated from mice called the (29) _____ [p.133] gene. It may be that some (30) _____ [p.133] people inherit malfunctioning appetite controls. Permanent weight loss requires combining a moderate reduction in (31) _____ [p.133] intake with increased physical (32) _____ [p.133].

Matching

33. _____ kwashiorkor [p.134]

34. _____ iron deficiency [p.135]

35. _____ malnutrition [p.134]

36. _____ marasmus [p.135]

37. _____ iodide deficiency [p.135]

38. _____ protein-energy malnutrition [p.134]

39. _____ undernutrition or starvation [p.134]

40. _____ xerophthalmia [p.135]

41. _____ scurvy [p.135]

42. _____ rickets [p.135]

43. _____ beriberi [p.135]

A. Child consumes near-normal amounts of calories but has chronic protein deficiency; swollen abdomen, sickly, many infections

B. Body-wasting disease in very young children; both protein and food calories extremely low; skin and hair dry; growth and mental development retarded

C. Vitamin C deficiency

D. A state in which body functions or development suffer as a result of inadequate or unbalanced food intake

E. Lack of vitamin A is responsible for this form of preventable blindness

F. A major factor contributing to high death rates from intestinal and respiratory effects in many countries; the most common nutritional deficiency among U.S. children

G. Vitamin B_1 deficiency

H. The individual lacks food and thus does not obtain sufficient kilocalories or nutrients to sustain proper growth, body functioning, and development

I. Causes weakness, weight loss, impaired immunity, and other symptoms in adults; especially devastating to infants and children

J. Vitamin D deficiency

K. Lack of this key component of thyroid hormones results in goiter as well as retarded physical and mental development

Self-Quiz

Multiple Choice

_____ 1. The process that moves nutrients into the blood or lymph is _____. [p.117]
   a. ingestion
   b. absorption
   c. assimilation
   d. digestion
   e. none of the above

_____ 2. The enzymatic digestion of proteins begins in the _____ . [p.120]
   a. mouth
   b. stomach
   c. liver
   d. pancreas
   e. small intestine

_____ 3. The enzymatic digestion of starches begins in the _____ . [p.118]
   a. mouth
   b. stomach
   c. liver
   d. pancreas
   e. small intestine

_____ 4. A bolus moves from the pharynx to the _____ . [p.119]
   a. trachea
   b. larnyx
   c. glottis
   d. oral cavity
   e. esophagus

_____ 5. Water moves through the membranes of the small intestine by _____ . [p.123]
   a. peristalsis
   b. osmosis
   c. diffusion
   d. active transport
   e. bulk flow

_____ 6. Digestion of fats is aided by bile and _____ . [p.121]
   a. lecithin
   b. cholesterol
   c. pancreatic enzymes
   d. pigments
   e. *Escherichia coli*

_____ 7. Which one of the following does *not* apply to the large intestine? [p.125]
   a. It contains large populations of bacteria.
   b. It is divided into the duodenum, jejunum, and ileum.
   c. It concentrates undigested matter.
   d. It stores undigested matter.
   e. It absorbs water.

_____ 8. Of the following, _____ has (have) the highest amounts of all eight essential amino acids. [p.129]
   a. sunflower seeds
   b. cream cheese
   c. eggs
   d. black-eyed peas
   e. mushrooms

_____ 9. Males tend to burn more kilocalories than females because men _____ . [p.133]
   a. are more active
   b. are taller
   c. weigh more
   d. have more muscle
   e. have a higher BMR

_____ 10. _____ is an eating disorder in which an individual purposely starves and overexercises. [p.133]
   a. Beriberi
   b. Marasmus
   c. Bulimia
   d. Anorexia nervosa
   e. Xerophthalmia

Matching

11. _____ gallbladder [p.121,124]

12. _____ large intestine [p.125]

13. _____ liver [p.121,124]

14. _____ oral cavity [p.118]

15. _____ pancreas [p.121]

16. _____ small intestine [p.122]

17. _____ stomach [p.120]

A. Secretes bile
B. Produces at least one digestive enzyme for each major food category
C. Where most digestion and absorption occur
D. Where water and salts are absorbed; where indigestible food is concentrated and stored
E. Stores bile
F. Where pepsins work and alcohol is first absorbed
G. Where salivary amylase works

Chapter Objectives/Review Questions

This section lists general and detailed chapter objectives that can be used as review questions. You can make maximum use of these items by writing answers on a separate sheet of paper. To check for accuracy, compare your answers with information given in the chapter or glossary.

1. List all specialized regions (in order) of the human gastrointestinal tract through which food actually passes. Then list the accessory structures that contribute one or more substances to the digestive process. [pp.118–125]
2. Define and distinguish among *mechanical processing and motility, secretion, digestion, absorption,* and *elimination.* [p.117]
3. Briefly describe the four-layered wall of the gastrointestinal tract. [p.117]
4. What can cause "heartburn" and stomach ulcers? [p.120]
5. Describe the role of the stomach's acidity in protein digestion. [p.120]
6. Describe how the digestion and absorption of fats differ from the digestion and absorption of carbohydrates and proteins. [pp.121–123]
7. List the enzyme(s) that act in (a) the oral cavity, (b) the stomach, and (c) the small intestine. Then tell where each enzyme was originally produced. [p.121]
8. Describe the cross-sectional structure of the small intestine and explain how its structure is related to its function. [pp.121–122]
9. List the molecules that leave the digestive system and enter the circulatory system during the process of absorption. [p.123]
10. Tell which foods undergo digestion in each of the following parts of the human digestive system and state what kinds of simple biological molecules they are broken into: oral cavity, stomach, small intestine, large intestine. [pp.118–125]
11. Summarize the processes that occur in the colon. [p.125]
12. Explain how, during digestion, food is mechanically broken down. Then explain how it is chemically broken down. [pp.118–123]
13. Explain how the human body manages to meet the energy needs of various body parts even though the person may be feasting sometimes and fasting at other times. [p.127]
14. Compare the contributions of carbohydrates, proteins, and fats to human nutrition with the contributions of vitamins and minerals. [pp.128–131]
15. Distinguish vitamins from minerals. [p.130]
16. Name four minerals that are important in human nutrition and state the specific role of each. [p.131]

Integrating and Applying Key Concepts

Based on the current worldwide distribution of basic foods versus human populations, suggest ways to prepare for a near doubling of the world population within the next three decades. Include mention of how nutrition, not just food quantity, is relevant.

8

BLOOD

Interactive Exercises

CHAPTER INTRODUCTION [p. 139]

8.1. BLOOD: PLASMA, BLOOD CELLS, AND PLATELETS [pp.140–141]

8.2. HOW BLOOD TRANSPORTS OXYGEN [p.142]

8.3. LIFE CYCLE OF RED BLOOD CELLS [p.143]

Selected Words: albumin [p.140], *edema* [p.140], *hemoglobin* [p.138], neutrophils [p.141], eosinophils [p.141], basophils [p.141], monocytes [p.141], lymphocytes [p.141], "blood doping" [p.143],

Boldfaced, Page-Referenced Terms

[p.140] blood _____

_____

[p.140] plasma _____

_____

[p.140] red blood cells _____

_____

[p.140] stem cell _____

_____

[p.141] white blood cells

_____

[p.141] granulocytes _____

_____

[p.141] agranulocytes _____

_____

[p.141] platelets _____

_____

[p.142] oxyhemoglobin _____

_____

[p.143] cell count _____

_____

Fill-in-the-Blanks

For the average-sized female adult, blood volume is generally about (1) _____ [p.140] liters, which is 6 to 8 percent of body weight. About 55 percent of whole blood is (2) _____ [p.140], which is mostly (3) _____ [p.140] and serves as a transport medium for blood cells, platelets, and other substances. Two-thirds of all plasma proteins are molecules of (4) _____ [p.140], which plays an important role in the movement of (5) _____ [p.140] between the bloodstream and spaces within body tissues. (6) _____ [p.140] also contains hormones, immune system proteins, proteins involved in blood clotting, lipids, vitamins, ions, glucose and other simple sugars, amino acids, various communication molecules, and dissolved gases. The ions help maintain extracellular (7) _____ [p.140] and (8) _____ _____ [p.140].

(9) _____ _____ [p.140] cells or erythrocytes make up about 45 percent of whole blood. Each is a biconcave disk carrying the iron-containing protein (10) _____ [p.140]. These cells transport oxygen used in aerobic respiration as well as some carbon dioxide wastes. Red blood cells are derived from unspecialized (11) _____ _____ [p.140] located in the bone marrow.

Leukocytes, or (12) _____ _____ [p.141] cells, make up a tiny fraction of whole blood but have vital functions in day-to-day housekeeping and defense. They scavenge dead or worn-out cells, as well as any foreign material such as specific bacteria, viruses, or other disease agents. They circulate in the blood, but most squeeze out of blood vessels and do their work after they enter (13) _____ [p.141]. Leukocytes arise from stem cells located in the (14) _____ _____ [p.141]. There are five types of white blood cells divided into two major classes. The (15) _____ [p.141] contain various granules in the cytoplasm. This group includes (16) _____ [p.141], (17) _____ [p.141], and (18) _____ [p.141]. Leukocytes called (19) _____ [p.141] have no visible granules in the cytoplasm. The first of two types are called (20) _____ [p.141], which differentiate into macrophages that engulf invaders and cellular debris. The second type, (21) _____ [p.141] (B cells and T cells), carry out specific immune responses.

Some stem cells in bone marrow develop into "giant" cells that shed fragments of cytoplasm that become enclosed in a bit of plasma membrane. The fragments are called (22) _____ [p.141]. They release substances that initiate blood (23) _____ [p.141].

Complete the Table

24. Complete the following table, which describes the components of blood. [p.140]

Components	Relative Amounts	Functions
Plasma Portion (50%–60% of total volume):		
Water	a. _____	Solvent
b. _____ (albumin, globulins, fibrinogen, HDLs, LDLs, VLDLs, etc.)	7%–8%	Defense, clotting, lipid transport, roles in extracellular fluid volume, etc.
Ions, sugars, lipids, amino acids, hormones, vitamins, dissolved gases	1%–2%	Roles in extracellular fluid volume, pH, etc.
Cellular Portion (40%–50% of total volume, *per microliter*):		
White blood cells		
c. _____	3,000–6,750	Phagocytosis; inflammation
d. _____	1,000–2,700	Immunity
Monocytes (macrophages)	150–720	e. _____
Eosinophils	100–360	Inflammation
Basophils	25–90	Inflammation
f. _____ _____ cells	4,800,000–5,400,000	O_2, CO_2 transport
g. _____	250,000–300,000	Role in clotting

Interpreting diagrams

25. Study each diagram below in order to answer the questions that follow. [p.142]

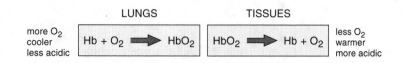

a. What is "Hb" in the diagram? _____

b. What is HbO_2? _____

c. Under what conditions (oxygen, temperature, pH) does oxygen tend to bind to hemoglobin?

    _____   _____   _____

d. Where in the body do these conditions exist? _____

e. Under what conditions (oxygen, temperature, pH) does hemoglobin tend to give up oxygen?

    _____   _____   _____

f. Where in the body do these conditions exist? _____

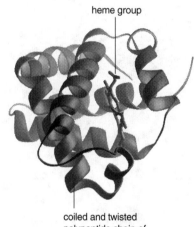

g. What is the molecule diagrammed above? _____
h. In what type of blood cells are molecules like this found? _____
i. How many polypeptide chains make up the globin component of this molecule? _____
j. What kind is found at the center of each heme group? _____
k. What is the function of this substance? _____
l. What is this molecule called when it binds oxygen? _____

Labeling-Matching

Identify the numbered cell types in the following illustration. Complete the exercise by matching and entering the letter of the appropriate function in the parentheses following the given cell types. You may use a letter more than once. [pp.140–141]

26. _____ ( )

27. _____ ( )

28. _____ ( )

29. _____ ( )

30. _____ ( )

31. _____ ( )

32. _____ ( )

A. Found in infected, inflamed, or damaged tissues
B. Engulf invading microbes
C. Involved in clotting
D. Specific immune responses
E. O_2, CO_2 transport

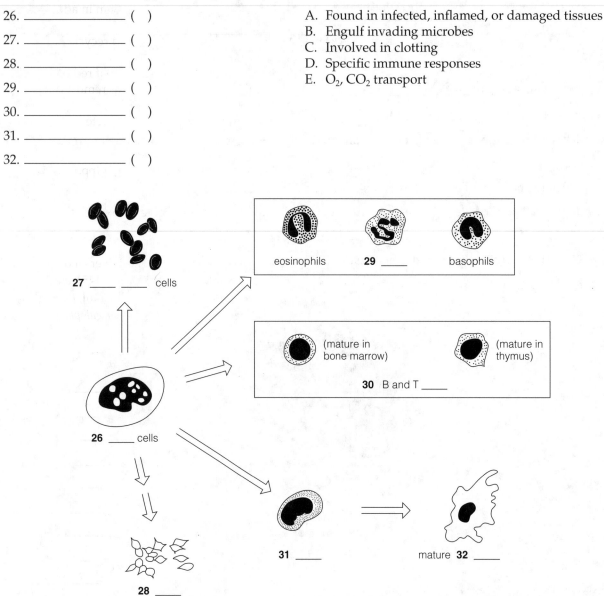

Matching

33. _____ erythropoietin
34. _____ bilirubin
35. _____ red marrow
36. _____ 120 days
37. _____ stem cell
38. _____ cell count
39. _____ "blood doping"
40. _____ spleen

A. found in skull, vertebrae, and sternum
B. average lifespan of red blood cell
C. "blank slate"
D. average per microliter is 5.4 million in adult male and 4.8 million in adult female
E. where macrophages remove and recycle old or damaged red blood cells
F. hormone that signals production of red blood cells
G. orangish remnant of heme group removed from bloodstream by liver
H. injection of stored blood of an athlete

8.4. BLOOD TYPES — GENETICALLY DIFFERENT RED BLOOD CELLS [pp.144–145]

8.5. HEMOSTASIS AND BLOOD CLOTTING [p.146]

8.6. BLOOD DISORDERS [p. 147]

Selected Words: antibodies [p.144], antigen [p.144], "self" [p.144], "nonself" [p.144], "universal donors" [p.144], "universal recipients" [p.144], "Rh factor" [p.144], *hemolytic disease of the newborn* [p.145], *cross-matching* [p.145], "intrinsic" clotting mechanism [p.146], "extrinsic" clotting mechanism [p.146], *thrombus* [p.146], *thrombosis* [p.146], *embolus* [p.146], *embolism* [p.146], *iron-deficiency anemia* [p.147], *megaloplastic anemia* [p.147], *aplastic anemia* [p.147], *hemolytic anemias* [p.147]

Boldfaced, Page-Referenced Terms

[p.144] agglutination _____

_____

[p.144] Rh blood typing _____

_____

[p.146] hemostasis _____

_____

[p.147] anemias _____

_____

[p.147] infectious mononucleosis _____

_____

[p.147] leukemias _____

_____

Matching

Choose the most appropriate description for each term.

1. _____ antibodies [p.144]
2. _____ antigen [p.144]
3. _____ ABO [p.144]
4. _____ Rh factor [p.144]
5. _____ hemolytic disease of the newborn [p.145]
6. _____ cross-matching [p.145]
7. _____ agglutination [p.144]
8. _____ Type AB [p.144]
9. _____ Type O [p.144]
10. _____ Types A and B [p.144]

A. A person whose blood type is positive carries this marker
B. Blood typing based on presence or absence of glycoproteins A and B
C. Theoretical "universal recipient"
D. Theoretical "universal donor"
E. Blood types with one marker from ABO group
F. May occur in offspring produced by an Rh^- woman and an Rh^+ man
G. Method used to identify hundreds of human blood marker forms
H. Occurs when different blood types are commingled
I. Immune system proteins that recognize and organize an attack on most foreign entities
J. Molecule with "nonself" protein markers that prompt a defensive attack by antibodies

Sequence

The following lettered stages of the "intrinsic" clotting mechanism occur in a specific sequence. Place the letter of the first event in the blank beside number 11, and continue until the last event is placed in the blank beside number 16.

11. _____ [p.146]
12. _____ [p.146]
13. _____ [p.146]
14. _____ [p.146]
15. _____ [p.146]
16. _____ [p.146]

A. Fibrinogen proteins stick together, forming insoluble net of fibrin.
B. Blood vessel ruptures.
C. Platelets aggregate and form a temporary plug.
D. An activated blood protein triggers reactions that form thrombin.
E. Smooth muscle in a damaged blood vessel contracts, constricting the vessel and slowing blood flow.
F. Blood cells and platelets become entangled in the fibrin net, forming a clot.

Short Answer

17. How does the "extrinsic" clotting mechanism differ from the sequence above? [p.146] _____

_____

_____

_____

18. How are a thrombus and an embolus the same? How are they different? Which is more dangerous? [p.146] _____

_____

_____

_____

Matching

11. _____ [p.146]

19. _____ iron-deficiency anemia [p.147]

20. _____ aplastic anemia [p.147]

21. _____ infectious mononucleosis [p.147]

22. _____ hemolytic anemia [p.147]

23. _____ megaloplastic anemia [p.147]

24. _____ leukemia [p.147]

A. Runaway multiplication of abnormal white blood cells
B. Premature destruction of red blood cells
C. Hemoglobin cannot be formed due to low iron supply
D. Destruction of red bone marrow by radiation or toxins
E. Deficiency of folic acid or vitamin B12
F. Overproduction of agranulocytes, caused by Epstein-Barr virus

Self-Quiz

_____ 1. Most of the oxygen in human blood is transported by _____. [p.142]
   a. plasma
   b. serum
   c. platelets
   d. hemoglobin
   e. leukocytes

_____ 2. Of all the different kinds of white blood cells, the two kinds of _____ are the ones that respond to specific invaders. [p.141]
   a. basophils
   b. eosinophils
   c. monocytes
   d. neutrophils
   e. lymphocytes

_____ 3. Oxygen becomes bound to _____ in hemoglobin molecules. [p.142]
   a. iron in heme groups
   b. polypeptides
   c. plasma
   d. globin
   e. carbon dioxide molecules

_____ 4. How does aspirin help to prevent blood clots? [p.146]
   a. prevents the production of fibrinogen
   b. blocks the activity of thrombin
   c. stops blood clots from breaking free of where they form
   d. causes more water to enter the plasma, thus thinning the blood
   e. reduces the aggregation of platelets

_____ 5. Oxyhemoglobin contains _____. [p.142]
   a. carbon dioxide
   b. red blood cells
   c. plasma
   d. calcium
   e. oxygen

_____ 6. Megakaryocytes shed millions of _____ into the blood that last about a week. [p.141]
   a. platelets
   b. red blood cells
   c. lymphocytes
   d. macrophages
   e. neutrophils

_____ 7. Too little albumin can cause swelling called _____. [p.140]
   a. thrombosis
   b. edema
   c. embolism
   d. hemostasis
   e. granuloma

_____ 8. What combination can lead to hemolytic disease of the newborn? [p.145]
   a. Rh⁻ mother, Rh⁻ fetus
   b. Rh⁺ mother, Rh⁺ fetus
   c. Rh⁺ mother, Rh⁻ fetus
   d. Rh⁻ mother, Rh⁺ fetus
   e. any of these

9. Red blood cells and white blood cells develop from _____ cells in the bone marrow. [pp.140–141]
   a. stem
   b. hemolytic
   c. megakaryocyte
   d. oxyhemoglobin
   e. albumin

10. A low red blood cell count could result from _____ . [pp.143,147]
   a. megaloplastic anemia
   b. hemolytic anemia
   c. iron-deficiency anemia
   d. aplastic anemia
   e. any of these

Chapter Objectives/Review Questions

This section lists general and detailed chapter objectives that can be used as review questions. You can make maximum use of these items by writing answers on a separate sheet of paper. Fill in answers where blanks are provided. To check for accuracy, compare your answers with information given in the chapter or glossary.

1. Name and describe the composition of the two major components of human blood, using percentages of volume. [p.140]
2. Name the most common plasma protein and describe its functions. [p.140]
3. State where erythrocytes, leukocytes, and platelets are produced. [p.140–141]
4. Contrast the two main types of leukocytes in terms of cell structure. Then name and state the general functions of the three types of granulocytes and the two types of agranulocytes. [p.141]
5. A hormone produced by the kidneys, _____ , stimulates certain stem cells to produce red blood cells. [p.143]
6. Describe what happens to red blood cells when they are old or damaged . [p.143]
7. Describe how blood is typed for the ABO blood group and for the Rh factor. [pp.144–145]
8. List in sequence the chemical events that occur in the formation of a blood clot. [p.146]
9. When red blood cells contain a less-than-normal amount of hemoglobin, it is termed _____ . [p.147]

Integrating and Applying Key Concepts

For each blood cell marker, there is a gene that codes for it. A baby is type AB+. The mother is type B−. Which blood types would be possible for the father? Which would not? What other means could be used to determine the identity of the father?

9

CIRCULATION — THE HEART AND BLOOD VESSELS

CHAPTER INTRODUCTION

THE CARDIOVASCULAR SYSTEM — MOVING BLOOD THROUGH THE BODY
>   The heart and blood vessels make up the cardiovascular system
>   The cardiovascular system is linked to the lymphatic system

THE HEART: A DOUBLE PUMP
>   Heart structure: two halves, four chambers
>   In a "heartbeat," the heart's chambers contract and relax

THE TWO CIRCUITS OF BLOOD FLOW
>   The pulmonary circuit: Blood picks up oxygen in the lungs
>   The systemic circuit: Blood travels to and from tissues
>   A detour through the liver

Science Comes to Life: HEART-SAVING DRUGS

HOW THE HEART CONTRACTS
>   Electrical impulses drive the heart's contractions
>   The nervous system adjusts heart activity

BLOOD PRESSURE
>   Blood exerts pressure against the walls of blood vessels
>   Pressure drops as blood circulates

THE STRUCTURE AND FUNCTIONS OF BLOOD VESSELS
>   Arteries: Pipelines for speedy transport of oxygenated blood
>   Arterioles are control points for blood flow
>   Capillaries are specialized for diffusion
>   Venules and veins: Return routes to the heart
>   Vessels help control blood pressure

Focus on Your Health: HEART-HEALTHY EXERCISE

EXCHANGES AT CAPILLARIES
>   Blood flows at different rates in different parts of the system
>   Substances enter and leave capillaries by four routes

CARDIOVASCULAR DISORDERS

THE MULTIPURPOSE LYMPHATIC SYSTEM
>   The lymph vascular system: Drainage, delivery, and disposal
>   Lymphoid organs and tissues are specialized for body defense

---

Interactive Exercises

CHAPTER INTRODUCTION [p.149]

9.1. THE CARDIOVASCULAR SYSTEM — MOVING BLOOD THROUGH THE BODY [pp.150–151]

9.2. THE HEART: A DOUBLE PUMP [pp.152–153]

9.3. THE TWO CIRCUITS OF BLOOD FLOW [pp.154–155]

9.4. *Science Comes to Life:* HEART-SAVING DRUGS [p.155]

9.5. HOW THE HEART CONTRACTS [p.156]

Selected Words: *electrocardiogram* [p.149], *capillary beds* [p.150], *tricuspid valve* [p.152], *bicuspid valve* or *mitral valve* [p.152], chordae tendineae [p.152], "coronary circulation" [p.152], "heartbeat" [p.152], *renal arteries* [p.154], *superior vena cava* [p.154], *inferior vena cava* [p.154], "true capillaries" [p.155], "thoroughfare channels" [p.155], *hepatic portal vein* [p.155], *hepatic vein* [p.155], *hepatic artery* [p.155], *statins* [p.155], *intercalated discs* [p.156], *Purkinje fibers* [p.156]

Boldfaced, Page-Referenced Terms

[p.149] cardiovascular system _____

_____

[p.150] arteries _____

_____

[p.150] arterioles _____

_____

[p.150] capillaries _____

_____

[p.150] venules _____

_____

[p.150] veins _____

_____

[p.152] heart _____

_____

[p.152] myocardium _____

_____

[p.152] septum _____

_____

[p.152] atrium (plural: atria) _____

_____

[p.152] ventricle _____

_____

[p.152] atrioventricular (AV) valve _____

_____

[p.152] semilunar valve _____

_____

[p.152] coronary arteries _____

_____

[p.152] aorta _____

_____

[p.153] systole _____

_____

[p.153] diastole _____

_____

[p.153] cardiac cycle _____

_____

[p.154] pulmonary circuit _____

_____

[p.154] systemic circuit _____

_____

[p.155] capillary beds _____

_____

[p.156] cardiac conduction system _____

_____

[p.156] sinoatrial (SA) node _____

_____

[p.156] atrioventricular (AV) node _____

_____

[p.156] cardiac pacemaker _____

_____

Fill-in-the-Blanks

The (1) _____ [p.150] system is the body's rapid-transport system in that its basic task is to circulate blood to the immediate neighborhood of every living cell in the body. Cells depend on circulating (2) _____ [p.150] to make pickups and deliveries of a diverse range of substances. The two main elements in the system are the (3) _____ [p.150] and (4) _____ _____ [p.150], which are tubes of different diameters. The heart pumps blood into large-diameter (5) _____ [p.150]. From there the blood flows into smaller, muscular (6) _____ [p.150], which branch into even-smaller-diameter (7) _____ [p.150]. Blood flows from capillaries into small (8) _____ [p.150], then into large-diameter (9) _____ [p.150] that return blood to the heart. Blood flow slows in (10) _____ [p.150] beds, where it fans out through vast numbers of capillaries. As the heart's pumping keeps pressure throughout the cardiovascular system, some water and proteins are forced out of the vast network of capillaries and become part of the interstitial fluid. A network called the (11) _____ [p.150] system picks up excess interstitial fluid and reclaimable solutes, then returns them to the cardiovascular system.

Labeling

In the following illustration, color in red all vessels that carry oxygen-rich blood (including all parts indicated by "aorta" or "artery," except for the pulmonary arteries, which carry oxygen-poor blood from the heart to the lungs). Then fill in all the blanks on the right side of the diagram. Next, color in blue all vessels that carry oxygen-poor blood (including all parts indicated by "vena cava" or "vein," except for the pulmonary veins, which return oxygen-rich blood to the heart). Then fill in all the blanks on the left side of the diagram. [p.151]

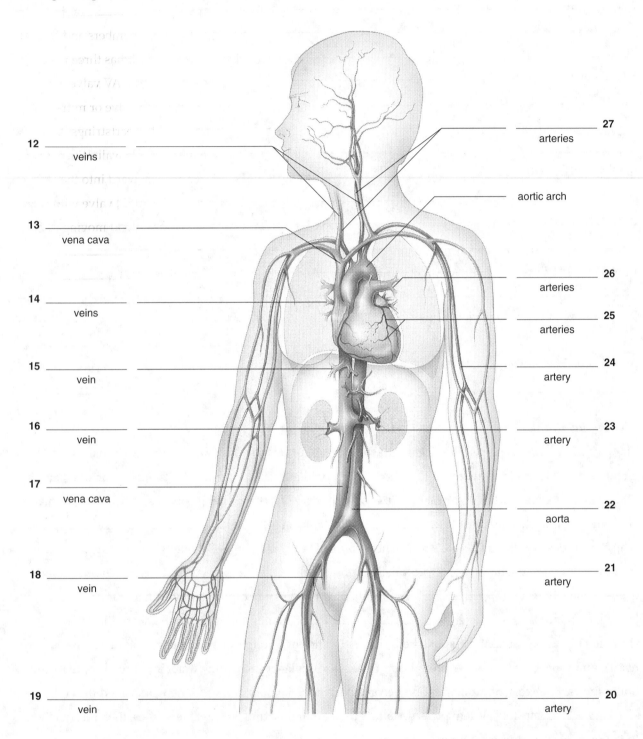

12 _____ veins

13 _____ vena cava

14 _____ veins

15 _____ vein

16 _____ vein

17 _____ vena cava

18 _____ vein

19 _____ vein

27 _____ arteries

_____ aortic arch

26 _____ arteries

25 _____ arteries

24 _____ artery

23 _____ artery

22 _____ aorta

21 _____ artery

20 _____ artery

Fill-in-the-Blanks

The heart is mostly cardiac muscle tissue, the (28) _____ [p.152]. The (29) _____ [p.152] is a tough, fibrous sac that surrounds, protects, and lubricates the heart. The inner chambers of the heart have a smooth lining, the (30) _____ [p.152], composed of connective tissue and a layer of epithelial cells, the (31) _____ [p.152]. The (32) _____ [p.152] is a thick wall that divides the heart into right and left halves. Each half has two chambers: a(n) (33) _____ [p.152] located above a(n) (34) _____ [p.152]. Membranous flaps separate the two chambers and serve as a one-way (35) _____ [p.152] valve between them. This valve in the right half has three flaps that come together in pointed cusps and is called the (36) _____ [p.152] valve. The AV valve in the heart's left half consists of just two flaps and is called the (37) _____ [p.152] valve or mitral valve. Tough, collagen-reinforced strands, (38) _____ _____ [p.152] ("heartstrings"), connect the AV flaps to cone-shaped papillary muscles that extend out from the ventricle wall. When a blood-filled ventricle contracts, this arrangement prevents the flaps from opening backward into the (39) _____ [p.152]. Each half of the heart also has a(n) (40) _____ [p.152] valve within each artery leading away from it. During a heartbeat, it opens and closes in ways that keep blood moving in one direction through the body. The coronary circulation services the heart muscle. Two (41) _____ [p.152] arteries lead to a capillary bed for the cardiac muscle cells. They branch off the (42) _____ [p.152], the major artery carrying oxygenated blood away from the heart.

Labeling

Identify each indicated part of the accompanying illustration. [p.152]

43. _____ _____ _____

44. _____ _____ _____

45. _____ _____ _____

46. _____ _____ _____

47. _____ _____ _____

48. _____

49. _____ _____

50. _____ _____ _____

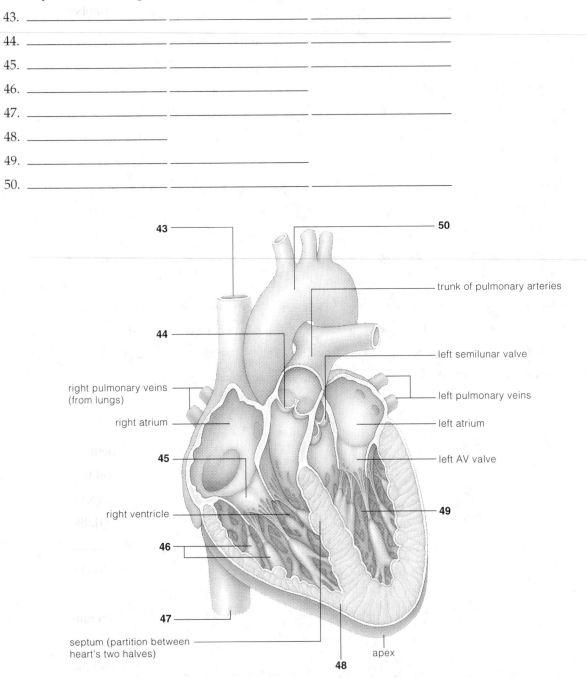

Sequencing

Show the correct sequence of the path of blood flow through the pulmonary and systemic circuits by placing the letter of the part of the heart that receives blood from body tissues in the blank beside number 51, then continuing the sequence until you are back at the starting point.

51. _____

52. _____

53. _____

54. _____

55. _____

56. _____

57. _____

58. _____

59. _____

60. _____

61. _____

62. _____

63. _____

64. _____

65. _____

A. capillary beds of lungs
B. aorta
C. main pulmonary artery
D. right atrium (use twice)
E. left atrium
F. right ventricle
G. left ventricle
H. right atrium
I. right AV valve
J. left AV valve
K. pulmonary veins
L. torso (systemic circulation)
M. right and left pulmonary arteries
N. right semilunar valve
O. left semilunar valve

Fill-in-the-Blanks

As the aorta descends into the torso, major (66) _____ [p.154] branch off it, funneling blood to (67) _____ [p.154] and tissues. In both the pulmonary and systemic circuits, blood travels through (68) _____ [p.154], (69) _____ [p.154], (70) _____ [p.154], and (71) _____ [p.154], and finally returns to the heart in (72) _____ [p.154]. Blood from the head, arms, and chest arrives through the (73) _____ _____ _____ [p.154], and the (74) _____ _____ _____ [p.154] collects blood from the lower body. The actual exchange of substances between blood and tissues occurs in (75) _____ _____ [p.155]. Blood flow into "true capillaries" is controlled by collars of smooth muscle cells called precapillary (76) _____ [p.155]. When the CO_2 level rises above a set point, the sphincter (77) _____ [p.155] so that blood flows through the capillary, delivering oxygen and picking up the excess CO_2. When the CO_2 level falls, the sphincter (78) _____ [p.155].

After a meal, blood passing through capillary beds in the GI tract detours through the (79) _____ _____ [p.155] to the liver. The liver removes (80) _____ [p.155] and processes absorbed substances. Blood leaves the liver through a(n) (81) _____ [p.155] vein. (The hepatic artery delivers (82) _____ [p.155] blood to the liver.)

Part of the processing that occurs in the liver synthesizes cholesterol. In the 1980s, drugs called (83) _____ [p.155] were shown to greatly reduce the amount of "bad" or (84) _____ [p.155] cholesterol in the blood. Other experiments showed that these drugs also raise the blood level of "good" cholesterol, called (85) _____ [p.155], and lower the levels of (86) _____ [p.155]. The effects of these drugs translate into dramatically reduced risks of (87) _____ _____ [p.155] and (88) _____ _____ [p.155], both of which can be caused when blood flow is blocked by the buildup of fatty, cholesterol-rich plaques in blood vessels.

Dichotomous Choice

Circle one of two possible answers given between parentheses in each statement.

89. The (cardiac conduction/intercalated disc) system produces the electrical impulses that stimulate contraction of the heart.
90. A signal to contract spreads rapidly through the heart because of communication junctions between abutting cells called (Purkinje fibers/intercalated discs).
91. Contraction follows a wave of excitation that begins at a mass of cells in the upper wall of the right atrium called the (atrioventricular/sinoatrial) node.
92. After causing both atria to contract, the wave of excitation reaches the (atrioventricular/sinoatrial) node in the septum between the two atria, where it slows down momentarily to give the atria time to finish contracting.
93. The ventricles then contract when the wave continues through conducting bundles that extend from the AV node and make contact with the muscle cells of each ventricle by means of conducting cells called (Purkinje fibers/SA nodes).
94. An artificial pacemaker is implanted when the cardiac pacemaker or (AV/SA) node malfunctions.

9.6. BLOOD PRESSURE [p.157]

9.7. THE STRUCTURE AND FUNCTIONS OF BLOOD VESSELS [pp.158–159]

9.8. *Focus on Your Health:* HEART-HEALTHY EXERCISE [p.160]

9.9. EXCHANGES AT CAPILLARIES [pp.160–161]

9.10. CARDIOVASCULAR DISORDERS [pp.162–163]

9.11. THE MULTIPURPOSE LYMPHATIC SYSTEM [pp.164–165]

Selected Words: *systolic pressure* [p.157], *diastolic pressure* [p.157], *hypertension* [p.157], *hypotension* [p.157], *varicose vein* [p.157], *stroke volume* [p.160], *arteriosclerosis* [p.162], *angina pectoris* [p.162], *angiography* [p.162], *coronary bypass surgery* [p.162], *laser angioplasty* [p.162], *balloon angioplasty* [p.162], *low-density lipoproteins* [p.163], *high-density lipoproteins* [p.163], *homocysteine* [p.163], *bradycardia* [p.163], *tachycardia* [p.163], *ventricular fibrillation* [p.163], *pulp* [p.165]

Boldfaced, Page-Referenced Terms

[p.157] blood pressure _____

_____

[p.158] pulse _____

_____

[p.159] vasodilation _____

_____

[p.159] vasoconstriction _____

_____

[p.159] baroreceptor reflex _____

_____

[p.159] carotid arteries _____

_____

[p.161] ultrafiltration _____

_____

[p.161] reabsorption _____

_____

[p.162] atherosclerosis _____

_____

[p.162] atherosclerotic plaque _____

_____

[p.163] LDLs _____

_____

[p.163] HDLs _____

_____

[p.163] arrhythmias _____

_____

[p.164] lymphatic system _____

_____

[p.164] lymph _____

_____

[p.165] lymph vascular system _____

_____

[p.165] lymph nodes _____

_____

[p.165] spleen _____

_____

[p.165] thymus _____

_____

Fill-in-the-Blanks

Heart contractions generate (1) _____ _____ [p.157], the fluid pressure blood exerts against vessel walls. Consider an adult with a blood pressure of 120/80; 120 is the (2) _____ [p.157] pressure, which is the peak of pressure in the (3) _____ [p.157], when the heart's left ventricle pushes blood into it. The number 80 is the (4) _____ [p.157] pressure, which is the lowest blood pressure in the (5) _____ [p.157], when the heart is relaxed and blood is flowing out of the (6) _____ [p.157]. Elevated blood pressure, or (7) _____ [p.157], is associated with atherosclerosis and kidney disease. Low blood pressure is called (8) _____ [p.157] and is usually not a cause for worry.

Choice

For questions 9–22, choose from the following:

                a. arteries   b. arterioles   c. capillaries   d. venules   e. veins

9. _____ Those near the body surface provide a "pulse." [p.158]

10. _____ These merge into venules. [p.159]

11. _____ The walls have rings of smooth muscle over a single layer of elastic fibers. [p.158]

12. _____ These serve as diffusion zones for exchanges between blood and interstitial fluid. [p.158]

13. _____ Their bulging walls keep blood flowing on through the system. [p.158]

14. _____ These offer more resistance to blood flow than other vessels do. [p.158]

15. _____ These present less total resistance to flow than do the arterioles leading into them; the total drop in blood pressure is more gradual in this region. [p.158]

16. _____ These serve as blood volume reservoirs. [p.159]

17. _____ Red blood cells must squeeze through them single-file. [p.158]

18. _____ Some, mainly in the limbs, have valves. [p.159]

19. _____ These branch into arterioles. [p.158]

20. _____ Weak valves lead to pooled blood and a varicose condition. [p.159]

21. _____ Contractions of smooth muscle in the thin walls occur when blood must circulate faster. [p.159]

22. _____ These merge into veins. [p.159]

Labeling

Identify each vessel and the parts indicated by asterisks in the accompanying illustrations. [p.158]

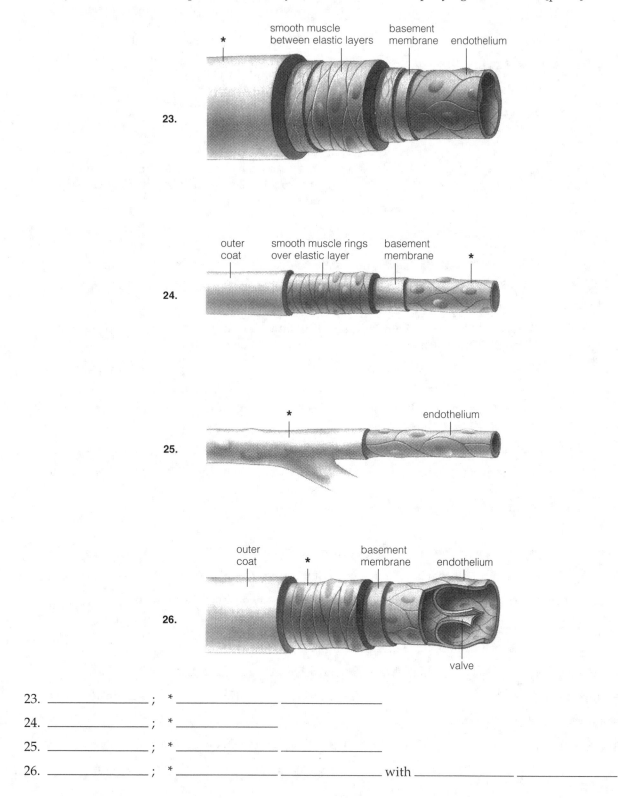

23. _____ ; * _____ _____

24. _____ ; * _____

25. _____ ; * _____ _____

26. _____ ; * _____ _____ with _____ _____

Matching

27. _____ vasodilation [p.159]

28. _____ vasoconstriction [p.159]

29. _____ baroreceptors [p.159]

30. _____ medulla oblongata [p.159]

31. _____ stroke volume [p.160]

A. Pressure receptors in the carotid arteries, the aorta, and elsewhere that monitor changes in mean arterial pressure

B. Part of the brain that coordinates the rate and strength of heartbeats with changes in the diameter of arterioles and veins

C. A decrease in blood vessel diameter brought about by brain centers when an abnormal decrease in blood pressure is detected

D. The amount of blood pumped with a heart contraction; increases with physical conditioning

E. An increase in blood vessel diameter brought about by brain centers when an abnormal increase in blood pressure is detected

True/False

If the statement is true, write a "T" in the blank. If the statement is false, make it correct by writing the word(s) in the blank that should take the place of the underlined word(s).

_____ 32. Blood flow is at its slowest in the arterioles. [p.160]

_____ 33. Because capillaries come within 0.01 millimeter of nearly all your living cells, most solutes move from bloodstream to cells by diffusion. [p.161]

_____ 34. Cholesterol contained in vesicles moves between capillaries and cells by endocytosis or exocytosis. [p.161]

_____ 35. Certain ions as well as white blood cells probably pass through pores in the capillary walls. [p.161]

_____ 36. Ultrafiltration and reabsorption are both examples of pressure-driven movements.

_____ 37. The excess fluid that moves from capillaries into surrounding tissues is returned to the blood by the cardiovascular system. [p.161]

Fill-in-the-Blanks

More than 40 million Americans have cardiovascular disorders. (38) _____ [p.162] is a term for a sustained high blood pressure. A progressive thickening of the arterial wall and progressive narrowing of the arterial lumen is known as (39) _____ [p.162]. These two disorders cause most

(40) _____ _____ [p.162] and (41) _____ [p.162]. Risk factors for cardiovascular disorders include (42) _____ [p.162], lack of regular (43) _____ [p.162], (44) _____ [p.162], high levels of blood (45) _____ [p.162], an inherited predisposition to (46) _____ _____ [p.162], and (47) _____ [p.162]. Two more contributing factors are (48) _____ [p.162] and (49) _____ [p.162]. Prior to menopause, (50) _____ [p.162] production in females provides some protection.

(51) _____ [p.162] results from gradually increasing resistance to flow of blood through small arteries. In time, blood pressure stays (52) _____ [p.162], even when a person is resting.

The effect is an overworked and ineffective heart as well as "hardening" of (53) _____ [p.162] walls. This condition is referred to as a(n) (54) "_____ _____" [p.162] because affected people may have no outward symptoms.

In (55) _____ [p.162], arteries thicken and lose elasticity. This condition worsens in atherosclerosis as cholesterol and other lipids build up and cause the arterial lumen to narrow. When plaque-clogged arteries become narrowed to one-quarter of their former diameter, the resulting symptoms can range from mild chest pain, (56) (_____ [p.162] *pectoris*), to a full-scale heart attack. Diagnosis of atherosclerosis can be accomplished by a stress (57) _____ [p.162], which records electrical activity of the cardiac cycle during treadmill exercise. It can also be diagnosed by (58) _____ [p.162], a procedure involving injection of a dye that causes plaques to show up on X rays. Severe blockage may require surgery. In (59) _____ _____ [p.162] surgery, a section of a large vessel taken from the chest is stitched to the aorta and to the coronary artery below the affected region. A technique using laser beams to vaporize the plaques is known as *laser* (60) _____ [p.162]. In (61) _____ [p.162] angioplasty, a small balloon is inflated within a blocked artery to flatten a plaque and so increase the arterial diameter. In some people, cells remove too little (62) _____ [p.163] from the blood. Those not removed, with their bound cholesterol and triglycerides, infiltrate the walls of arteries, leading to the formation of a(n) (63) _____ [p.163] plaque bulging into the artery lumen. Bony slivers covering the plaque shred the endothelium, setting in motion the formation of a blood (64) _____ [p.163]. As plaques grow, inflammation leads to further damage. A clot that breaks loose can cause a life-threatening (65) _____ [p.163]. Artery damage may result from high levels of (66) _____ [p.163] in the blood. Taking (67) _____ [p.163] vitamins can remove this amino acid before it causes damage.

(68) _____ [p.163] are irregular heart rhythms revealed by ECGs. Endurance athletes may have a below-average cardiac rate, a condition called (69) _____ [p.163], which is a normal adaptation to ongoing strenuous exercise. A cardiac rate above 100 beats per minute is called (70) _____ [p.163] and occurs normally during exercise or stressful situations. Coronary artery disease or some other disorders may cause abnormal rhythms that can degenerate rapidly into an extreme medical emergency called (71) _____ _____ [p.163].

Identification — Fill-in-the-Blanks

Identify each numbered part in the accompanying figure.

72. _____ [p.164]

73. _____ _____ [p.164]

74. _____ _____ [p.164]

75. _____ [p.164]

76. _____ _____ [p.164]

77. _____ _____ [p.164]

78. organized arrays of _____ [p.165]

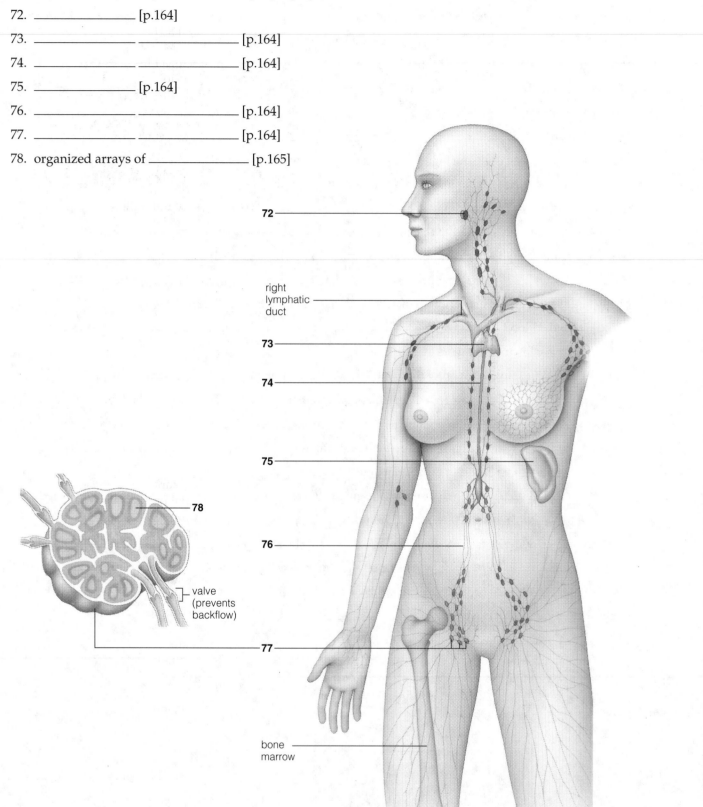

right
lymphatic
duct

72

73

74

75

76

77

78

valve
(prevents
backflow)

bone
marrow

Matching

Choose the most appropriate description for each term.

79. _____ red pulp [p.165]

80. _____ lymph nodes [p.165]

81. _____ thymus [p.165]

82. _____ white pulp [p.165]

83. _____ spleen [p.165]

84. _____ lymph vessels [p.165]

A. Located at intervals along lymph vessels; battlegrounds of lymphocyte armies and foreign agents

B. Largest lymphoid organ; a filtering station for blood and a holding station for lymphocytes

C. Reservoir of red blood cells and macrophages inside the spleen

D. Have a larger diameter than lymph capillaries, smooth muscle in the wall, and valves that prevent backflow

E. Site where lymphocytes multiply, differentiate, and mature into fighters of specific disease agents; also produces hormones that influence these events

F. Portion of the spleen filled with lymphocytes to destroy specific invaders

Short Answer

85. Briefly explain the "drainage, delivery, and disposal" functions of the lymph vascular system.

_____

_____

_____

_____

Self-Quiz

_____ 1. _____ valves may be either bicuspid or tricuspid. [p.152]
   a. Sinoatrial
   b. Semilunar
   c. Atrioventricular
   d. Lymphatic
   e. Venous

_____ 2. The "lub-dup" sound of the heart comes from _____ . [p.153]
   a. the SA valves closing followed by the AV valves closing
   b. the AV valves closing followed by the semilunar valves closing
   c. the semilunar valves closing followed by the SA valves closing
   d. the contraction of the atria followed by the contraction of the ventricles
   e. electrical impulses from the SA node

_____ 3. Arterioles and venules are connected by _____ . [p.155]
   a. thoroughfare channels
   b. interstitial passageways
   c. hepatic portals
   d. true capillaries
   e. renal portals

_____ 4. Blood pressure in capillaries, especially at the arteriole end, forces water out of the capillaries in a process called _____ . [p.161]
   a. diffusion
   b. reabsorption
   c. exocytosis
   d. endocytosis
   e. ultrafiltration

_____ 5. Fluid carried by lymph vessels is returned to collecting ducts that drain into veins in the _____ . [p.165]
a. kidneys
b. lower neck
c. right atrium
d. left atrium
e. liver

_____ 6. The pacemaker of the human heart is the _____ . [p.156]
a. sinoatrial node
b. semilunar valve
c. inferior vena cava
d. superior vena cava
e. atrioventricular node

_____ 7. During systole, _____ . [p.153]
a. the aorta contracts
b. both ventricles contract
c. the entire heart relaxes
d. only the left atrium and left ventricle contract
e. only the right atrium and right ventricle contract

_____ 8. _____ are blood reservoirs in which resistance to flow is low. [p.159]
a. Arteries
b. Arterioles
c. Capillaries
d. Venules
e. Veins

_____ 9. Begin with a red blood cell located in the superior vena cava and travel with it in proper sequence as it goes through the following structures. Which will be last in sequence? [p.154]
a. aorta
b. left atrium
c. pulmonary artery
d. right atrium
e. right ventricle

_____ 10. The lymphatic system is the principal avenue in the human body for transporting _____ . [p.165]
a. oxygen
b. carbon dioxide
c. wastes
d. amino acids
e. interstitial fluids

Chapter Objectives/Review Questions

This section lists general and detailed chapter objectives that can be used as review questions. You can make maximum use of these items by writing answers on a separate sheet of paper. Fill in answers where blanks are provided. To check for accuracy, compare your answers with information given in the chapter or glossary.

1. Describe the functional links that the circulatory system has with the lymphatic system. [p.150]
2. Describe the structure of the heart, including linings, chambers, and valves. [p.152]
3. The _____ [p.154] circuit receives blood from the body tissues and circulates it through the lungs for gas exchange; the _____ [p.154] circuit transports blood to and from tissues.
4. The contraction phase of the cardiac cycle is called _____ [p.147] and the relaxation phase is called _____ [p.153].
5. Trace the path of blood in the human body. Begin with the aorta and name all major components of the circulatory system through which all blood passes before it returns to the aorta. [p.153]
6. Describe the role of capillary beds in the cardiovascular system. Where are these structures found in the body? [pp.150, 154–155]
7. Explain what causes a heart to beat. [p.156]
8. Explain a blood pressure such as 118/76 in terms of systolic and diastolic pressure. Is this a healthy blood pressure? What medical terms refer to elevated blood pressure and low blood pressure? Which is more dangerous and why? [p.157]
9. Describe how the structures of arteries, arterioles, and capillaries differ; describe how the structures of venules and veins differ. [pp.158–159]
10. List and briefly describe the four routes by which substances enter and leave capillaries. [pp.160–161]

11. Distinguish between hypertension and atherosclerosis. What factors increase the risk of these cardiovascular disorders? What can either lead to? [p.162]
12. State the significance of high- and low-density lipoproteins to cardiovascular disorders. [p.163]
13. Describe the components and function of the lymphatic system. [pp.164–165]

Integrating and Applying Key Concepts

You are observing that some people appear as though fluid has accumulated in their lower legs and feet. Their lower extremities resemble those of elephants. You inquire about what is wrong and are told that the condition is caused by a parasite injected by the bite of a mosquito. Construct a testable hypothesis that would explain (1) why the fluid was not being returned to the torso as normal, and (2) how the parasite might prevent the return of fluid.

10

IMMUNITY

Interactive Exercises

CHAPTER INTRODUCTION [p.169]

10.1. THREE LINES OF DEFENSE [p.170]

10.2. COMPLEMENT PROTEINS: PARTNERS IN OTHER DEFENSES [p.171]

10.3. INFLAMMATION — RESPONSES TO TISSUE DAMAGE [pp.172–173]

Selected Words: "vaccination" [p.169], *pasteurization* [p.169], <u>Bacillus anthracis</u> [p.169], *athlete's foot* [p.170], *Lactobacillus* [p.170], *nonspecific response* [p.170], *specific response* [p.170], *chemotaxins* [p.173], *lactoferrin* [p.173], *endogenous pyrogen* [p.173]

Boldfaced, Page-Referenced Terms

[p.170] pathogen _____

_____

[p.170] lysozyme _____

_____

[p.170] immune response _____

_____

[p.171] complement system _____

_____

[p.171] membrane attack complexes _____

_____

[p.171] lysis _____

_____

[p.172] neutrophils _____

_____

[p.172] eosinophils _____

_____

[p.172] basophils _____

_____

[p.172] macrophages _____

_____

[p.172] inflammation _____

_____

[p.172] mast cells _____

_____

[p.172] histamine _____

_____

[p.173] interleukin _____

_____

[p.173] fever _____

_____

Fill-in-the-Blanks

The human body requires protection from a constant bombardment of (1) _____ [p.170], disease-causing organisms such as viruses, bacteria, fungi, protozoans, and parasitic worms. The first line of defense is at the body's (2) _____ [p.170] and includes the upper layer of dead, dry cells in the skin, competition from harmless bacteria on skin and mucus membranes, sticky mucus in the respiratory passageways, lysozyme in many body fluids, low pH in the urine, and diarrhea to rid the GI tract of pathogens. Our second line of defense includes (3) _____ [p.170] responses, in which white blood cells and plasma proteins react to (4) _____ _____ [p.170] in general, not to specific pathogens. The body's third line of defense is a(n) (5) _____ [p.170] response, in which lymphocytes recognize a specific invading pathogen.

About 20 kinds of proteins circulating in the blood plasma make up the (6) "_____ _____" [p.171]. This set of proteins takes part in both specific and nonspecific defense responses. The normally inactive system can be activated in two ways. A complement protein called (7) _____ [p.171] can bind to an antibody that has attached to an invader (an antigen). Alternatively, a different complement protein can interact directly with carbohydrate (8) _____ [p.171] on the surfaces of various pathogens. When a few molecules of a complement protein have become activated in either way, they trigger a (9) _____ [p.171] of reactions. Activation of one kind of complement protein causes the activation of another kind, which activates another, and so on. Some complement proteins unite to form (10) _____ _____ _____ [p.171], which become inserted into the plasma membrane of a pathogen, forming pores that result in (11) _____ [p.171]— disintegration of the cell. Some of these complexes create pores in bacterial cell walls through which molecules of (12) _____ [p.171] diffuse and digest essential structural elements of the cell. Other activated complement proteins promote (13) _____ [p.171], a nonspecific defense response. Their cascades attract phagocytic (14) _____ _____ _____ [p.171] to the irritated or damaged tissue. Activated complement proteins can also bind directly to pathogens. The "complement-coated" invader sticks to a (15) _____ [p.171], which ingests and kills it.

White blood cells develop from (16) _____ _____ [p.172] in the bone marrow. Each of the three types of white blood cells reacts quickly but only lives a few (17) _____ [p.172] or days. (18) _____ [p.172] are the most abundant kind of white blood cells. They (19) _____ [p.172] and kill bacteria. (20) _____ [p.172] defend us against parasitic worms, as well as phagocytizing foreign proteins and helping to control (21) _____ [p.172] responses. (22) _____ [p.172] secrete substances such as histamine that keep (23) _____ [p.172] going once it starts. Slower to act but living for months, (24) _____ [p.172] engulf and digest nearly any foreign agent, as well as helping clean up damaged tissue.

Dichotomous Choice

Circle one of two possible answers given between parentheses in each statement.

25. When cells of any tissue region are damaged or killed, (vasoconstriction/inflammation) occurs, in which the area becomes red, warm, swollen, and painful. [p.172]
26. Fast-acting phagocytes, complement proteins, and some plasma proteins escape from blood (veins/capillaries). [p.172]
27. Mast cells release (histamine/lysozyme), which triggers vasodilation of arterioles (resulting in warmth).
28. The capillaries in the region become (constricted/"leaky"), resulting in edema and pain. [p.172]
29. Within hours of the tissue damage, (eosinophils/neutrophils) are squeezing through capillary walls and starting to work. [p.173]
30. (Macrophages/Basophils) arrive and engulf pathogens and debris. [p.173]
31. Macrophages secrete chemotaxins that attract more (phagocytes/pathogens). [p.173]
32. Endogenous pyrogens secreted by macrophages trigger the release of (interleukin/prostaglandins), leading to fever, which slows down pathogens.
33. The plasma proteins that leak out of capillaries include complement proteins and (blood-thinning/clotting) factors that wall off the inflamed area to prevent the spread of invaders and toxins.

10.4. AN IMMUNE SYSTEM OVERVIEW [pp.174–175]

10.5. HOW LYMPHOCYTES FORM AND DO BATTLE [pp.176–177]

10.6. ANTIBODIES IN ACTION [pp.178–179]

10.7. CELL-MEDIATED RESPONSES — COUNTERING THREATS INSIDE CELLS [pp.180–181]

10.8. *Focus on Your Health:* ORGAN TRANSPLANTS: BEATING THE (IMMUNE) SYSTEM [p.181]

Selected Words: "nonself" markers [p.174], *effector cells* [p.174], *cell-mediated* responses [p.174], *antibody-mediated responses* [p.174], *variable regions* [p.174], *clonal selection* [p.176], *immunological memory* [p.176], *IgM* [p.178], *IgD* [p.179], *IgG* [p.179], *IgA* [p.179], *IgE* [p.179], "touch-kill" [p.181], perforins [p.181], *xenotransplantation* [p.181]

Boldfaced, Page-Referenced Terms

[p.174] immune system _____

_____

[p.174] self/nonself recognition _____

_____

[p.174] specificity _____

_____

[p.174] diversity _____

_____

[p.174] memory _____

_____

[p.174] antigen _____

_____

[p.174] memory cells _____

_____

[p.174] recognition _____

_____

[p.174] rounds of cell division _____

_____

[p.174] specialization _____

_____

[p.174] MHC markers _____

_____

[p.174] antigen-MHC complexes _____

_____

[p.174] helper T cells _____

_____

[p.174] cytotoxic T cells _____

_____

[p.174] natural killer (NK) cells _____

_____

[p.174] B cells _____

_____

[p.174] antibody _____

_____

[p.175] suppressor T cells _____

_____

[p.176] TCRs (T-Cell Receptors) _____

_____

[p.177] tonsils _____

_____

[p.178] immunoglobulins (Igs) _____

_____

[p.180] interferons _____

_____

[p.181] apoptosis _____

_____

Short Answer

1. Which line of defense are B and T cells a part of? When does the body call upon this line of defense? [p.174] _____

_____

_____

_____

2. List and briefly define the four features of the immune system. [p.174] _____

_____

_____

_____

Sequencing

Sequence the events listed and write their letters in the appropriate blanks to show the order in which they occur. [p.174]

A. Repeated division of activated B and T cells

B. Specialization of B and T lymphocytes as memory cells for future attacks of the same invader

C. Specialization of B and T lymphocytes as effector cells to fight the invader

D. Recognition of "nonself" marker by B and T lymphocytes

3. _____

4. _____

5. _____   6. _____

Choice

Choose one of the white blood cell types (a–e) for each of the following functions.

a. effector cytotoxic T cells and natural killer cells   b. B cells
c. antigen-presenting cells   d. helper T cells   e. suppressor T cells

7. _____ take part in an antibody-mediated response [p.174]

8. _____ sends out a chemical signal that activates T cells [p.174]

9. _____ attack infected body cells, tumor cells, and organ transplant cells [p.174]

10. _____ binds to antigen-MHC complexes displayed on antigen-presenting cell [p.174]

11. _____ secrete huge numbers of antibodies that can bind to a specific invader that will then be destroyed by a phagocytic white blood cell [p.174]

12. _____ produces chemical signals that help shut down an immune response [p.175]

13. _____ ingests antigen, then links some antigen fragments to MHC markers for display [p.174]

14. _____ take part in a cell-mediated response [p.174]

15. _____ types include macrophages, B cells, and dendritic cells [p.174]

16. _____ signals B and T cells sensitive to a specific antigen to divide repeatedly, producing effector and memory cells [p.174]

17. _____ destroy cells by releasing chemicals that form lethal pores in a target's plasma membrane [p.174]

True/False

If the statement is true, write a "T" in the blank. If the statement is false, make it correct by writing the word(s) in the blank that should take the place of the underlined word(s).

_____ 18. Antigen-presenting cells display ingested <u>antibody</u> fragments linked with MHC markers. [p.174]

_____ 19. Antigen-presenting cells send out chemical signals that activate <u>B cells</u>.

_____ 20. After binding to antigen-MHC complexes, <u>helper T cells</u> produce chemical signals that will cause T and B cells to multiply.

_____ 21. Cytotoxic T cells and natural killer cells attack invader cells as part of an <u>antibody-mediated</u> response.

_____ 22. Antibodies are released from <u>NK cells.</u>

_____ 23. Antibodies bind to specific <u>antigens.</u>

_____ 24. NK cells, macrophages, and neutrophils will destroy cells with <u>antibodies</u> attached to them.

Fill-in-the-Blanks

All of the antigen receptors on a particular B or T cell are identical, consisting of polypeptide chains. Certain parts of each chain are called (25) _____-_____ [p.176]. In each new T and B cell, genetic instructions for building these antigen-binding regions of receptor molecules are (26) _____ [p.176] around, producing unique polypeptides. As a B cell matures, it makes copies of its genetically unique, typically Y-shaped (27) _____ [p.176] molecule. The copies move to the B cell membrane and the antigen-binding arms (28) _____ [p.176] from the cell. This cell is a (29) "_____" [p.176] B cell, as it has not yet encountered the specific antigen it is programmed to detect. As T cells form in the bone marrow, they migrate to the (30) _____ [p.176], where they mature and acquire unique antigen-binding receptors called (31) _____ [p.176]. Once released into the bloodstream, these molecules can recognize only (32) _____-_____ [p.176] complexes. In the armies of B and T cells that form during an immune response, each one bears only a single type of (33) _____ _____ [p.176], all specific for a detected invader. These armies arise from a mechanism called (34) _____ _____ , [p.176] in which an antigen binds only to the B or T cell that is displaying a receptor specific for it. Once this first cell is activated, repeated rounds of cell division produce a clone of genetically identical cells. (35) _____ _____ [p.176] refers to the body's capacity to make a secondary immune response to a later encounter with the same type of antigen that provoked the primary response. (36) _____ _____ [p.176] that form during the primary response circulate for years or decades. Because there are many of these already on patrol, an infection by a previously encountered invader is (37) _____ [p.176] before a person gets sick.

    In the skin, mucous membranes, and internal organs, the (38) _____ [p.177] and dendritic cells pick up antigens by phagocytosis or endocytosis. They then migrate to the (39) _____ [p.177] nodes all over the body. There, they present the antigen to (40) _____ and

_____ [p.177] cells. Any antigens that have made their way into tissue fluid enter the

(41) _____ [p.177] vascular system. Some may enter the bloodstream, but (42) _____

_____ [p.177] trap most of them. Antigens that enter the bloodstream are intercepted

by the (43) _____ [p.177]. In the lymph nodes, antigens move through the region occupied

by B cells, macrophages, and dendritic cells. The invaders are captured, processed, and presented to

(44) _____ _____ [p.177] cells, thereby activating them. At this point, a full-blown

(45) _____ _____ [p.177] gets under way.

Sequencing

Write the letter of the first of the following events to occur beside number 46, with the other letters following in sequence. [p.178]

46. _____

47. _____

48. _____

49. _____

50. _____

51. _____

52. _____

A. Effector B cells produce huge numbers of antibodies that flag invaders for destruction by phagocytes.

B. The clonal descendants of the activated B cell specialize as effector and memory B cells.

C. An antigen binds with antibodies on the B cell and links them together.

D. The TCRs of a helper T cell bind to the antigen-MHC complex displayed on a B cell, allowing an exchange of signals, after which the cells disengage.

E. The antibodies of a B cell bind to a second, unprocessed antigen. This, along with interleukins secreted from helper T cells, causes B cell cloning.

F. A naive B cell is produced, bristling with many copies of a unique antigen.

G. Endocytosis occurs and antigen-MHC complexes form, which are displayed on the surface of the B cell.

Matching

53. _____ IgG [p.179]

54. _____ IgA [p.179]

55. _____ IgM [p.178]

56. _____ IgE [p.179]

57. _____ IgD [p.179]

A. Binds to basophils and mast cells, prompting histamine release

B. Ten binding sites, first antibody made by newborns

C. Commonly found bound to naive B cells; helps activate helper T cells

D. Found in tears, saliva, breast milk, and mucus

E. Most efficient at turning on complement proteins; neutralized many toxins

Fill-in-the-Blanks

Pathogens such as viruses, some bacteria, some fungi, protozoans, and tumor cells hide inside body

(58) _____ [p.180]. The body's weapons against such dangers are (59) _____-

_____ [p.180] immune responses. Helper T cells and cytotoxic T cells respond to particular

(60) _____ [p.180]. NK cells, macrophages, neutrophils, and eosinophils make more

(61) _____ [p.180] responses. Chemical signals such as interleukins and defensive proteins called

(62) _____ [p.180] attract defenders and stimulate them to divide, specialize, and attack.

Interferons also spur the making of (63) _____ [p.180] molecules, so some cells display antigens

more effectively.

Cytotoxic T cells produce molecules used to (64) "_____-_____" [p.181] infected

and abnormal body cells. Examples are (65) _____ [p.181] proteins, which form pores in a

target's plasma membrane. Cytotoxic T cells also secrete chemicals that cause the programmed death, or

(66) _____ [p.181], of a target cell, as well as contributing to the (67) _____ [p.181] of

transplanted tissues and organs. Natural killer (NK) cells act when infections trigger a flood of interferons

and (68) _____ [p.181]. They are the (69) _____ [p.181] immune cells called into

battle; they buy time while others are mobilizing.

Organ transplants are risky because the organ recipient's immune system will perceive donated tissues

as (70) _____ [p.181] and attempt to reject them. Transplants usually succeed only when the

donor and recipient share at least (71) _____ [p.181] percent of their MHC markers. Because of

this, the best donors are close (72) _____ [p.181] of the recipient. After surgery, the organ

recipient receives drugs that (73) _____ [p.181] the immune system. This means that the patient

must take large doses of (74) _____ [p.181] to control infections. Some researchers are trying to

genetically alter (75) _____ [p.181] to create varieties bearing common human MHC markers.

Transplantation of organs from one species to another is called (76) _____ [p.181]. Tissues of the

(77) _____ [p.181] and (78) _____ [p.181] are examples of exceptions to the "rule"

that transplanted tissues evoke immune defenses, thus making transplants of structures such as the cornea

simpler.

10.9. PRACTICAL APPLICATIONS OF IMMUNOLOGY [pp.182–183]

10.10. *Science Comes to Life:* A CAN'T WIN PROPOSITION? [p.183]

10.11. ABNORMAL, HARMFUL, OR DEFICIENT IMMUNE RESPONSES [pp.184–185]

Selected Words: *active immunization* [p.182], "booster shot" [p.182], "attenuated" pathogens [p.182], "transgenic" viruses [p.182], *passive* immunization [p.182], hybridoma cells [p.182], *gamma* interferon [p.183], *beta* interferon [p.183], *multiple sclerosis* [p.183], *lymphokines* [p.183], LAK cells [p.183], *influenza* virus [p.183], "gene swapping" [p.183], *allergens* [p.183], *hay fever* [p.183], *anaphylactic shock* [p.183], antihistamines [p.184], *rheumatoid arthritis* [p.185], *diabetes mellitus* [p.185], *systemic lupus erythematosus* (SLE) [p.185], *severe combined immune deficiency* (SCID) [p.185], human immunodeficiency virus (HIV) [p.185]

Boldfaced, Page-Referenced Terms

[p.182] vaccine _____

_____

[p.182] monoclonal antibodies _____

_____

[p.184] allergy _____

_____

[p.185] autoimmune response _____

_____

[p.185] AIDS (acquired immunodeficiency syndrome) _____

_____

Matching

1. _____ monoclonal antibodies

2. _____ gamma interferon

3. _____ active immunization

4. _____ passive immunization

5. _____ LAK cells

6. _____ attenuated

7. _____ booster shot

8. _____ beta interferon

9. _____ transgenic

A. injection of antigen-containing vaccines [p.182]
B. elicits a secondary response that results in more effector and memory cells being made [p.182]
C. being used to treat multiple sclerosis [p.183]
D. genetically engineered viruses used to make vaccines [p.182]
E. Injections of purified antibody molecules [p.182]
F. B cells cloned from a single antibody-producing cell; used in research to find very small quantities of a substance [pp.182–183]
G. produced by T cells; calls NK cells into action and enhances activity of macrophages [p.183]
H. killed or extremely weakened pathogens used in making vaccines [p.182]
I. tumor-infiltrating lymphocytes produced by research [p.183]

Short Answer

10. How do the three strains of influenza virus differ from each other? In addition to this variation, name one way in which a strain can mutate to become a new variant of the flu. [p.183] _____

_____

_____

_____

Fill-in-the-Blank

A(n) (11) _____ [p.184] is an altered secondary response to a normally harmless substance. Such a substance is called a(n) (12) _____ [p.184]. Some people are genetically (13) _____ [p.184] to develop allergies. When an allergic person is first exposed to certain antigens, IgE (14) _____ [p.184] are secreted and bind to (15) _____ [p.184] cells. These secrete prostaglandins, (16) _____ [p.184], and other substances that cause inflammation. They also cause the person's airways to (17) _____ [p.184]. If an allergen spreads and promotes wide-ranging inflammation, a life-threatening condition called (18) _____ _____ [p.184] results. The emergency treatment is an injection of the hormone (19) _____ [p.184].

(20) _____ [p.184] are anti-inflammatory drugs that are used to relieve short-term allergy symptoms. In a desensitization program, doses of allergens are administered so that a person's body produces more circulating (21) _____ [p.184] molecules and memory cells, thus blocking allergic inflammation.

A(n) (22) _____ [p.185] response is a disorder in which the body mobilizes its forces against normal body cells or proteins. (23) _____ _____ [p.185] is an example of this kind of disorder, in which skeletal joints are chronically inflamed. Another example is type 1 (24) _____ [p.185], in which the immune system destroys the insulin-secreting cells of the pancreas. In the autoimmune disease (25) _____ _____ _____ [p.185], the affected person develops antibodies to her or his own DNA and other "self" components. Deficient immune responses include disorders in which the body does not have enough functioning (26) _____ [p.185]. Both T and B cells are in short supply in (27) _____ _____ _____ _____ [p.185] (SCID), an inherited life-threatening disorder. Infection by the (28) _____ _____ _____ [p.185] (HIV) causes AIDS — (29) _____ _____ _____ [p.185]. HIV is transmitted when (30) _____ _____ [p.185] of an infected person enter another person's tissues. The virus cripples the immune system by attacking (31) _____ _____ [p.185] cells and (32) _____ [p.185]. The body then becomes dangerously susceptible to opportunistic infections and to some otherwise rare forms of (33) _____ [p.185].

Self-Quiz

Multiple Choice

_____ 1. All the body's white blood cells are derived from stem cells in the _____ . [p.172]
   a. spleen
   b. liver
   c. thymus
   d. bone marrow
   e. thyroid

_____ 2. The plasma proteins that are activated when they contact a microorganism or virus are collectively known as the _____ system. [p.171]
   a. shield
   b. complement
   c. IgG
   d. MHC
   e. HIV

_____ 3. The lymphocytes known as _____ attack only specific nonself cells or substances. [p.174]
   a. macrophages
   b. B and T cells
   c. mast cells
   d. basophils
   e. interleukins

_____ 4. _____ produce and secrete antibodies that set up bacterial invaders for subsequent destruction by macrophages. [pp.174–175]
   a. B cells
   b. Phagocytes
   c. T cells
   d. Mast cells
   e. Thymus cells

_____ 5. Antibody molecules are shaped like the letter _____ . [p.176]
   a. *Y*
   b. *W*
   c. *Z*
   d. *H*
   e. *E*

_____ 6. The markers for every cell in the human body are referred to by the letters _____ . [p.174]
   a. HIV
   b. MBC
   c. RNA
   d. DNA
   e. MHC

_____ 7. B and T lymphocytes that enter a resting stage are called _____ . [p.174]
   a. effector cells
   b. macrophages
   c. mast cells
   d. helper cells
   e. memory cells

_____ 8. The great variety of naive B and T cells differ in _____ regions of polypeptide chains. [p.176]
   a. shuffled
   b. constant
   c. clonal
   d. mutated
   e. variable

_____ 9. When an allergic person is first exposed to certain antigens, IgE antibodies cause _____ . [p.184]
   a. inflammation
   b. clonal cells to be produced
   c. B cell division
   d. the immune response to be suppressed
   e. an autoimmune disorder to develop

_____ 10. The clonal selection theory explains _____ . [p.174]
   a. how all the B or T cells responding to an invasion bear a single type of antigen receptor
   b. how B cells differ from T cells
   c. how so many different kinds of antigen-specific receptors can be produced by lymphocytes
   d. how memory cells are set aside from effector cells
   e. how antigens differ from antibodies

Matching

Choose the most appropriate description for each term.

11. _____ autoimmune response [p.185]

12. _____ antibody [p.176]

13. _____ antigen [p.174]

14. _____ macrophage [p.172]

15. _____ clone [p.176]

16. _____ complement [p.171]

17. _____ histamine [p.172]

18. _____ MHC marker [p.174]

19. _____ effector cells [p.174]

20. _____ T cell [p.176]

A. Begins its development in bone marrow but matures in the thymus gland
B. A population of genetically identical cells that descended from a selected T or B cell
C. A potent chemical that causes blood vessels to dilate and let protein leak through the vessel walls
D. Y-shaped immunoglobulin produced by a B cell
E. A nonself marker that triggers the formation of lymphocytes
F. Produced by activated B or T lymphocytes and specialized to destroy an invader
G. A group of about 20 proteins that participate in nonspecific and specific defenses
H. A disorder in which the body's immune system attacks its own cells and proteins
I. Proteins sticking out of the plasma membrane of body cells that identify "self"
J. "Big eaters" that live for months and phagocytize foreign agents

Chapter Objectives/Review Questions

This section lists general and detailed chapter objectives that can be used as review questions. You can make maximum use of these items by writing answers on a separate sheet of paper. To check for accuracy, compare your answers with information given in the chapter or glossary.

1. Describe typical external barriers that organisms such as humans present to invading organisms. [p.170]
2. List and discuss three nonspecific defense responses that serve to exclude microbes from the body. [p.170]
3. Explain how the complement system is involved in destroying invaders and fanning inflammation. [p.171]
4. Describe what occurs in the course of an acute inflammation and how these events stop the spread of a pathogen. [p.172]
5. List the four features that define the immune system. [p.174]
6. List the three basic steps of an immune response. [p.174]
7. Distinguish between the cell-mediated response and the antibody-mediated response. [pp.174–175]
8. Describe the clonal selection theory and relate this concept to immunological memory. [pp.174–175]
9. Explain how helper T cells and cytotoxic T cells are involved in countering threats inside cells. [pp.180–181]
10. Contrast active and passive immunization in the areas of procedure and effectiveness. [p.182]
11. Distinguish allergy from autoimmune disease. [pp.184–185]
12. Describe how AIDS specifically interferes with the human immune system. [p.185]

Integrating and Applying Key Concepts

Vaccines can be developed against many pathogens, but not all. Diseases like the common cold and AIDS cannot be prevented by vaccination. Others, like influenza, must be vaccinated against every year. Humans are not capable of developing lasting immunity to any of the pathogens causing these diseases. Offer a possible genetic explanation for this variation among pathogens.

11

THE RESPIRATORY SYSTEM

Interactive Exercises

CHAPTER INTRODUCTION [p.189]

11.1. THE RESPIRATORY SYSTEM — BUILT FOR GAS EXCHANGE [pp.190–191]

Selected Words: "altitude sickness" [p.189], *edema* [p.189], "respiration" [p.189], *nasal cavity* [p.190], "Adam's apple" [p.190], *epiglottis* [p.191], *glottis* [p.191], *laryngitis* [p.191], *pleural sac* [p.191], *intrapleural space* [p.189], *intrapleural fluid* [p.191], *pleurisy* [p.191], "bronchial trees" [p.191], *alveolar sac* [p.191]

Boldfaced, Page-Referenced Terms

[p.190] respiratory system _____

_____

[p.190] respiration _____

_____

[p.190] pharynx _____

_____

[p.190] larynx _____

_____

[p.191] trachea _____

_____

[p.191] bronchus _____

_____

[p.191] vocal cords _____

_____

[p.191] lungs _____

_____

[p.191] diaphragm _____

_____

[p.191] pleurae _____

_____

[p.191] bronchioles _____

_____

[p.191] respiratory bronchioles _____

_____

[p.191] alveolus (plural: alveoli) _____

_____

Labeling-Matching

Identify each of the components of the human respiratory system in the accompanying illustration by entering the correct names in the numbered blanks provided. Complete the exercise by matching and entering the letter of the correct function of each component in the parentheses that follow the labels. [p.190]

1. _____ _____ ( )
2. _____ _____ ( )
3. _____ _____ ( )
4. _____ ( )
5. _____ _____ ( )
6. _____ ( )
7. _____ ( )
8. _____ ( )
9. _____ ( )
10. _____ ( )
11. _____ _____ ( )
12. _____ ( )
13. _____ _____ ( )
14. _____ ( )
15. _____ _____ ( )
16. _____ _____ ( )

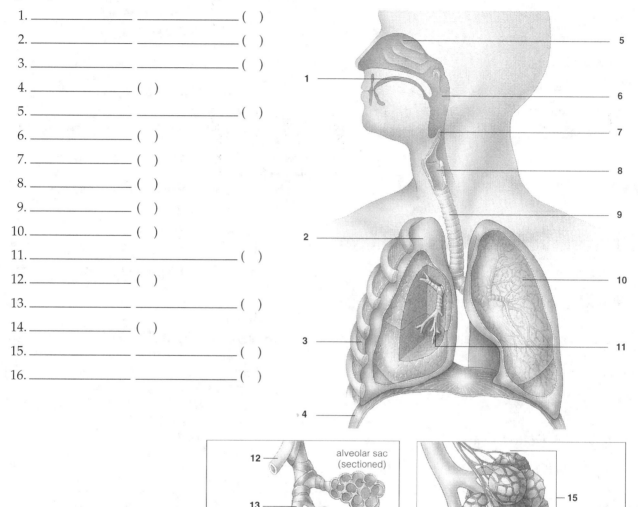

alveolar sac (sectioned)

A. Airway where sound is produced; closed off while swallowing
B. Muscle sheet between chest cavity and abdominal cavity with roles in breathing
C. Increasingly branched airways between two bronchi and alveoli
D. Closes off larynx during swallowing
E. Chamber in which air is warmed, moistened, and filtered, and in which sounds resonate
F. Ribcage muscles with roles in breathing
G. Airway that connects larynx with two bronchi
H. Supplemental airway when breathing is labored

I. Thin-walled air sacs where gases are exchanged between lungs and pulmonary capillaries
J. Membranes that separate lungs from other organs; fluid-filled cavity between has roles in breathing
K. Airway that connects nasal cavity and mouth with larynx; enhances sounds; also connects with esophagus
L. Lobed, elastic organ of breathing that enhances gas exchange between the internal environment and the outside air

Matching

Choose the most appropriate description for each term.

17. _____ edema [p.189]

18. _____ pleural sac [p.191]

19. _____ intrapleural space [p.191]

20. _____ intrapleural fluid [p.191]

21. _____ pleurisy [p.191]

22. _____ diaphragm [p.191]

23. _____ glottis [p.191]

A. A sheet of muscle between the thoracic and abdominal cavities

B. Inflammation of pleurae, causing membranes to dry out and rub against each other, which makes breathing painful, or to oversecrete fluid, which hampers breathing movements

C. A gap between the vocal cords that is forced open with each exhalation

D. Formed by a thin, double membrane of epithelium covering each lung

E. A thin layer of lubricating substance filling the intrapleural space

F. Swelling of tissues due to an accumulation of plasma fluid

G. A very narrow space between the two pleural membranes

11.2. THE "RULES" OF GAS EXCHANGE [p.192]

11.3. *Focus on Our Environment:* BREATHING AS A HEALTH HAZARD [p.193]

11.4. BREATHING — AIR IN, AIR OUT [pp.194–195]

Selected Words: pressure gradients [p.192], "partial pressure" [p.192], *hypoxic* [p.192], *hyperventilation* [p.192], *the bends* [p.192], *decompression sickness* [p.192], *nitrogen narcosis* [p.192], *carbon monoxide poisoning* [p.192], *bronchitis* [p.193], *emphysema* [p.193], "secondhand smoke" [p.193], *asthma* [p.193], *intrapulmonary pressure* [p.194], *negative pressure gradient* [p.195], *inspiratory reserve volume* [p.195], *expiratory reserve volume* [p.195], *residual volume* [p.195]

Boldfaced, Page-Referenced Terms

[p.192] respiratory surface _____

_____

[p.194] respiratory cycle _____

_____

[p.194] inspiration _____

_____

[p.194] expiration _____

_____

[p.195] tidal volume _____

_____

[p.195] vital capacity _____

_____

Fill-in-the-Blanks

(1) _____ _____ [p.192] relies on the tendency of oxygen and carbon dioxide to diffuse down their concentration gradients, or as we say in the case of gases, their (2) _____ [p.192] gradients. The partial pressure of oxygen at sea level is (3) _____ [p.192] percent of 760 mm Hg (atmospheric pressure at sea level), which equals 160 mm Hg. The partial pressure of carbon dioxide is about 0.3 mm Hg, or about 0.04 percent of 760 mm Hg. Large animals such as humans must be capable of efficient gas exchange. Gases enter and leave the body by crossing a thin, moist (4) _____ [p.192] surface of epithelium. The surface must be moist because gases cannot diffuse across membranes unless they are dissolved in fluid. The larger the surface area and the steeper the partial (5) _____ _____ [p.192], the faster diffusion occurs. In healthy human lungs, millions of thin-walled (6) _____ [p.192] provide a huge surface area for gas exchange. Gas exchange is facilitated by the (7) _____ [p.192] in red blood cells. In the lungs, each hemoglobin molecule binds with up to (8) _____ [p.192] oxygen molecules. When blood carries red blood cells into tissues where oxygen concentration is low, hemoglobin (9) _____ [p.192] oxygen. By carrying oxygen away from the respiratory surface, hemoglobin helps maintain the required pressure (10) _____ [p.192] that helps draw oxygen into the lungs — and into blood in lung capillaries.

Dichotomous Choice

Circle one of two possible answers given between parentheses in each statement.

11. Partial pressure of oxygen (decreases/increases) as altitude increases. [p.192]
12. Higher than 2,400 meters, or about 8,000 feet, brain respiratory centers compensate for oxygen deficiency by triggering faster and deeper breathing called (hypoxia/hyperventilation). [p.192]
13. To compensate for increased pressure underwater, a diver may inhale pressurized air, which (decreases/increases) the total pressure of gases in the lungs. [p.192]
14. If a diver ascends too rapidly, the decrease in pressure may result in the formation of bubbles of (O_2/N_2), causing a painful condition called the bends, or decompression sickness. [p.192]
15. Deeper than 150 meters, the high partial pressure of N_2 may cause a euphoric feeling called (nitrogen narcosis/decompression sickness). [p.192]
16. In carbon monoxide poisoning, tiny amounts of the gas can tie up half of the body's hemoglobin, resulting in tissues not receiving enough oxygen, a condition called (hypoxia/hyperventilation). [p.192]

Choice

For questions 17–26, choose from the following:

a. bronchitis   b. emphysema   c. asthma

17. _____ A chronic, incurable condition of the respiratory tract in which breathing can become so difficult so quickly that the victim feels in imminent danger of suffocating [p.193]

18. _____ Scar tissue from chronic bronchitis builds up, the bronchi become clogged with mucus, and a reduced number of enlarged alveoli skew the balance between air flow and blood flow [p.193]

19. _____ Can be brought on by air pollution that increases mucus secretions and interferes with ciliary action in the lungs [p.193]

20. _____ Many sufferers lack a functional gene coding for antitrypsin, a protein that inhibits tissue-destroying enzymes produced by bacteria [p.193]

21. _____ Factors that trigger an attack actually cause the smooth muscle to contract in strong spasms, so that the bronchi and bronchioles suddenly narrow [p.193]

22. _____ Cilia are lost from the lining and mucus-secreting cells multiply as the body fights against the accumulating debris [p.193]

23. _____ Among other triggers, allergens can set off an attack [p.193]

24. _____ The disease can develop over 20 or 30 years, but by the time it is detected, lung tissue is permanently damaged [p.193]

25. _____ Triggering substances may include pollen, dairy products, shellfish, flavorings and other foods, pet hairs or dandruff — even mite dung in house dust [p.193]

26. _____ With this disorder, the lungs are so distended and inelastic that gases cannot be exchanged efficiently; running, walking, even exhaling can be difficult [p.193]

Analyzing Diagrams

To better understand the mechanisms of ventilation, study each indicated part of the following illustration. Then answer the accompanying questions. As you do the exercise, it aids understanding if you are conscious of the same parts of your own anatomy presently undergoing the ventilation cycle. [p.194]

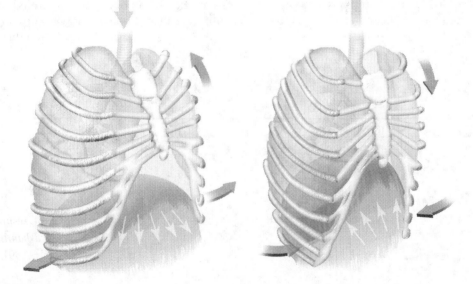

_____ 27. Which diagram, left or right, shows inspiration?

_____ 28. Which diagram, left or right, shows expiration?

_____ 29. Which muscles contract and relax to cause the rib cage movement indicated by the three larger arrows on each diagram?

_____ 30. The small arrows on each diagram show the contraction and relaxation of which muscle?

_____ 31. On the left-hand diagram, in which direction does the diaphragm move?

_____ 32. On the left-hand diagram, in which directions does the ribcage move?

_____ 33. What effect do the movements on the left-hand diagram have on lung volume?

_____ 34. On the right-hand diagram, in which direction does the diaphragm move?

_____ 35. On the right-hand diagram, in which directions does the ribcage move?

_____ 36. What effect do the movements on the right-hand diagram have on the lungs?

Dichotomous Choice

Circle one of two possibilities given between parentheses in each statement.

37. The air movements of ventilation result from rhythmic increases and decreases in the volume of the (lungs/thoracic cavity). [p.194]
38. The cohesiveness of water molecules in the fluid inside the (pleural sac/thoracic cavity) helps keep the lungs close to the thoracic wall. [p.195]
39. When the air pressure in alveolar sacs is lower than atmospheric pressure, air flows down its gradient and (enters/exits) the alveoli. [p.195]
40. The normally passive part of the respiratory cycle is (inspiration/expiration). [p.195]
41. When the muscles that cause inhalation relax, air flows (into/out of) the lungs. [p.195]
42. The amount of air taken in by a person in a normal breath, about 500 ml, is known as the (vital capacity/tidal volume). [p.195]
43. The maximum volume of air that can move out of the lung after a person inhales as deeply as possible is called the (vital capacity/tidal volume). [p.95]
44. In addition to the air taken in as tidal volume, a person can forcibly inhale roughly 3,100 ml of air, called the (expiratory/inspiratory) reserve volume. [p.195]
45. Forcibly exhaling, you can expel an additional (expiratory/inspiratory) reserve volume of about 1,200 ml. [p.195]
46. Even at the end of your deepest exhalation, your lungs still cannot be completely emptied of air; roughly another 1,200 ml of (expiratory reserve/residual) volume remains. [p.195]
47. People rarely take more than (one-half/three-fourths) of their vital capacity, even when they breathe deeply during strenuous exercise. [p.195]
48. About (150/350) ml of inhaled air is actually available for gas exchange. [p.195]

11.5. HOW GASES ARE EXCHANGED AND TRANSPORTED [pp.196–197]

11.6. CONTROLS OVER GAS EXCHANGE [pp.198–199]

11.7. *Focus on Your Health:* TOBACCO AND OTHER THREATS TO THE RESPIRATORY SYSTEM [pp.200–201]

Selected Words: respiration [p.196], *external* respiration [p.196], *internal* respiration [p.196], *respiratory membrane* [p.196], *pulmonary surfactant* [p.196], *infant respiratory distress syndrome* [p.196], *carbaminohemoglobin* ($HbCO_2$) [p.197], breathing *rhythm* [p.198], breathing *magnitude* [p.198], *cerebrospinal fluid* [p.199], *apnea* [p.199], *sleep apnea* [p.199], "secondhand smoke" [p.200]

Boldfaced, Page-Referenced Terms

[p.196] oxyhemoglobin (HbO$_2$) _____

_____

[p.197] carbonic anhydrase _____

_____

[p.199] carotid bodies _____

_____

[p.199] aortic bodies _____

_____

Matching

Choose the most appropriate description for each term.

1. _____ aortic bodies [p.199]

2. _____ carbaminohemoglobin (HbCO$_2$) [p.197]

3. _____ carbonic anhydrase [p.197]

4. _____ carotid bodies [p.199]

5. _____ external respiration [p.196]

6. _____ infant respiratory distress syndrome [p.196]

7. _____ internal respiration [p.196]

8. _____ oxyhemoglobin (HbO$_2$) [p.196]

9. _____ respiration [p.196]

10. _____ respiratory centers [p.199]

11. _____ pulmonary surfactant [p.196]

12. _____ ventilation [p.196]

A. Hemoglobin with bound oxygen
B. Enzyme in red blood cells mediating the chemical reaction that forms and dissociates carbonic acid
C. Phase of respiration in which oxygen moves from blood into tissues and carbon dioxide moves from tissues into blood
D. Sensory receptors located where carotid arteries branch to the brain; detect arterial changes in carbon dioxide and oxygen levels as well as pH
E. Provides the body as a whole with oxygen for aerobic respiration in cells and disposes of carbon dioxide
F. Stimulated by decreasing pH of cerebrospinal fluid
G. Life-threatening breathing disorder in premature infants whose partially developed lungs do not yet have functional surfactant-secreting cells
H. Sensory receptors in arterial walls near the heart that detect changes in carbon dioxide, oxygen, and pH levels
I. Draws oxygen into the lungs and moves carbon dioxide out
J. Carbon dioxide bound with hemoglobin
K. Phase of respiration in which oxygen moves from alveoli to blood and carbon dioxide moves from blood to alveoli
L. Secreted by certain cells of the alveolar epithelium; reduces surface tension of the watery fluid film between alveoli

Dichotomous Choice

Circle one of two possibilities given between parentheses in each statement.

13. Hemoglobin in red blood cells enables blood to carry (17/70) times more oxygen than it otherwise would. [p.196]
14. The higher the partial pressure of oxygen around the alveoli, the (less/more) oxygen will be picked up by hemoglobin. [p.196]
15. Oxygen-binding to hemoglobin weakens as temperature rises, or as pH (increases/decreases). [p.197]
16. The (more/less) DPG bound to hemoglobin, the less affinity hemoglobin has for binding oxygen, and thus more oxygen can be available to tissues. [p.197]
17. About 70 percent of the body's carbon dioxide is transported in plasma in the form of (carbonic acid/bicarbonate). [p.197]
18. Bicarbonate forms after carbon dioxide combines with (water/carbonic acid) in plasma. [p.197]
19. Carbonic anhydrase mediates reactions that convert unbound carbon dioxide to carbonic acid and its dissociation products; the blood level of carbon dioxide then (rises/falls) rapidly. [p.197]
20. In the alveoli, the partial pressure of carbon dioxide is (higher/lower) than it is in the surrounding capillaries. [p.197]
21. Driven by its partial pressure gradient, (carbon dioxide/oxygen) diffuses from alveoli, through interstitial fluid, and into lung capillaries. [p.197]

Fill-in-the-Blanks

Gas exchange is most efficient when breathing is regulated so that the rate of air flow is matched by the rate of blood flow. The (22) _____ [p.198] system acts to balance the flow rates. It controls breathing (23) _____ [p.198], which is the rate of breathing, and breathing (24) _____ [p.198], which is the depth of breathing. Breathing rhythm involves nervous system control of the diaphragm and rib cage muscles. One group of cells of the reticular formation in the brain stem coordinates signals calling for (25) _____ [p.198]; another cell cluster coordinates the signals for (26) _____ [p.198]. The resulting rhythmic contractions are fine-tuned by another part of the brain stem, the (27) _____ [p.198], which can stimulate or inhibit both cell clusters.

The body's control over breathing magnitude involves monitoring the level of (28) _____ [p.199] in the blood. However, the nervous system is more sensitive to levels of (29) _____ _____ [p.199]. Both oxygen and carbon dioxide levels are monitored in blood flowing through (30) _____ [p.199]. When conditions warrant, nervous system signals adjust contractions of the (31) _____ [p.199] and muscles in the chest wall, and so adjust the rate and depth of breathing. Sensory receptors in the (32) _____ [p.199] of the brain detect rising carbon dioxide levels. Corresponding increases of H^+ in the blood affect the pH of (33) _____ [p.199] fluid, which bathes the medulla. The shift in pH stimulates the receptors that signal changes to the brain's (34) _____ [p.199] centers. In addition, the brain receives input from sensory receptors such as the (35) _____ [p.199] bodies located where the carotid arteries branch to the brain, and the (36) _____ [p.199] bodies in arterial walls near the heart. Both types of receptors can detect changes in carbon dioxide and oxygen levels in arterial blood as well as changes in blood pH. The brain responds by increasing the (37) _____ [p.199] rate, so more oxygen can be delivered to tissues.

Fill-in-the-Blanks

Cigarette smoke — including (38) "_____ [p.200] smoke" inhaled by a nonsmoker — causes lung

cancer and contributes to various other ills. Cigarette smoking causes at least (39) _____ [p.200]

percent of all lung cancer deaths.

Consider Your Own Health

40. A considerable amount of scientific research has been reported that firmly establishes the devastating effects of smoking on human beings. Study the Focus on Your Health essay and table on page 200 closely and then complete the following table by first checking in the left column only those risks associated with smoking that *you personally are willing to take* as a smoker. Be prepared to seriously discuss with your friends, relatives, and classmates the reasons you have for accepting the risks you checked. Then, in the right column, check those risks to which you are willing to subject nonsmokers among your family and close friends through secondhand smoke. [p.200]

Personal Risks	Risks Associated with Smoking	Risks to Others
	a. shortened life expectancy	
	b. chronic bronchitis, emphysema	
	c. lung cancer	
	d. cancer of the mouth	
	e. cancer of the larynx	
	f. cancer of the esophagus	
	g. cancer of the pancreas	
	h. cancer of the bladder	
	i. coronary heart diseases	
	j. effects on your offspring	
	k. impaired immune system	
	l. slow bone healing (about 30% longer)	

Self-Quiz

_____ 1. Most forms of life depend on _____ [p.192] down concentration gradients to obtain oxygen and eliminate carbon dioxide.
   a. active transport
   b. bulk flow
   c. diffusion
   d. osmosis
   e. muscular contractions

_____ 2. _____ [p.192] is the most abundant gas in Earth's atmosphere.
   a. Water vapor
   b. Oxygen
   c. Carbon dioxide
   d. Hydrogen
   e. Nitrogen

_____ 3. _____ [p.199] are involved in local chemical controls over air flow operation in the lungs.
   a. Smooth muscles in bronchiole walls
   b. Nerve cell clusters in the pons and medulla
   c. The vagus nerve and its associated stretch receptors
   d. Carotid bodies
   e. Aortic bodies

_____ 4. The amount of a gas diffusing across a respiratory surface does _not_ depend on the _____ . [p.192]
   a. amount of surface area of the membrane involved
   b. level of glucose in the blood
   c. differences in partial pressure of a gas across the membrane involved
   d. presence and amount of hemoglobin
   e. whether or not a respiratory membrane is moist

_____ 5. Immediately before reaching the alveoli, air passes through the _____ [p.191].
   a. bronchioles
   b. glottis
   c. larynx
   d. pharynx
   e. trachea

_____ 6. During inhalation, _____ [p.195].
   a. air pressure in the alveolar sacs is lower than atmospheric pressure
   b. fresh air follows a gradient from the alveoli up to the trachea
   c. the diaphragm moves upward and becomes more curved
   d. the thoracic cavity volume decreases
   e. all of the above

_____ 7. Hemoglobin _____ [pp. 196–197].
   a. releases oxygen more readily in active tissues
   b. tends to release oxygen in places where the temperature is lower
   c. tends to hold on to oxygen when the pH of the blood drops
   d. tends to give up oxygen in regions where partial pressure of oxygen is higher than in the blood
   e. all of the above

_____ 8. Because mechanisms for sensing carbon dioxide and oxygen levels become less effective, along with loss of lung elasticity with age, older people are more affected by _____ [p.199].
   a. sleep apnea
   b. emphysema
   c. asthma
   d. bronchitis
   e. edema

_____ 9. The wall of an alveolus separated from lung capillaries by a thin film of interstitial fluid constitutes a _____ [p.196].
   a. bronchial tree
   b. pleural sac
   c. respiratory membrane
   d. pulmonary surfactant
   e. respiratory center

_____ 10. Inflammation of the mucous lining of the vocal cords results in _____ [p.191].
   a. emphysema
   b. bronchitis
   c. hypoxia
   d. laryngitis
   e. asthma

Chapter Objectives/Review Questions

This section lists general and detailed chapter objectives that can be used as review questions. You can make maximum use of these items by writing answers on a separate sheet of paper. Fill in answers where blanks are provided. To check for accuracy, compare your answers with information given in the chapter or glossary.

1. List all the principal parts of the human respiratory system and explain how each structure contributes to transporting oxygen from the external world to the bloodstream. [pp.190–191]
2. Describe the behavior of gases and the type of respiratory surface that participates in gas exchange in humans. [pp.192, 196–197]
3. Describe the relationship of the human lung to the pleural sac and to the thoracic cavity. [p.195]
4. Distinguish bronchitis from emphysema. [p.193]
5. The two phases of ventilation are _____ and _____ . [p.194]
6. For the human lung, distinguish tidal volume from vital capacity. [p.195]
7. Distinguish respiration from ventilation. [p.196]
8. Explain why oxygen diffuses from the bloodstream into the tissues far from the lungs. Then explain why carbon dioxide diffuses into the bloodstream from the same tissues. [pp.196–197]
9. Explain why oxygen diffuses from alveolar air spaces, through interstitial fluid, and across capillary epithelium. Then explain why carbon dioxide diffuses in the reverse direction. [pp.196–197]
10. Hemoglobin with oxygen bound to it is called _____ [p.196]; when carbon dioxide binds to hemoglobin, it is called _____ [p.197].
11. Describe what happens to carbon dioxide when it dissolves in water under conditions normally present in the human body. [p.197]
12. List the structures involved in detecting carbon dioxide levels in the blood and in regulating the rate of breathing. Name the location of each structure. [pp.198–199]
13. List some of the things that go awry with the respiratory system, and describe the characteristics of the breakdown. [p.198]
14. How long after quitting smoking does it take smokers to return to the life expectancy of nonsmokers?

Integrating and Applying Key Concepts

Explain the need for an extensive and efficient circulatory system to go along with the human respiratory system, including the huge number of capillaries, efficient heart, and presence of hemoglobin.

12

THE URINARY SYSTEM

Interactive Exercises

CHAPTER INTRODUCTION [p.205]

12.1. THE CHALLENGE: SHIFTS IN EXTRACELLULAR FLUID [pp.206–207]

12.2. *Focus on Our Environment:* IS YOUR DRINKING WATER POLLUTED? [p.207]

12.3. THE URINARY SYSTEM — BUILT FOR FILTERING AND WASTE DISPOSAL [pp.208–209]

Selected Words: "insensible" water losses [p.206], *uric acid* [p.207], *gout* [p.207], *ammonia* [p.207], "deamination" reactions [p.207], *urea* [p.207], "potential human carcinogen" [p.207], *cortex* [p.208], *medulla* [p.208], *renal capsule* [p.208], *collecting ducts* [p.208], *renal pelvis* [p.208], *afferent arteriole* [p.209], *efferent arteriole* [p.209]

Boldfaced, Page-Referenced Terms

[p.206] extracellular fluid (ECF) _____

[p.206] intracellular fluid _____

[p.206] urinary system _____

[p.206] urinary excretion _____

[p.207] electrolytes _____

[p.208] kidney _____

[p.208] nephrons _____

[p.208] ureter _____

[p.208] urinary bladder _____

[p.208] urethra _____

[p.209] renal corpuscle _____

[p.209] glomerulus _____

[p.209] Bowman's (glomerular) capsule _____

[p.209] proximal tubule _____

[p.209] loop of Henle _____

[p.209] distal tubule _____

[p.209] peritubular capillaries _____

Short Answer

1. What is the basic task of the urinary system? Use the terms extracellular fluid and intracellular fluid in your answer. [p.206] _____

_____

_____

_____

Complete the Table

2. Complete the table by categorizing the movements of water and solutes. [pp.206–207]

Water Gain	Water Loss	Solute Gain	Solute Loss
a.	c.	g.	k.
b.	d.	h.	l.
----------	e.	i.	m.
----------	f.	j.	----------

True/False

If the statement is true, write a "T" in the blank. If the statement is false, make it correct by writing the word(s) in the blank that should take the place of the underlined word(s).

_____ 3. When there is a water deficit in body tissues, the <u>kidneys</u> compel(s) us to seek out water. [p.206]

_____ 4. Of the ways that water is lost, the body exerts the most control over <u>sweating</u>. [p.206]

_____ 5. Water loss that a person is not aware of, such as through evaporation from lungs or skin, is called "<u>insensible</u>". [p.206]

_____ 6. The <u>excretory</u> system brings oxygen into the blood, and respiring cells add carbon dioxide to it. [p.206]

_____ 7. The most abundant metabolic waste is <u>carbon dioxide</u>. [p.207]

_____ 8. All metabolic wastes besides carbon dioxide leave in the <u>feces</u>. [p.207]

_____ 9. <u>Uric acid</u> forms when cells break down nucleic acids. [p.207]

_____ 10. Ammonia is a product of <u>carbohydrate</u> metabolism. [p.207]

_____ 11. The liver combines ammonia and carbon dioxide to make <u>urine</u>. [p.207]

_____ 12. The kidneys regulate the balance of ions, such as sodium, potassium, and calcium, that are called <u>electrolytes</u>. [p.207]

Labeling

Identify each indicated part of the accompanying illustrations. [p.208]

13. _____

14. _____

15. _____ _____

16. _____

17. _____ _____

18. _____ _____

19. _____

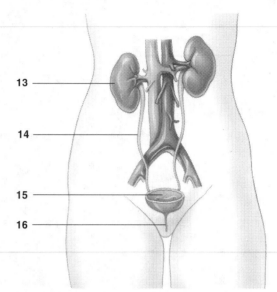

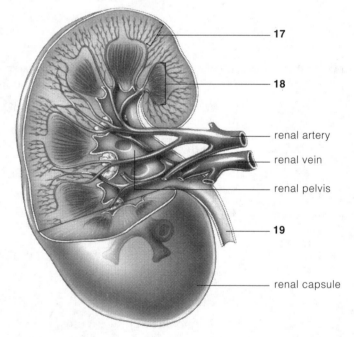

Fill-in-the-Blanks

Urine formation occurs in a pair of (20) _____ [p.208], the central components of the urinary system. Each kidney contains blood vessels and slender tubes called (21) _____ [p.208], which filter water and (22) _____ [p.208] from the (23) _____ [p.208]. More than a (24) _____ [p.208] nephrons are packed inside each kidney. A(n) (25) _____ [p.209] arteriole delivers blood to each nephron, and filtration starts at the (26) _____ _____

[p.209]. Here, the nephron balloons around a cluster of capillaries called the (27) _____ [p.209].
Its wall region, the (28) _____ [p.209] capsule, receives water and solutes filtered from the
blood. This filtrate moves into the (29) _____ [p.209] tubule, then through the loop of
(30) _____ [p.209] and into the (31) _____ [p.209] tubule, which empties into a
collecting duct.

The capillaries inside a glomerular (Bowman's) capsule converge to form a(n) (32) _____
[p.209] arteriole. This branches into a set of (33) _____ [p.209] capillaries that weave around the
nephron's tubules. Water and essential solutes (34) _____ [p.209] from the tubule system enter
these capillaries and are returned to the bloodstream.

Labeling

Identify each indicated part of the accompanying illustration. [p.209]

35. _____ _____

36. _____ _____

37. _____ _____

38. _____ _____

39. _____ _____

40. _____ _____

41. _____ _____

42. _____ _____

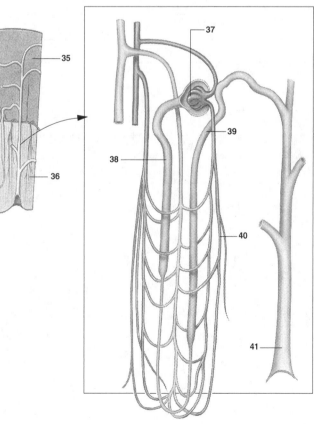

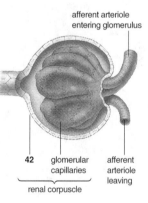

afferent arteriole
entering glomerulus

42   glomerular   afferent
      capillaries   arteriole
                        leaving

renal corpuscle

12.4. HOW URINE FORMS [pp.210–211]

12.5. KEEPING WATER AND SODIUM IN BALANCE [pp.212–213]

12.6. *Focus on Your Health:* DRINK TO THE HEALTH OF YOUR URINARY TRACT [p.213]

12.7. ADJUSTING REABSORPTION: HORMONES ARE THE KEY [p.214]

Selected Words: urination [p.205], *internal urethral sphincter* [p.205], *external urethral sphincter* [p.205], *stress incontinence* [p.205], *kidney stones* [p.205], *lithotripsy* [p.205], sodium "pumps" [p.206], *cystitis* [p.207], *pyelonephritis* [p.207], *chlamydia* [p.207], *urinalysis* [p.207], *diuretic* [p.208], *renin* [p.209], "thirst" [p.209]

Boldfaced, Page-Referenced Terms

[p.204] urine _____

_____

[p.204] filtration _____

_____

[p.204] reabsorption _____

_____

[p.204] secretion _____

_____

[p.206] proximal tubule _____

_____

[p.208] ADH (antidiuretic hormone) _____

_____

[p.208] aldosterone _____

_____

[p.209] juxtaglomerular apparatus _____

_____

[p.209] thirst center _____

_____

Complete the Table

1. Complete the table below by categorizing events that occur in the kidneys into the three steps of urine formation. [p.210]

Step 1: Filtration	Step 2: Reabsorption	Step 3: Secretion
a. occurs at the _____	d. occurs along tubular parts of _____	g. occurs along tubes of _____
b. moves _____ and small _____ out of blood	e. most of filtrate moves _____ of nephron	h. wastes and ions enter nephron by diffusion out of _____
c. forces substances into _____ capsule, then into _____ tubule of nephron	f. returns most of filtrate into neighboring _____	i. substances are secreted into forming _____

Dichotomous Choice

Circle one of two possible answers given between parentheses in each statement.

2. Drug testing relies on the use of (urinalysis/dialysis), which shows which substances have been secreted in the urine. [pp.210–211]
3. Urination is a reflex response signaled by tension across the walls of the (kidney/urinary bladder). [p.212]
4. During urination, the internal urethral sphincter (relaxes/contracts) while the bladder walls force urine through the urethra. [p.212]
5. A person can exert control over the (internal/external) urethral sphincter. [p.212]
6. (Gall/Kidney) stones are deposits of substances that have settled out of the urine. [p.212]

Fill-in-the-Blanks

Of all the water and sodium the kidneys filter, about two-thirds is quickly reabsorbed at the part of the nephron nearest the glomerulus, called the (7) _____ _____ [p.212]. The epithelial cells here have transport proteins at their outer surface that function as (8) _____ "_____" [p.212]. Sodium ions are actively transported from the inside of the tubule into the (9) _____ _____ [p.212] outside. Glucose and (10) _____ _____ [p.212] are reabsorbed by active transport linked to sodium reabsorption. The wall of the proximal tubule is quite permeable to water, so water follows its gradient and flows (11) _____ [p.212] of the tubule into the (12) _____ _____ [p.212], returning to the bloodstream.

The hairpin-shaped loop of Henle descends into the kidney (13) _____ [p.212]. The (14) _____ [p.212] limb of the loop is permeable to water, so water moves (15) _____ [p.212] and is reabsorbed. In the ascending limb, water cannot cross the tubule wall, but (16) _____ [p.213] ions are actively transported out of the nephron. The filtrate that finally reaches the distal tubule in the kidney cortex is quite (17) _____ [p.213], with a low sodium concentration.

Two hormones, ADH (antidiuretic hormone) and (18) _____ [p.214], adjust the reabsorption of water and (19) _____ [p.214] along the distal tubules and collecting ducts. The hypothalamus triggers (20) _____ [p.214] secretion from the pituitary gland when the concentration of (21) _____ [p.214] in extracellular fluid rises above a certain point or when (22) _____ _____ [p.214] falls. ADH makes the walls of distal tubules and collecting ducts more permeable to (23) _____ [p.214], so additional water is reabsorbed from the dilute urine inside the distal tubule of nephrons. In the presence of ADH, only a(n) (24) _____ [p.214] volume of very concentrated urine is excreted. Excess water inhibits ADH secretion, so more water is excreted in the urine. A(n) (25) _____ [p.214] is any substance that promotes the loss of water, such as alcohol, which acts by (26) _____ [p.214] the release of ADH. When sensory receptors in the walls of kidney blood vessels detect a decrease in sodium and water with it, kidney cells in the juxtaglomerular apparatus secrete an enzyme called (27) _____ [p.214]. Renin activity results in the production of (28) _____ [p.215] II, which stimulates cells of the adrenal cortex to secrete (29) _____ [p.215]. This causes cells of the distal tubules and collecting ducts to reabsorb (30) _____ [p.215] faster, so that less is excreted in the urine. Too much sodium in the extracellular fluid inhibits aldosterone secretion, so that (31) _____ [p.215] sodium is reabsorbed and more is excreted. As well as prompting the release of ADH, when the hypothalamus detects an increase in solute concentration in the extracellular fluid, its (32) _____ _____ [p.214] is stimulated. This inhibits saliva production, and the resulting "thirst" stimulates a person to seek fluids.

12.8. MAINTAINING THE BODY'S ACID–BASE BALANCE [p.216]
12.9. KIDNEY DISORDERS [p.217]
12.10. MAINTAINING THE BODY'S CORE TEMPERATURE [pp.218–219]

Selected Words: *polycystic kidney disease* [p.217], *nephritis* [p.217], *glomerulonephritis* [p.217], *"dialysis"* [p.217], *hemodialysis* [p.217], *peritoneal dialysis* [p.217], *nonshivering heat production* [p.218], *"brown fat"* [p.218], *hypothermia* [p.218], *frostbite* [p.218], *hyperthermia* [p.219], *heat exhaustion* [p.219], *heat stroke* [p.219], *"thermostat"* [p.219]

Boldfaced, Page-Referenced Terms

[p.216] acid–base balance _____

_____

[p.216] bicarbonate–carbon dioxide buffer system _____

_____

[p.218] core temperature _____

_____

[p.218] endotherms _____

_____

[p.218] peripheral vasoconstriction _____

_____

[p.218] pilomotor response _____

_____

[p.213] peripheral vasodilation _____

Fill-in-the-Blanks

Normal pH of extracellular body fluids ranges from (1) _____ to _____ [p.216].

Buffer systems have only a temporary effect in neutralizing excess H$^+$, but only the urinary system can

(2) _____ [p.216] excess H$^+$ and thus restore the buffer. H$^+$ formed in cells is secreted into the

(3) _____ [p.216], where it may combine with bicarbonate ions, phosphate ions, or ammonia and

be excreted in the (4) _____ [p.216]. Adjustments in acid–base balance and other kidney

functions are essential to maintaining (5) _____ [p.216].

Matching

6. _____ polycystic kidney disease [p.217]

7. _____ peritoneal dialysis [p.217]

8. _____ nephritis [p.217]

9. _____ glomerulonephritis [p.217]

10. _____ hemodialysis [p.217]

A. Fluid is put into the abdominal cavity and later drained out
B. Inflammation of the kidneys; interferes with blood filtering
C. Damage to kidneys from hypertension, diabetes, etc.
D. Blood pumped from bloodstream is cleaned and returned
E. Masses form in and destroy kidney tissue

Fill-in-the-Blanks

Normal human core temperature is about (11) _____ °C or _____ °F [p.218]. Humans

are endotherms with body temperature controlled mainly by (12) _____ [p.218] activity and

homeostatic controls. Responses to cold stress are governed by the (13) _____ [p.218]. When

thermoreceptors at the body surface detect a drop in temperature, smooth muscle in the walls of arterioles

in the skin are commanded to (14) _____ [p.218]. The resulting (15) _____

_____ [p.218] reduces blood flow to the body surface, so that heat is retained. In the (16)

_____ _____ [p.218] to falling temperature, body hair "stands on end." This traps

still air close to the body, thereby reducing heat loss. When these responses are not enough to counter

cold stress, the hypothalamus causes low-level skeletal muscle contractions, and you start to (17)

_____ [p.218], which produces heat. Prolonged or severe exposure to cold leads to a hormonally

induced elevation of cellular metabolic rate called (18) _____ _____ _____

[p.218], especially notable in "brown fat." (19) _____ [p.218] is a condition in which the body

core temperature falls below the normal range. Survival of extreme hypothermia due to immersion in cold

water may be due to the (20) _____ [p.218] reflex of mammals, in which heart rate slows and

blood is shunted to vital organs.

When core temperature rises above a set point, the hypothalamus signals blood vessels in the skin to dilate in a response called (21) _____ _____ [p.219], resulting in heat loss from more blood flowing to the skin. (22) _____ _____ _____ [p.219] occurs when the hypothalamus activates (23) _____ [p.219] glands. Heat is dissipated with evaporation of the water in sweat. Heavy sweating may bring about (24) _____ [p.219] losses that can alter the internal environment to the extent that the affected person faints. When heat stress cannot be adequately countered, the result is (25) _____ [p.219]. A person can suffer (26) _____ _____ [p.219] when vasodilation and water loss cause a drop in blood pressure. When the body's temperature controls completely fail, (27) _____ _____ [p.219] occurs, which can be lethal.

Self-Quiz

_____ 1. The most toxic waste product of metabolism is _____ . [p.207]
   a. water
   b. uric acid
   c. urea
   d. ammonia
   e. carbon dioxide

_____ 2. The functional subunit of a kidney that filters blood as well as restoring solute and water balance is called a _____ . [p.208]
   a. glomerulus
   b. loop of Henle
   c. nephron
   d. ureter
   e. none of the above

_____ 3. In humans, the thirst center is located in the _____ . [p.215]
   a. adrenal cortex
   b. thymus
   c. heart
   d. adrenal medulla
   e. hypothalamus

_____ 4. Of all the water and sodium that leave the bloodstream at the glomerulus, about two-thirds is promptly reabsorbed across the _____ . [p.212]
   a. loop of Henle
   b. proximal tubule
   c. ureter
   d. Bowman's capsule
   e. collecting duct

_____ 5. During reabsorption, sodium ions cross the proximal tubule walls into the interstitial fluid principally by means of diffusion or _____ . [p.210]
   a. phagocytosis
   b. countercurrent multiplication
   c. bulk flow
   d. active transport
   e. all of the above

_____ 6. Filtration of the blood in the kidney takes place in the _____ . [p.210]
   a. loop of Henle
   b. proximal tubule
   c. distal tubule
   d. glomerulus
   e. all of the above

_____ 7. _____ primarily controls the concentration of sodium in urine. [pp.212–213]
   a. Insulin
   b. Glucagon
   c. Antidiuretic hormone
   d. Aldosterone
   e. Epinephrine

_____ 8. Hormonal control over excretion primarily affects the _____ . [p.214]
   a. Bowman's capsule
   b. distal tubules
   c. proximal tubules
   d. urinary bladder
   e. loop of Henle

9. The last portion of the excretory system passed by urine before it is eliminated from the body is the _____. [p.208,211]
   a. renal pelvis
   b. bladder
   c. ureter
   d. collecting ducts
   e. urethra

10. _____ is the principal waste product of protein breakdown, a combination of ammonia and carbon dioxide produced by the liver. [p.207]
   a. Urea
   b. Uric acid
   c. Creatine
   d. Amino acid
   e. ADH

Chapter Objectives/Review Questions

This section lists general and detailed chapter objectives that can be used as review questions. You can make maximum use of these items by writing answers on a separate sheet of paper. To check for accuracy, compare your answers with information given in the chapter or glossary.

1. List some of the factors that can change the composition and volume of body fluids. [pp.206–207]
2. List three electrolytes that the kidneys help to maintain correct levels of in the body. [p.207]
3. List successively the parts of the human urinary system that constitute the path of urine formation and excretion. [pp.208–209]
4. Locate the processes of filtration, reabsorption, and secretion along a nephron and tell what makes each process happen. [pp.210–211]
5. Discuss the factors that allow the kidneys to effectively filter large amounts of blood fairly quickly. [p.211]
6. State explicitly how the hypothalamus, pituitary, and tubules of the nephrons are interrelated in regulating water and solute levels in body fluids. [pp.212–213]
7. Why are urinary tract infections more common in women than in men? Why are they more of a problem for men as they get older? [p.213]
8. Describe the role of the kidneys in maintaining the pH of the extracellular fluids between 7.37 and 7.43. [p.216]
9. List three kidney disorders and explain what can be done if kidneys become too diseased to work properly. [p.217]
10. Explain how humans and other endotherms maintain their body temperature when environmental temperatures fall. [p.218]
11. Define *hypothermia* and state the situations in which a human might experience the disorder. [p.218]
12. Explain how humans and other endotherms maintain their body temperature when environmental temperatures rise 3 to 4 degrees Fahrenheit above standard body temperature. [p.219]
13. Discuss what can happen as a result of hyperthermia. [p.219]
14. What happens in the body at the onset of a fever as well as when a fever "breaks"? [p.219]

Integrating and Applying Key Concepts

The hemodialysis machine used in hospitals is expensive and time consuming. So far, artificial kidneys capable of allowing people who have nonfunctional kidneys to purify their blood by themselves, without having to go to a hospital or clinic, have not been developed. Which aspects of the hemodialysis procedure do you think have presented the most problems for developing a method of home self-care? If you had an unlimited budget and were appointed head of a team to develop such a procedure and its instrumentation, what strategy would you pursue?

13

THE NERVOUS SYSTEM

Interactive Exercises

CHAPTER INTRODUCTION [p.223]

13.1. NEURONS — THE COMMUNICATION SPECIALISTS [pp.224–225]

13.2. A CLOSER LOOK AT ACTION POTENTIALS [pp.226–227]

13.3. CHEMICAL SYNAPSES: COMMUNICATION JUNCTIONS [pp.228–229]

Selected Words: Parkinson's disease (PD) [p.223], "glia" [p.223], *excitable* neuron [p.224], "input zones" [p.224], "conducting zone" [p.224], "trigger zone" [p.224], *axon hillock* [p.224], "output zones" [p.224], *voltage* [p.224], "nerve impulse" [p.224], "gates" [p.225], *electric* gradient [p.225], *graded* signal [p.226], *local* signals [p.226], *all-or-nothing* events [p.226], *presynaptic cell* [p.228], *post*synaptic cell [p.228], *myasthenia gravis* [p.228], *serotonin* [p.228], *norepinephrine* [p.228], *dopamine* [p.228], *GABA* [p.228], *substance P* [p.228], *nitric oxide* (NO) [p.228], Alzheimer's disease [p.229], *endorphins* [p.229], *depolarize* [p.229], *hyperpolarize* [p.229], *summation* [p.229]

Boldfaced, Page-Referenced Terms

[p.223] sensory neurons _____

[p.223] interneurons _____

[p.223] motor neurons _____

[p.223] neuroglia _____

[p.223] Schwann cells _____

[p.224] axons _____

[p.224] dendrites _____

[p.224] resting membrane potential _____

[p.224] action potential _____

[p.225] sodium–potassium pumps _____

[p.225] threshold level of stimulation _____

[p.228] neurotransmitters _____

[p.228] chemical synapses _____

[p.228] acetylcholine (ACh) _____

[p.229] neuromodulators _____

[p.229] synaptic integration _____

Matching

1. _____ Schwann cells
2. _____ axon
3. _____ action potential
4. _____ interneurons
5. _____ sodium–potassium pump
6. _____ sensory neurons
7. _____ neuroglia
8. _____ resting membrane potential
9. _____ dendrite
10. _____ motor neurons

A. Physically support and protect neurons [p.223]
B. A "nerve impulse" [p.224]
C. Carrier protein; moves Na^+ and K^+ across membrane [p.225]
D. Respond to a stimulus and relay information to the brain [p.223]
E. Insulate axons of sensory and motor neurons [p.223]
F. Steady charge difference across neuron cell membrane [p.224]
G. Receive and integrate input, then signal other neurons [p.223]
H. Neuron's "input zone"; receives incoming signals [p.224]
I. Relay messages from brain to muscles or glands [p.223]
J. Neuron's "conducting zone"; carries outgoing signals [p.224]

Sequence

Write the letter of the first of the following events to occur beside number 11, with the other letters following in sequence. [p.223]

11. _____
12. _____
13. _____
14. _____
15. _____
16. _____

A. motor neuron
B. stimulus (input)
C. receptor (sensory neuron)
D. response (output)
E. effector (muscle, gland)
F. integrator (interneurons)

Labeling [p.224]

17. _____

18. _____ _____

19. _____ _____

20. _____

21. _____ _____

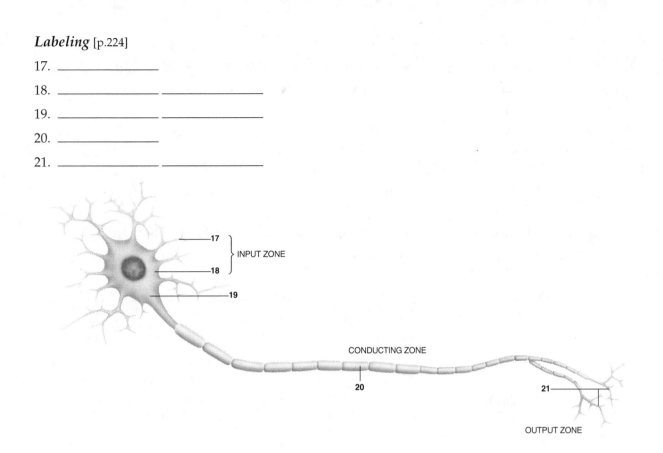

INPUT ZONE

CONDUCTING ZONE

20

21

OUTPUT ZONE

Fill-in-the-Blanks

When a neuron is resting, there is a steady difference in electric charge of about −70 millivolts across its

plasma membrane, with the cytoplasm side being more (22) _____ [p.224] compared to the

interstitial fluid just outside the neuron. This charge difference is called the (23) _____

_____ [p.224] and has the potential to do work. If a strong signal arrives at the resting neuron's

(24) _____ _____ [p.224], it may cause the voltage difference to reverse for an instant.

This "nerve impulse" or (25) _____ _____ [p.224] is a neuron's communication signal.

When the charge difference is reversed, that bit of membrane cannot receive another signal until its

(26) _____ _____ _____ [p.224] is restored. Two factors aid in this process.

The first is that the membrane prevents charged substances such as K$^+$ and Na$^+$ from passing through it, so

that differences in ion (27) _____ [p.225] can build up across the membrane. Second, the cell can

control the flow of ions through channel (28) _____ [p.225] that span the bilayer. Some stay open

so that ions can (29) _____ [p.225] ("leak") through them. Others have (30) "_____"

[p.225], which open only when the neuron is adequately stimulated. In a resting motor neuron, the gated

sodium channels are (31) _____ [p.225]. A neuron can't respond to an incoming signal unless the

concentration and (32) _____ [p.225] gradients across its plasma membrane are in place. To prevent "leaks" of K^+ and Na^+, a neuron expends energy on a(n) (33) _____ _____ [p.225] mechanism to maintain the gradients. This involves channel proteins called (34) _____ _____ _____ , [p.225] which move potassium in and sodium out of the cell.

True/False

If the statement is true, write a "T" in the blank. If the statement is false, make it correct by writing the word(s) in the blank that should take the place of the underlined word(s).

_____ 35. If you feel a touch to your arm, it means that action potentials <u>have</u> been triggered. [p.226]

_____ 36. A <u>local</u> signal means that signals at an input zone vary in magnitude. [p.226]

_____ 37. When a stimulus is intense enough or long-lasting, graded signals spread into an <u>action potential</u> zone of the neuron. [p.226]

_____ 38. The minimum voltage change needed to produce an action potential is the <u>threshold</u> level of stimulation. [p.226]

_____ 39. An appropriate stimulus causes sodium ions to flow <u>out of</u> the neuron, resulting in even more gates opening. [p.226]

_____ 40. Every action potential spikes to the same level as an <u>all-or-nothing</u> event. [p.226]

_____ 41. When a spike ends, the gated sodium channels close, while <u>chloride</u> channels are open so that ions flow out and restore the original voltage difference across the membrane. [p.227]

_____ 42. Action potentials always propagate <u>toward</u> a trigger zone. [p.227]

Fill-in-the-Blanks

The arrival of a signal may prompt a neuron to release chemical messengers called (43) _____ [p.228]. These molecules diffuse across the (44) _____ _____ [p.228], a gap between the neuron's output zone and the input zone of the neighboring neuron. The (45) _____ [p.228] cell stores neurotransmitter molecules. When an action potential arrives, calcium ions flow into the cell, resulting in synaptic (46) _____ [p.228] fusing with the plasma membrane and releasing neurotransmitters. These diffuse across the synapse and bind with receptor proteins on the receiving, or (47) _____ [p.228], cell. This binding causes a channel to open so that ions can flow through and enter the receiving cell. Some neurotransmitters excite the receiving cell, while others (48) _____ [p.228] the cell's activity. Examples of neurotransmitters include acetylcholine (ACh), serotonin, norepinephrine, (49) _____ [p.229], and GABA.

Between 1,000 and 10,000 communication lines form (50) _____ [p.229] with a typical neuron in the brain, and the brain has at least 100 (51) _____ [p.229] neurons. At any moment, many signals are washing over the input zone of a receiving neuron. All are (52) _____ [p.229] potentials. Signals called (53) _____ [p.229] depolarize the membrane, bringing it closer to threshold. (54) _____ [p.229] will hyperpolarize the membrane and drive it away from threshold. (55) _____ _____ [p.229] tallies up competing signals in a process called

(56) _____ [p.229]. (57) _____ [p.229] occurs when neurotransmitter molecules from more than one presynaptic cell reach a neuron's input zone at the same time.

The flow of signals through the nervous system depends on the rapid, controlled (58) _____ [p.229] of neurotransmitters from synapses. Some diffuse out, others are cleaved by (59) _____ [p.229] such as acetylcholinesterase, and some are pumped back into the presynaptic cells. Some drugs, such as cocaine, inhibit the reuptake of neurotransmitters. Antidepressants like Prozac alter mood by blocking the reuptake of (60) _____ [p.229].

Matching

61. _____ norepinephrine

62. _____ acetylcholine

63. _____ myasthenia gravis

64. _____ substance P

65. _____ nitric oxide

66. _____ serotonin

A. Conveys information about pain [p.228]
B. Affects emotional states, dreaming, and awaking [p.228]
C. Gas that controls blood vessel dilation [p.228]
D. Excites or inhibits muscles, glands, the brain, and the spinal cord [p.228]
E. Affects sleeping, sensory perception, body temperature, and emotional states [p.228]
F. Produces drooping eyelids, muscle weakness, and fatigue [p.228]

Labeling

Label the parts of the accompanying illustration. [p.229]

67. _____ _____

68. _____

69. _____ _____
    _____

70. _____

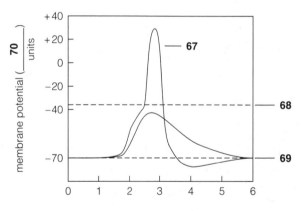

13.4. INFORMATION PATHWAYS [pp.230–231]

13.5. THE NERVOUS SYSTEM: AN OVERVIEW [pp.232–233]

13.6. *Focus on Our Environment:* AN ENVIRONMENTAL ASSAULT ON THE NERVOUS SYSTEM [p.233]

Selected Words: "reverberating" circuits [p.230], *somatic* subdivision [p.232], *autonomic* subdivision [p.233], *ganglia* (singular: ganglion) [p.233]

Boldfaced, Page-Referenced Terms

[p.230] nerve _____

_____

[p.230] myelin sheath _____

_____

[p.231] reflex _____

_____

[p.231] reflex arcs _____

_____

[p.232] central nervous system (CNS) _____

_____

[p.232] peripheral nervous system (PNS) _____

_____

Fill-in-the-Blanks

Blocks of hundreds or thousands of interneurons are parts of (1) _____ [p.230]. In the brain, some circuits diverge, some converge into just a few signals, and others synapse back on themselves, including the (2) "_____" [p.230] circuits that make your eye muscles twitch rhythmically as you sleep. A(n) (3) _____ [p.230] consists of the long axons of sensory neurons, motor neurons, or both. Each axon has a(n) (4) _____ _____ [p.230] that speeds the rate at which (5) _____ _____ [p.230] propagate. The sheath consists of glia called (6) _____ [p.230] cells. A(n) exposed (7) _____ [p.230] separates each Schwann cell from the next one. Action potentials (8) _____ [p.230] from node to node in what is called saltatory conduction. The (9) _____ _____ _____ [p.230] has no Schwann cells. There, glia called (10) _____ [p.230] sheath myelinated axons.

Sensory and motor neurons participate in a path known as the (11) _____ _____ [p.231]. In the simplest (12) _____ _____ [p.231], sensory neurons synapse directly on motor neurons. The stretch reflex (13) _____ [p.231] a muscle involuntarily when gravity or some other load has caused the muscle to stretch.

Labeling

Label the parts of the illustrated nerve. [p.230]

14. _____

15. _____ _____

16. _____ _____

17. _____ _____

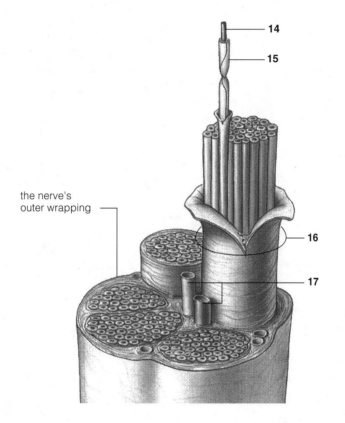

the nerve's
outer wrapping

Matching

Match the lettered choices with the correct number in the diagram. [p.231]

18. _____

19. _____

20. _____

21. _____

22. _____

23. _____

24. _____

A. Muscle cell plasma membrane stimulated to contract (response)
B. Action potentials generated in motor neuron and propagated along its axon toward muscle
C. Axon ending of motor neuron synapse with muscle cells
D. Spinal cord
E. Receptor endings of sensory neurons generate action potential toward spinal cord
F. Muscle spindle stretches (stimulus)
G. Axon endings of sensory neurons synapse with motor neuron

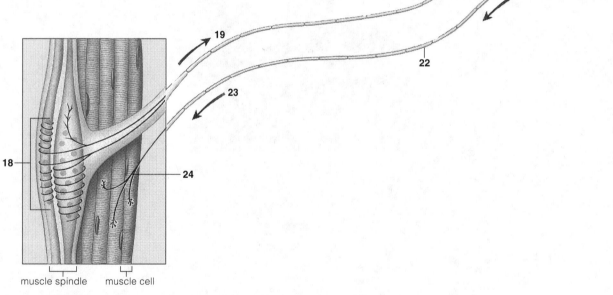

muscle spindle   muscle cell

Short Answer

25. Distinguish between the central and peripheral nervous systems, including subdivisions. [p.232] _____

_____

_____

_____

Labeling

Label each numbered part of the accompanying illustration. [p.232]

26. _____ nerves

27. _____ _____

28. _____ nerves

29. _____ nerves

30. _____ nerves

31. _____ nerves

32. _____ nerves

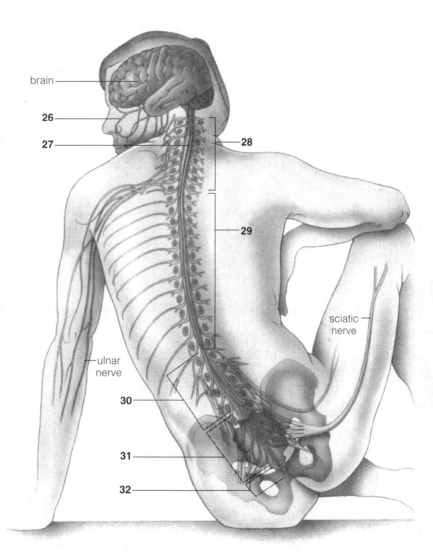

13.7. MAJOR EXPRESSWAYS: PERIPHERAL NERVES AND THE SPINAL CORD [pp.234–235]

13.8. THE BRAIN — COMMAND CENTRAL [pp.236–237]

13.9. A CLOSER LOOK AT THE CEREBRUM [pp.238–239]

13.10. MEMORY AND STATES OF CONSCIOUSNESS [pp.240–241]

Selected Words: rebound effect [p.235], *white matter* [p.235], *gray matter* [p.235], *meninges* [p.229], *spinal reflexes* [p.235], *autonomic reflexes* [p.235], *dura mater* [p.236], *olfactory bulbs* [p.237], *nuclei* [p.237], *basal nuclei* [p.237], *ventricles* [p.237], *motor* areas [p.238], *sensory* areas [p.238], *association* areas [p.238], *short-term* storage [p.240], *long-term* storage [p.240], *facts* [p.240], *skills* [p.240], *amnesia* [p.240]

Boldfaced, Page-Referenced Terms

[p.234] somatic nerves _____

[p.234] autonomic nerves _____

[p.234] parasympathetic nerves _____

[p.235] sympathetic nerves _____

[p.235] fight–flight response _____

[p.235] spinal cord _____

[p.236] brain _____

[p.236] meninges _____

[p.236] cerebrospinal fluid _____

[p.236] brain stem _____

[p.236] medulla oblongata _____

[p.236] cerebellum _____

[p.236] pons _____

_____

[p.237] cerebrum _____

_____

[p.237] thalamus _____

_____

[p.237] hypothalamus _____

_____

[p.237] reticular formation _____

_____

[p.237] cerebrospinal fluid _____

_____

[p.237] blood–brain barrier _____

_____

[p.238] cerebral hemispheres _____

_____

[p.238] cerebral cortex _____

_____

[p.239] limbic system _____

_____

[p.240] memory _____

_____

Choice

Choose "A" or "S" to match each statement about the peripheral nervous system.

A. autonomic nerves   S. somatic nerves

1. _____ Signals travel to and from internal organs and other structures [p.234]
2. _____ Sensory axons carry information from receptors in skin, skeletal muscles, and tendons [p.234]
3. _____ Signals concern moving the head, trunk, and limbs [p.234]
4. _____ Includes preganglionic and postganglionic neurons [p.234]
5. _____ Motor axons carry messages to smooth muscle, cardiac muscle, and glands [p.234]
6. _____ Motor axons deliver commands to skeletal muscles [p.234]
7. _____ Includes parasympathetic and sympathetic nerves [p.234]

Dichotomous Choice

Circle one of two possible answers given between parentheses in each statement.

8. (Sympathetic/Parasympathetic) nerves cause the pupils to constrict and heart rate to decrease, as well as increasing stomach and intestinal movements. [p.234]
9. (Sympathetic/Parasympathetic) nerves cause glandular secretions in the airways to decrease and salivary gland secretions to thicken. [p.234]
10. (Sympathetic/Parasympathetic) nerves tend to slow down the body when there is not much outside stimulation. [p.234]
11. (Sympathetic/Parasympathetic) nerves tend to speed up the body during heightened awareness, excitement, or danger. [p.235]
12. The release of (endorphins/norepinephrine) primes the body in what is called the fight–flight response. [p.235]
13. The (gray matter/white matter) is found on the inside of the spinal cord. [p.235]
14. The (gray matter/white matter) contains dendrites, cell bodies of neurons, interneurons, and neuroglial cells. [p.235]
15. The spinal cord is protected by the vertebral column, as well as three layers of coverings called the (intervertebral disks/meninges). [p.235]
16. Spinal reflexes (do/do not) require direct input from the brain. [p.235]
17. The spinal cord deals with (somatic/autonomic) reflexes that deal with internal organ functions such as emptying the bladder. [p.235]

Labeling

Identify the numbered parts of the accompanying illustration. [p.235]

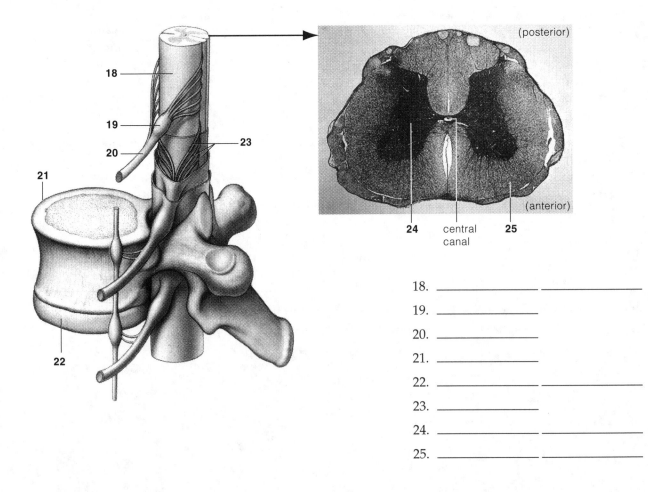

18. _____ _____
19. _____
20. _____
21. _____
22. _____ _____
23. _____
24. _____ _____
25. _____ _____

Fill-in-the-Blanks

The spinal cord merges with the (26) _____ [p.236], a master control center. The brain is

protected by the bones of the (27) _____ [p.236] and by the three (28) _____ [p.236].

Folds in the tough, outermost (29) _____ [p.236] mater separate the brain into right and left

(30) _____ [p.236]. Spaces in the brain are filled with (31) _____ _____

[p.236], which cushions and helps nourish the brain. The tissue of the brain that controls basic reflexes

is the (32) _____ _____ [p.236]. Through evolution, expanded layers of

(33) _____ _____ [p.236] developed over this structure. The most recent layers

are correlated with humans' increasing reliance on the nose, ears, and eyes.

Matching

To the left of each number, indicate what division of the brain includes the structure by writing "H" for hindbrain, "M" for midbrain, or "F" for forebrain. Then, match the named part with the letter of the phrase that describes its function in the blank to the right of each number.

_____ 34. _____ cerebrum [p.237]

_____ 35. _____ hypothalamus [p.237]

_____ 36. _____ cerebellum [p.236]

_____ 37. _____ medulla oblongata [p.236]

_____ 38. _____ tectum [p.236]

_____ 39. _____ olfactory bulbs [p.237]

_____ 40. _____ pons [p.236]

_____ 41. _____ thalamus [p.237]

A. Monitors internal organs; influences behaviors related to thirst, hunger, sexual behavior, and emotional expression
B. Information is processed and sensory input and motor responses are integrated
C. Relays and coordinates sensory signals to cerebrum through basal nuclei
D. Deal with sensory information about smell
E. Directs signal traffic between cerebellum and forebrain
F. Coordinates reflex responses to sights and sounds; sensory input converges on gray matter roof
G. Coordinates nerve signals for movement and balance
H. Site of reflex centers involved in respiration and cardiovascular function

Fill-in-the-Blanks

The mesh of interneurons extending from the uppermost spinal cord, through the brain stem, and into the

cerebral cortex is called the (42) _____ _____ [p.237]. It helps govern muscle activity

associated with balance, (43) _____ [p.237], and muscle tone. It can also activate centers in

the (44) _____ _____ [p.237] and so help govern the entire nervous system.

The brain and spinal cord are surrounded by transparent (45) _____ [p.237] fluid. It is

secreted from specialized capillaries inside cavities called (46) _____ [p.237]. The fluid is also

found in the central canal of the spinal cord and in the space between the innermost layer of the

(47) _____ [p.237] and the brain itself, helping to cushion the brain and spinal cord from

jarring movements.

The structure of brain capillaries forces substances to pass through endothelial cells, not between

them, in order to reach the brain. This (48) _____ _____ _____ [p.237] helps control which substances reach the brain's neurons. The "loophole" in the system allows

(49) _____-soluble [p.237] substances through, including caffeine, nicotine, alcohol, and

many illegal drugs.

True/False

If the statement is true, write a "T" in the blank. If the statement is false, make it correct by writing the word(s) in the blank that should take the place of the underlined word(s).

_____ 50. The <u>cerebellum</u> is divided into right and left hemispheres. [p.238]

_____ 51. The cerebral cortex is a thin, outer layer of <u>gray</u> matter. [p.238]

_____ 52. The <u>left</u> hemisphere deals with visual–spatial relationships, music, and other creative activities. [p.206]

_____ 53. The <u>left</u> hemisphere dominates the right hemisphere in most people. [p.238]

_____ 54. The corpus callosum is a band of <u>connective tissue</u> between the hemispheres. [p.238]

_____ 55. Each hemisphere is divided into four regions called <u>lobes</u>. [p.238]

_____ 56. Everything people comprehend, communicate, remember, and voluntarily act upon arises in the <u>brain stem</u>. [p.238]

_____ 57. <u>Motor areas</u> in the primary motor cortex of the frontal lobe control coordinated movements of skeletal muscles. [p.238]

_____ 58. The premotor cortex deals with <u>instinctive</u> behaviors. [p.238]

_____ 59. Broca's area, used in speech, as well as the eye field controlling voluntary eye movements, are in the <u>temporal</u> lobe of each hemisphere. [p.238]

_____ 60. The main receiving center for sensory input from the skin and joints is in the <u>frontal</u> lobe. [p.238]

_____ 61. Taste is perceived in the parietal lobe, sight in the occipital lobe, and sound and smell in the <u>temporal</u> lobes. [p.238]

_____ 62. <u>Association</u> areas in all parts of the cortex integrate, analyze, and respond to many inputs. [p.239]

_____ 63. The limbic system governs emotions and has roles in <u>memory</u>. [p.239]

_____ 64. The limbic system correlates organ activities with <u>reasoning</u> behaviors. [p.239]

Labeling

Identify each numbered part of the accompanying illustration. [p.238]

65. _____

66. _____

67. _____ _____

68. _____

69. _____

70. _____ _____

71. _____

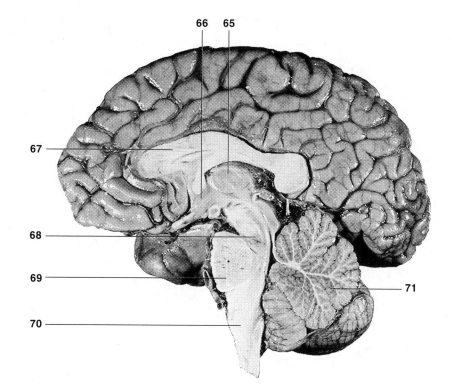

Fill-in-the-Blanks

The first stage in forming memories is (72) _____ _____ [p.240] storage, which lasts a few seconds to a few hours and is limited to bits of sensory information. In (73) _____ _____ [p.240] storage, a great amount of information is kept more or less permanently. While facts are often forgotten or filed in long-term storage, (74) _____ [p.240] are gained by practicing specific motor activities. A memory circuit leading to fact memory flows from the sensory cortex to the (75) _____ [p.240] and (76) _____ [p.240] in the limbic system. From there, information flows to the (77) _____ [p.240] cortex and to the basal nuclei, which send it back to the cortex for reinforcement. Skill memory flows from the sensory cortex to the (78) _____ _____ [p.241], which promotes motor responses. The circuit extends to the (79) _____ [p.241], which coordinates motor activity. (80) _____ [p.241] is the loss of fact memory, but it does not affect a person's capacity to learn new (81) _____ [p.241]. (82) _____ [p.241] disease destroys basal nuclei and (83) _____ [p.241] ability, although skill memory remains. (84) _____ [p.241] disease results in people remembering long-standing information but having trouble remembering very recent events.

EEGs and (85) _____ [p.241] scans show brain activity. Part of the (86) _____ _____ [p.241] promotes chemical changes that influence whether you stay awake or fall asleep. One of its sleep centers releases (87) _____ [p.241], which triggers drowsiness and sleep.

13.11. NERVOUS SYSTEM DISORDERS [p.242]
13.12 THE BRAIN ON DRUGS [pp.242–243]

Selected Words: *meningitis* [p.242], *encephalitis* [p.242], *concussion* [p.242], *epilepsy* [p.242], *seizure disorders* [p.242], *grand mal* seizure [p.242], *multiple sclerosis* (MS) [p.242], *Alzheimer's disease* [p.242], *Parkinson's diseases* [p.242], *"speed"* [p.242], *cocaine* [p.242], *blood alcohol concentration* (BAC) [p.243], *analgesic* [p.242], *tolerance* [p.243], *habituation* [p.243], *synergistic* interaction [p.243], *antagonistic* [p.243], *potentiating* [p.243]

Boldfaced, Page-Referenced Terms

[p.242] drug _____

_____

[p.242] drug abuse _____

_____

[p.242] psychoactive drugs _____

_____

[p.242] stimulants _____

_____

Matching

1. _____ multiple sclerosis
2. _____ epilepsy (seizure disorders)
3. _____ meningitis
4. _____ concussion
5. _____ grand mal seizure
6. _____ encephalitis
7. _____ Parkinson's disease
8. _____ Alzheimer's disease

A. Degenerative brain disease affecting lower rear of brain [p.242]
B. Inflammation of the brain, usually due to viral infection [p.242]
C. Autoimmune disease causing destruction of the myelin sheaths of the central nervous system [p.242]
D. Brain's normal electric activity becomes chaotic [p.242]
E. Results from violent blow to head or neck [p.242]
F. Degeneration of brain neurons and buildup of amyloid protein, leading to loss of memory and intellect [p.242]
G. Uncontrollable jerking with loss of consciousness [p.242]
H. Often-fatal inflammation of the meninges caused by viral or bacterial infection [p.242]

Fill-in-the-Blanks

A(n) (9) _____ [p.242] is a substance introduced into the body to provoke a specific physiological response. (10) _____ [p.242] drugs act on the central nervous system by binding to (11) _____ [p.242] in the neuron plasma membrane normally occupied by (12) _____ [p.242] molecules that transmit chemical messages among neurons. Psychoactive drugs act on parts of the brain that govern states of consciousness and (13) _____ [p.242], and may influence physiological events. Many affect a(n) (14) _____ [p.242] center in the (15) _____ [p.242]. Caffeine, nicotine, cocaine, and amphetamines are all (16) _____ [p.242] that increase alertness initially but then lead to depression. Nicotine mimics the neurotransmitter (17) _____ [p.242] and thus increases the metabolic rate. Amphetamines stimulate the pleasure center in the hypothalamus, but over time an unhealthy (18) _____ [p.242] develops. Cocaine blocks the reabsorption of (19) _____ [p.242] and other signaling molecules. It harms the cardiovascular system and weakens the (20) _____ [p.242] system. (21) _____ [pp.242–243] is a drug that dampens brain activity and produces disorientation, uncoordinated motor functions, and diminished judgment. Morphine is a(n) (22) _____ [p.243], or painkiller, derived from opium poppy seeds. Like its cousin (23) _____ [p.243], it blocks the transmission of pain signals.

When the body develops (24) _____ [p.243] to a drug, more is required to produce the same effect. The level of detoxifying liver enzymes (25) _____ [p.243] in response to the continuing presence of the drug in the blood, so more of the substance is needed to stay ahead of the rate of breakdown. Habituation, or (26) _____ [p.243] drug dependence, occurs when a user begins to crave the sensations associated with using a particular drug. Both habituation and tolerance are evidence that the user is (27) _____ [p.243] to the substance.

When different psychoactive drugs are taken simultaneously, they sometimes cause a(n)
(28) _____ [p.243] effect if the two drugs used together have a much more powerful effect
than they would if used separately. When one drug blocks the effect of another, they are said to be
(29) _____ [p.243] to each other. Other combinations of drugs are (30) _____ [p.243],
which means that one enhances the effect of the other, as when alcohol deepens the drowsiness effect of
antihistamines.

Matching

Choose the most appropriate category for each drug.

31. _____ amphetamine [p.242]

32. _____ caffeine [p.242]

33. _____ cocaine [p.242]

34. _____ alcohol [pp.242–243]

35. _____ heroin [p.243]

36. _____ LSD [p.242]

37. _____ inhalants [p.242]

38. _____ nicotine [p.242]

39. _____ morphine [p.236]

A. Analgesic
B. Psychedelic or hallucinogenic drug
C. Drug that reduces brain activity
D. Stimulant
E. Deliriant

Self-Quiz

_____ 1. The conducting zone of a neuron is the
_____ . [p.224]
   a. axon
   b. axon ending
   c. cell body
   d. dendrite

_____ 2. An action potential is brought about by
_____ . [p.226]
   a. a sudden membrane impermeability
   b. the movement of negatively charged proteins through the neuronal membrane
   c. the movement of lipoproteins to the outer membrane
   d. a local change in membrane permeability caused by a greater-than-threshold stimulus

_____ 3. The resting membrane potential
_____ . [p.224]
   a. exists as long as a charge difference sufficient to do work exists across a membrane
   b. occurs because there are more potassium ions outside the neuronal membrane than there are inside
   c. occurs because of the unique distribution of receptor proteins located on the dendrite exterior
   d. is brought about by a local change in membrane permeability caused by a greater-than-threshold stimulus

____ 4. The phrase *all-or-nothing* used in conjunction with discussion about an action potential means that _____ . [p.226]
   a. a resting membrane potential has been received by the cell
   b. it will always spike totally once stimulated past threshold
   c. the membrane either achieves total equilibrium or remains as far from equilibrium as possible
   d. propagation along the neuron is saltatory

____ 5. _____ are responsible for integration within the nervous system. [p.223]
   a. Interneurons
   b. Schwann cells
   c. Motor neurons
   d. Sensory neurons

____ 6. _____ nerves generally dominate internal events when environmental conditions permit normal body functioning. [p.234]
   a. Ganglia
   b. Pacemaker
   c. Sympathetic
   d. Parasympathetic

____ 7. The center of consciousness, memory, and intelligence is the _____ . [p.237]
   a. hindbrain
   b. reticular formation
   c. midbrain
   d. brain stem
   e. forebrain

____ 8. For which of these is the left cerebral hemisphere responsible? [p.238]
   a. music
   b. mathematics
   c. spatial relationships
   d. abstract abilities
   e. artistic ability

____ 9. The part of the brain that controls the basic responses necessary for maintaining life processes (breathing, heartbeat) is the _____ . [p.236]
   a. cerebral cortex
   b. cerebellum
   c. corpus callosum
   d. medulla oblongata

____ 10. The center for balance and coordination in the human brain is the _____ . [p.236]
   a. cerebrum
   b. pons
   c. cerebellum
   d. hypothalamus
   e. thalamus

Chapter Objectives/Review Questions

This section lists general and detailed chapter objectives that can be used as review questions. You can make maximum use of these items by writing answers on a separate sheet of paper. To check for accuracy, compare your answers with information given in the chapter or glossary.

1. Draw a neuron and label it according to its three general zones, its specific structures, and the specific function(s) of each structure. [p.224]
2. Explain the chemical basis of the action potential. Look at Figure 13.6 in the text and determine which part of the curve represents each of the following: [pp.226–227]
   a. the point at which the stimulus was applied
   b. the events prior to achievement of the threshold value
   c. the opening of the ion gates and the diffusing of the ions
   d. the change from net negative charge inside the neuron to net positive charge and back again to net negative charge
   e. the active transport of sodium ions out of and potassium ions into the neuron

3. Describe six major disorders/diseases of the human nervous system by naming the causes and symptoms of each. [p.242]
4. Explain what a reflex is by drawing and labeling a diagram and explaining how it functions. [p.231]
5. Define and contrast the central and peripheral nervous systems. [p.232]
6. Explain how parasympathetic nerve activity balances sympathetic nerve activity. List activities of the sympathetic and parasympathetic nerves in regulating pupil diameter, rate of heartbeat, activities of the gut, and elimination of urine. [p.234]
7. Compare the structures of the spinal cord and brain with respect to white matter and gray matter. [pp.235,238]
8. List the parts of the brain found in the hindbrain, midbrain, and forebrain and tell the basic functions of each. [pp.236–237]
9. Explain what an electroencephalogram is and what EEGs can tell us about the levels of conscious experience. [p.241]
10. Name eight drugs that are commonly abused and state the category in which each belongs. [pp.242–243]

Integrating and Applying Key Concepts

Suppose that anger is eventually determined to be caused by excessive amounts of specific transmitter substances in the brains of angry people. Also suppose that an inexpensive antidote that neutralizes these anger-producing transmitter substances is readily available. Can violent murderers now argue that they have been wrongfully punished because they were victimized by their brain's transmitter substances and could not have acted in any other way? Suppose an antidote is prescribed to curb violent temper in an easily angered person. Suppose also that the person forgets to take the pill and subsequently murders a family member. Can the murderer still claim to be victimized by transmitter substances?

14

SENSORY RECEPTION

Interactive Exercises

CHAPTER INTRODUCTION [p.247]

14.1. SENSORY RECEPTORS AND PATHWAYS — AN OVERVIEW [pp.248–249]

14.2. SOMATIC "BODY" SENSATIONS [pp.250–251]

Selected Words: "smell genes" [p.247], *sensory systems* [p.247], *compound sensations* [p.248], *somatic pain* [p.251], *visceral pain* [p.251], "referred pain" [p.251], *phantom pain* [p.251]

Boldfaced, Page-Referenced Terms

[p.248] sensation _____

_____

[p.248] perception _____

_____

[p.248] sensory receptors _____

_____

[p.248] mechanoreceptors _____

_____

[p.248] thermoreceptors _____

_____

[p.248] nociceptors _____

_____

[p.248] chemoreceptors _____

_____

[p.248] osmoreceptors _____

_____

[p.248] photoreceptors _____

_____

[p.249] sensory adaptation _____

_____

[p.249] somatic sensations _____

_____

[p.249] special senses _____

_____

[p.250] somatosensory cortex _____

_____

[p.250] free nerve endings _____

_____

[p.250] encapsulated receptors _____

_____

[p.250] pain _____

Short Answer

1. Distinguish between a sensation and a perception. [p.248] _____

_____

_____

_____

2. How does a compound sensation differ from a simple sensation? [p.248] _____

_____

_____

_____

3. List the three ways in which the brain determines the nature of a given stimulus. [pp.248–249] _____

_____

_____

_____

4. Name one sensation with which sensory adaptation occurs. Name one with which it does not. [p.249] __

_____

_____

_____

5. Distinguish between somatic sensations and special senses. [p.249] _____

_____

_____

_____

Matching

Choose the most appropriate phrase to match each term. [p.248]

6. _____ Vision is associated with _____

7. _____ Pain is associated with _____

8. _____ Odors are detected by _____

9. _____ Hearing is detected by _____

10. _____ Detects CO_2 concentration in the blood

11. _____ Detects environmental temperature

12. _____ Detects internal body temperature

13. _____ Touch is detected by _____

14. _____ Rods and cones

15. _____ Hair cells in organ inside the ear

16. _____ Olfactory receptors in the nose

17. _____ Any stimulus that causes tissue damage

A. chemoreceptors
B. mechanoreceptors
C. nociceptors
D. photoreceptors
E. thermoreceptors

Dichotomous Choice

Circle one of two possible answers given between parentheses in each statement.

18. Somatic sensations travel from body receptors to the spinal cord and then to the somatosensory cortex in the brain's (cerebellum/cerebrum). [p.250]
19. Interneurons in the somatosensory cortex form a "map" of the body's surface, with the largest areas corresponding to the body parts where the density of neurons is (least/greatest). [p.250]
20. Free nerve endings such as mechanoreceptors, thermoreceptors, and nociceptors adapt (quickly/slowly) when stimulated. [p.250]
21. You might feel a spider walking on your arm because of (chemoreceptors/mechanoreceptors). [p.250]
22. Encapsulated receptors that respond to low-frequency vibration are (Meissner's corpuscles/bulbs of Krause). [p.250]
23. Receptors that respond to steady touching and pressure, as well as to high temperature, are called (Pacinian corpuscles/Ruffini endings). [p.250]
24. Pain associated with internal organs is called (somatic pain/visceral pain). [p.251]
25. Sensing the pain of a heart attack along the left shoulder is an example of (referred pain/phantom pain). [p.251]

14.3. TASTE AND SMELL — CHEMICAL SENSES [pp.252–253]

14.4. *Science Comes to Life:* A TASTY MORSEL OF SENSORY SCIENCE [p.253]

14.5. HEARING: DETECTING SOUND WAVES [pp.254–255]

14.6. BALANCE: SENSING THE BODY'S NATURAL POSITION [pp.256–257]

14.7. *Focus on Our Environment:* NOISE POLLUTION: AN ATTACK ON THE EARS [p.257]

Selected Words: chemical senses [p.252], "tastes" [p.252], "smell" [p.252], vomeronasal organ or "sexual nose" [p.253], gustation [p.253], olfaction [p.253], "tastants" [p.253], *amplitude* [p.254], *frequency* [p.254], *outer ear* [p.254], *middle ear* [p.254], *inner ear* [p.254], *semicircular canals* [p.254], *malleus* [p.254], *incus* [p.254], *stapes* [p.254], *oval window* [p.254], *cochlear duct* [p.254], *scala vestibuli* [p.254], *scala tympani* [p.254], *basilar membrane* [p.255], "pitch" [p.255], *round window* [p.255], *Eustachian tube* [p.255], "equilibrium position" [p.256], *cupula* [p.257], *static equilibrium* [p.257], *otolith organ* [p.257], *motion sickness* [p.257], *decibels* [p.257]

Boldfaced, Page-Referenced Terms

[p.252] taste receptors _____

_____

[p.252] olfactory receptors _____

_____

[p.254] cochlea _____

_____

[p.254] tympanic membrane _____

_____

[p.255] organ of Corti _____

_____

[p.255] hair cells _____

_____

[p.255] tectorial membrane _____

_____

[p.256] vestibular apparatus _____

_____

[p.256] semicircular canals _____

_____

Fill-in-the-Blanks

With both taste and smell, (1) _____ [p.252] bind molecules that are dissolved in the fluid bathing them. Sensory information travels from the receptors through the (2) _____ [p.252] and on to the cerebral cortex. In the case of taste, these receptors are often part of sense organs called (3) _____ _____ [p.252], such as those scattered over the tongue. These distinguish only (4) _____ [p.252] basic types of flavors: sweet, sour, (5) _____ [p.252], and (6) _____ [p.252].

Animals smell substances by means of (7) _____ [p.252] receptors that detect water-soluble or volatile substances. Sensory nerve pathways lead from the nasal cavity to the region of the brain called the (8) _____ [pp.252–253] bulbs. From there, other neurons forward the message to a center in the (9) _____ _____ [p.252], which interprets it as a particular smell. About half an inch inside the nose is the (10) _____ [p.253] organ, or "sexual nose." It detects signal molecules called (11) _____ [p.253]. These affect the behavior of other individuals.

The loudness of a sound corresponds to the (12) _____ [p.254] of its wave form. The number of wave cycles per second is the sound's (13) _____ [p.254]. The sense of hearing starts with vibration-sensitive (14) _____ [p.254] deep in the ear, which respond to fluid motion in the ear. Vibrations from sound waves cause this motion, which bends the tips of (15) _____ [p.254] on the mechanoreceptors. Enough bending results in a(n) (16) _____ _____ [p.254] that ultimately reaches the brain as a(n) (17) "_____" [p.254].

Matching

18. _____ malleus
19. _____ incus
20. _____ stapes
21. _____ tympanic membrane
22. _____ cochlea
23. _____ basilar membrane
24. _____ Eustachian tube
25. _____ organ of Corti
26. _____ inner ear
27. _____ middle ear
28. _____ outer ear
29. _____ round window

A. Location of semicircular canals and cochlea [p.254]
B. "Anvil" of middle ear [p.254]
C. Pathway by which sound waves enter the ear [p.254]
D. "Hammer" of middle ear [p.254]
E. Its hair cells touch the tectorial membrane [p.255]
F. "Release valve" for force of sound waves [p.255]
G. Eardrum [p.254]
H. Equalizes air pressure of middle ear with outside pressure [p.255]
I. Location of three tiny bones that transmit sound waves [p.254]
J. "Stirrup" of middle ear [p.254]
K. Different sound frequencies cause different parts to vibrate [p.255]
L. Location of scala vestibuli, scala tympani, and cochlear duct [p.254]

Sequence

In the blanks beside the numbers, write the letters of the ear parts to show the pathway of sound vibrations through the ear.

30. _____
31. _____
32. _____
33. _____
34. _____
35. _____
36. _____
37. _____
38. _____
39. _____
40. _____

a. round window
b. basilar membrane
c. fluid of cochlear duct
d. malleus
e. organ of Corti hair cells pushing against tectorial membrane (trigger action potential)
f. oval window
g. incus
h. tympanic membrane
i. outer ear
j. fluid inside scala vestibuli and scala tympani
k. stapes

Labeling

Identify each indicated part of the accompanying illustrations.

41. _____ _____ [p.254]

42. _____ _____ _____ [p.254]

43. _____ _____ [p.254]

44. _____ _____ [p.254]

45. _____ _____ [p.254]

46. _____ _____ [p.255]

47. _____ _____ [p.255]

48. _____ _____ [p.255]

49. _____ _____ [p.255]

50. _____ _____ [p.255]

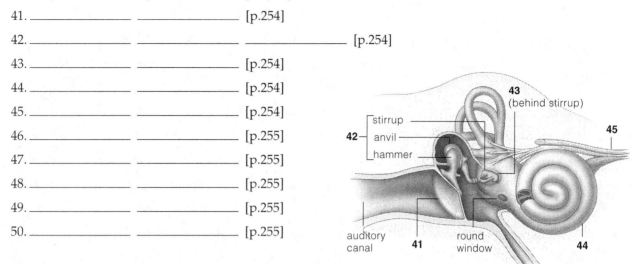

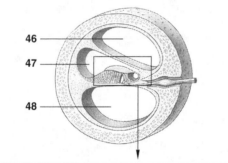

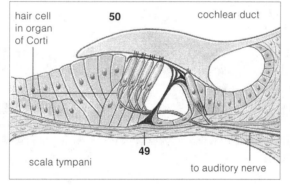

Fill-in-the-Blanks

The baseline for assessing any body displacement from its natural position is called the (51) "_____

_____" [p.256]. Our sense of balance relies partly on messages from (52) _____

[p.256] in the eyes, skin, and joints. Within the inner ear is the (53) _____ [p.256] apparatus,

which consists of three (54) _____ _____ [p.256]. These are positioned at right angles

to one another. Sensory receptors inside monitor (55) _____ _____ [p.256], while

other receptors in the apparatus monitor acceleration and (56) _____ [p.256]. Sensory hairs at the

base of each semicircular canal project into a jellylike (57) _____ [p.257] and respond to rotation

of the head. The head's position in space is called (58) _____ [p.257] equilibrium. The receptors

that detect this are two fluid-filled sacs in the (59) _____ [p.257] apparatus, called the utricle and

the saccule. Each contains a(n) (60) _____ _____ [p.257] with hairs embedded in a

jellylike membrane. Movements of the membrane and bits of calcium carbonate called (61) _____

[p.257] signal changes in the head's orientation. Action potentials from different parts of the vestibular

apparatus travel to reflex centers in the (62) _____ _____ [p.257]. The brain integrates

this information with other sensory information, then orders movements that help you keep your (63)

_____ [p.257]. (64) _____ _____ [p.257] develops when extreme motion

overstimulates hair cells in the balance organs or when conflicting messages about motion or position are

received from the eyes and ears.

14.8. VISION: AN OVERVIEW [pp.258–259]
14.9. FROM VISUAL SIGNALS TO "SIGHT" [pp.260–261]
14.10. DISORDERS OF THE EYE [pp.262–263]

Selected Words: "tunics" [p.258], *sclera* [p.258], *choroid* [p.258], "iris scans" [p.258], *pupil* [p.258], *ciliary body* [p.258], *aqueous humor* [p.258], *vitreous humor* [p.258], "blind spot" [p.258], *bipolar* interneurons [p.261], *ganglion cells* [p.261], *horizontal cells* [p.261], *amacrine cells* [p.261], "receptive fields" [p.261], "visual field" [p.261], *red–green color blindness* [p.262], *astigmatism* [p.262], *nearsightedness* (myopia) [p.262], *farsightedness* (hyperopia) [p.262], *histoplasmosis* [p.262], *herpes simplex* [p.262], *trachoma* [p.262], *cataracts* [p.263], *macular degeneration* [p.263], *glaucoma* [p.263], *retinal detachment* [p.263], *corneal transplant surgery* [p.263], "lasik" and "lasek" [p.263], *laser coagulation* [p.263].

Boldfaced, Page-Referenced Terms

[p.258] vision _____

_____

[p.258] eyes _____

_____

[p.258] cornea _____

_____

[p.258] iris _____

_____

[p.258] lens _____

_____

[p.258] retina _____

_____

[p.258] visual cortex _____

_____

[p.259] accommodation _____

_____

[p.260] rod cells _____

_____

[p.260] cone cells _____

_____

[p.260] rhodopsin _____

_____

[p.260] fovea _____

Fill-in-the-Blanks

(1) _____ [p.258] requires a system of (2) _____ [p.258] and (3) _____

[p.258] centers that can interpret images. The (4) _____ [p.258] are sensory organs that contain a

tissue with a dense array of photoreceptors. The outer layer of the eye consists of a sclera and transparent

(5) _____ [p.258]. The middle layer includes a choroid, ciliary body, and (6) _____

[p.258]. The key feature of the inner layer is the (7) _____ [p.258].

Matching

8. _H._ pupil

9. _K._ sclera

10. _A._ visual cortex

11. _B._ lens

12. _E._ vitreous humor

13. _G._ ciliary body

14. _J._ retina

15. _F._ aqueous humor

16. _D._ choroid

17. _I._ cornea

18. _C._ iris

A. Part of brain where signals are interpreted as sight [p.258]
B. Focuses incoming light onto the retina [p.258]
C. Pigmented ring behind cornea; prevents light from scattering [p.258] iris
D. Pigmented area beneath sclera [p.258] choroid
E. Jellylike substance in chamber behind lens [p.258]
F. Clear fluid bathing both sides of lens [p.258]
G. Smooth muscle that focuses light [p.258]
H. Entrance through which light enters eye [p.258] pupil
I. Transparent covering of iris and pupil [p.258] cornea
J. Layer of neural tissue at back of eye [p.258] retina
K. Fibrous "white" of eye [p.258] sclera

Labeling

Identify each indicated part of the accompanying illustration. [p.258]

19. _____

20. _____ _____

21. _____ _____

22. _____

23. _____

24. _____

25. _____

26. _____ _____

27. _____

28. _____

29. _____

30. _____ _____

31. _____ _____

Multiple Choice

_____ 32. The correct path of light waves and/or electrochemical impulses used in photoreception and vision is _____ [p.258]
   a. cornea → lens → retina → optic nerve → thalamus → visual cortex
   b. sclera → iris → retina → optic nerve → visual cortex → thalamus
   c. cornea → lens → retina → thalamus → optic nerve → visual cortex
   d. sclera → lens → retina → optic nerve → visual cortex → thalamus
   e. cornea → retina → lens → optic nerve → thalamus → visual cortex

_____ 33. What is the "blind spot" or optic disk of the retina? [p.258]
   a. where the optic nerve exits the eye
   b. an area with no photoreceptors
   c. area that can be scanned as a form of identification
   d. area most vulnerable to damage by strong, focused light
   e. both a and b

_____ 34. What causes "upside-down and backwards" imaging that must be corrected in the brain? [pp.258–259]
   a. density of the vitreous humor
   b. presence of a blind spot
   c. "double" focusing at both the lens and the retina
   d. curve of cornea causing light trajectories to bend
   e. contraction of ciliary muscles around iris

_____ 35. Which of these is *not* true concerning focusing? [p.259]
   a. The lens is adjusted so that light strikes the retina very precisely.
   b. Adjustments of the lens are called accommodation.
   c. Focusing is necessary because light rays strike the cornea at different angles.
   d. Ciliary muscle changes the size of the pupil.
   e. Muscle contractions cause the lens to bulge.

True/False

If the statement is true, write a "T" in the blank. If the statement is false, make it correct by writing the word(s) in the blank that should take the place of the underlined word(s).

_____ 36. Daytime vision and color perception are the job of <u>rods</u>. [p.260]

_____ 37. Rods are better than cones at detecting light <u>intensity</u>. [p.260]

_____ 38. Vitamin A is used in making the pigment <u>rhodopsin</u>. [p.260]

_____ 39. To start an action potential to allow sight in dim surroundings, <u>many</u> photons must be absorbed by rhodopsin. [p.260]

_____ 40. In the part of the retina called the fovea, visual acuity is <u>lower</u> than in the rest of the retina. [p.260]

_____ 41. The optic nerves form from the axons of <u>bipolar neurons</u>. [p.261]

_____ 42. Horizontal cells and amacrine cells play a role in processing <u>before</u> visual information is sent to the brain. [p.261]

_____ 43. The organization of the retina into receptive fields <u>leads to</u> the brain being bombarded with signals that cause confusion. [p.261]

_____ 44. The part of the outside world that a person actually sees is his or her "<u>receptive field</u>." [p.261]

_____ 45. The optic nerve leading out of each eye delivers signals from <u>both</u> side(s) of a person's visual field. [p.261]

Matching

Choose the most appropriate description for each term.

46. _____ astigmatism [p.262]

47. _____ cataracts [p.263]

48. _____ farsightedness (hyperopia) [p.262]

49. _____ glaucoma [p.263]

50. _____ *Herpes simplex* [p.262]

51. _____ nearsightedness (myopia) [p.262]

52. _____ red–green color blindness [p.262]

53. _____ retinal detachment [p.263]

54. _____ trachoma [p.262]

55. _____ histoplasmosis [p.262]

56. _____ macular degeneration [p.263]

A. Caused by a physical blow to the head or an illness that separates the inner layer of the eyeball from the choroid

B. Damaged eyeballs and conjunctiva caused by chlamydial bacteria; in North Africa and the Middle East

C. Objects close to the eye are focused in front of the retina; eyeball too long from front to back

D. Gradual clouding of the lens

E. Inherited abnormality; retina lacks a particular type of cone cell

F. Objects close to the eye are focused beyond the retina; eyeball too short from front to back

G. Sometimes causes ulcerated cornea

H. Excess aqueous humor accumulates inside the eyeball, causing neurons in the retina and optic nerve to die

I. Uneven curvature of the cornea; cannot bend incoming light rays to the same focal point

J. Portion of retina breaks down and is replaced by scar tissue

K. Occurs when fungal infection of lungs moves to the eye

Self-Quiz

Multiple Choice

_____ 1. What type of mechanoreceptors help sense limb motions and the body's position in space? [p.250]
   a. stretch receptors
   b. baroreceptors
   c. touch receptors
   d. balance receptors
   e. auditory receptors

_____ 2. The principal place in the human ear in which sound waves are amplified is the _____ . [p.254]
   a. pinna
   b. ear canal
   c. middle ear
   d. organ of Corti
   e. none of the above

_____ 3. The place in which vibrations are translated into patterns of nerve impulses is _____ . [p.255]
   a. the pinna
   b. the ear canal
   c. the middle ear
   d. the organ of Corti
   e. none of the above

_____ 4. Nearsightedness is caused by _____ . [p.262]
   a. eye structure that focuses an image in front of the retina
   b. uneven curvature of the lens
   c. eye structure that focuses an image posterior to the retina
   d. uneven curvature of the cornea
   e. none of the above

Choice

For questions 5–9, choose from the following:

   a. fovea   b. cornea   c. iris   d. retina   e. sclera

5. The white protective fibrous tissue of the eye is the _____ . [p.258]
6. Rods and cones are located in the _____ . [p.260]
7. The highest concentration of cones is in the _____ . [p.255]
8. The adjustable ring of contractile and connective tissues that controls the amount of light entering the eye is the _____ . [p.258]
9. The outer transparent protective covering of part of the eyeball is the _____ . [p.258]

Multiple Choice

_____ 10. Accommodation involves the ability to _____ . [p.259]
   a. change the sensitivity of the rods and cones by means of transmitters
   b. change the width of the lens by relaxing or contracting certain muscles
   c. change the curvature of the cornea
   d. adapt to large changes in light intensity
   e. all of the above

Chapter Objectives/Review Questions

This section lists general and detailed chapter objectives that can be used as review questions. You can make maximum use of these items by writing answers on a separate sheet of paper. To check for accuracy, compare your answers with information given in the chapter or glossary.

1. Define and distinguish among chemoreceptors, mechanoreceptors, photoreceptors, and thermoreceptors. Name one example of each type that appears in humans. [p.248]
2. Describe the function of nociceptors. [pp.250–251]
3. Explain how a taste bud works. [p.252]
4. Follow a sound wave from the outer ear to the organ of Corti; mention the name of each structure it passes and state where the sound wave is amplified and where the pattern of pressure waves is translated into nervous impulses. [pp.254–255]
5. State how low- and high-frequency sounds affect the basilar membrane and the organ of Corti. [p.255]
6. Explain the roles of the oval and round windows in hearing. [pp.254–255]
7. Explain how the three semicircular canals of the human ear detect changes of position and acceleration in a variety of directions. [pp.256–257]
8. Describe the structure of the human eye. [p.258]
9. Describe how the human eye perceives color and black-and-white. [p.260]
10. Explain the general principles that affect how light is detected by photoreceptors and changed into electrochemical messages. [pp.260–261]
11. Define nearsightedness and farsightedness and relate each to eyeball structure. [p.262]
12. Describe inherited eye problems. Name several eye diseases and give their causes. [pp.262–263]

Integrating and Applying Key Concepts

Discuss the benefits and problems associated with iris scans as a form of identification. Can you think of any other sensory structures or characteristics that are being or could be used for identification?

15

THE ENDOCRINE SYSTEM

Interactive Exercises

CHAPTER INTRODUCTION [p.267]

15.1. THE ENDOCRINE SYSTEM: HORMONES [pp.268–269]

15.2. HORMONE CATEGORIES AND SIGNALING [pp.270–271]

Selected Words: "peptide hormones" [p.270], *testicular feminization syndrome* [p.270]

Boldfaced, Page-Referenced Terms

[p.268] target cell _____

[p.268] hormones _____

[p.268] pheromones _____

[p.268] endocrine system _____

[p.268] opposing interaction _____

[p.268] synergistic interaction _____

[p.268] permissive interaction _____

[p.271] second messenger _____

Matching

Choose the most appropriate description for each term.

1. _____ hormones [p.268]
2. _____ opposing interaction [p.268]
3. _____ permissive interaction [p.268]
4. _____ target cells [p.268]
5. _____ synergistic interaction [p.268]
6. _____ endocrine systems [p.268]
7. _____ pheromones [p.268]

A. Sum total of the actions of two or more hormones necessary to produce the required effect on target cells
B. Group of glands that release hormones
C. Exocrine gland secretions; signaling molecules that act on animals of the same species to help integrate social behavior
D. Effect of one hormone works against the effect of another
E. Secretions from endocrine glands, endocrine cells, and some neurons; distributed by the bloodstream to nonadjacent target cells
F. Cells that have receptors for a given type of signaling molecule
G. One hormone exerts its effect only when a target cell has been "primed" to respond to that hormone

Complete the Table

8. Complete the following table by identifying the numbered components of the endocrine system shown in the illustration as well as the hormones produced by each gland. [p.269]

Gland Name	Number	Hormone(s) Produced
a. parathyroids (four)		
b. adrenal cortex		
c. pineal		
d. pancreatic islets		
e. ovaries		
f. hypothalamus		
g. pituitary, anterior lobe		
h. thyroid		
i. thymus		
j. testes		
k. pituitary, posterior lobe		
l. adrenal medulla		

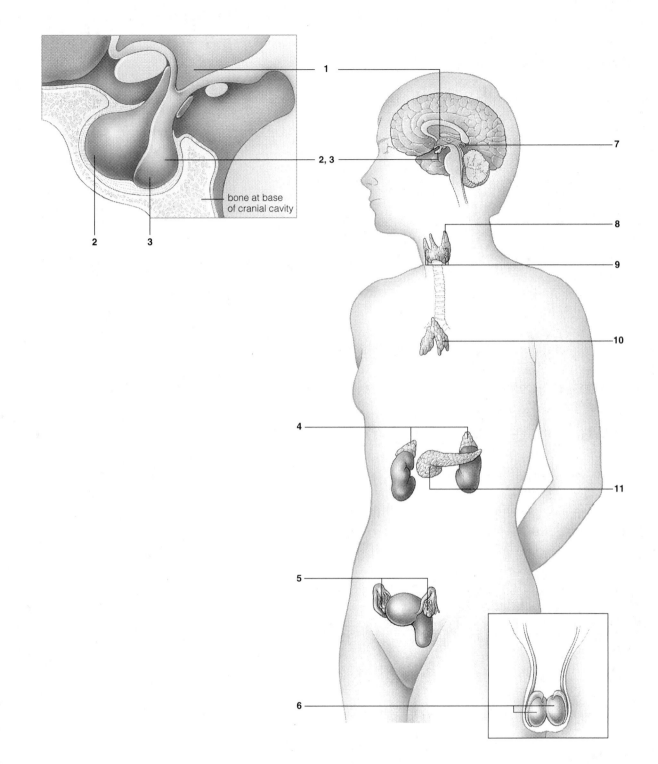

bone at base
of cranial cavity

Choice

For questions 9–15, choose from the following:

a. steroid hormones   b. nonsteroid hormones

9. _____ Lipid-soluble molecules synthesized from cholesterol; made in adrenal glands, ovaries, and testes [p.270]

10. _____ Includes amines, peptides, proteins, and glycoproteins [p.270]

11. _____ One example involves testosterone, defective receptors, and a condition called *testicular feminization syndrome* [p.270]

12. _____ Hormones that often activate second messengers [p.271]

13. _____ Peptide hormones that bind to receptors at the plasma membrane [p.271]

14. _____ Lipid-soluble hormones that move through a target cell's plasma membrane and bind to a protein receptor; the hormone–receptor complex moves into the nucleus and interacts with specific DNA regions to stimulate or inhibit transcription of mRNA [p.270]

15. _____ Involves molecules such as cyclic AMP that activate many enzymes in the cytoplasm which, in turn, cause alteration in some cell activity [p.271]

15.3. THE HYPOTHALAMUS AND PITUITARY GLAND — MAJOR CONTROLLERS [pp.272–273]

15.4. WHEN PITUITARY SIGNALS GO AWRY [p.274]

Selected Words: posterior lobe and *anterior* lobe (of the pituitary) [p.272], "metabolic hormone" [p.273], *gigantism* [p.274], *pituitary dwarfism* [p.274], *acromegaly* [p.274], *diabetes insipidus* [p.274]

Boldfaced, Page-Referenced Terms

[p.272] hypothalamus _____

_____

[p.272] pituitary gland _____

_____

[p.273] releasers _____

_____

[p.273] inhibitors _____

_____

Choice-Match

Label each hormone given in the following list with "A" if it is secreted by the anterior lobe of the pituitary and "P" if it is released from the posterior pituitary. Complete the exercise by entering the letter of the corresponding target and hormone action in the parentheses following each label.

1. _____ (   ) ACTH [p.272]

2. _____ (   ) ADH (vasopressin) [p.272]

3. _____ (   ) FSH [p.272]

4. _____ (   ) GH (STH) [pp.272–273]

5. _____ (   ) LH [pp.272–273]

6. _____ (   ) oxytocin [p.272]

7. _____ (   ) PRL [pp.272–273]

8. _____ (   ) TSH [pp.272–273]

A. Acts on ovaries and testes to produce gametes
B. Acts on mammary glands to stimulate and sustain milk production
C. Acts on ovaries and testes to release gametes; promotes testosterone secretion in males and formation of corpus luteum in females
D. Induces uterine contractions and milk movement into secretory ducts
E. Acts on the thyroid gland to stimulate release of thyroid hormones
F. Acts on the kidneys to induce water conservation and control extracellular fluid volume
G. Acts on the adrenal cortex to stimulate release of adrenal steroid hormones
H. Acts on most cells to promote growth in young; induces protein synthesis and cell division; plays roles in glucose and protein metabolism

Dichotomous Choice

Circle one of two possibilities given between parentheses in each statement.

9. The (hypothalamus/pituitary gland) monitors internal organs and activities related to their functioning, such as eating, sexual behavior, and body temperature; it also secretes some hormones. [p.272]
10. The (posterior/anterior) lobe of the pituitary stores and secretes two hypothalamic hormones, ADH and oxytocin. [p.272]
11. The (posterior/anterior) lobe of the pituitary produces and secretes its own hormones, which govern the release of hormones from other endocrine glands. [p.272]
12. Most hypothalamic hormones acting in the anterior pituitary lobe are (releaser/inhibitor) hormones and cause target cells to secrete their own hormones. [p.273]
13. Some hypothalamic hormones slow down hormone secretion from their targets; these are classed as (releaser/inhibitor) hormones. [p.273]
14. (ACTH/TSH) is an anterior pituitary hormone acting on the adrenal glands. [p.272]
15. In addition to LH, the anterior pituitary hormone having a role in reproduction is (FSH/TSH). [p.272]
16. (Pituitary dwarfism/Gigantism) results when not enough somatotropin is produced during childhood. [p.274]
17. Production of excessive amounts of somatotropin during childhood results in (pituitary dwarfism/ gigantism). [p.274]
18. Excess somatotropin production during adulthood results in thicker bone, cartilage, and connective tissues of hands, feet, jaws, and epithelia; this condition is known as (gigantism/acromegaly). [p.274]

15.5. SOURCES AND EFFECTS OF OTHER HORMONES [p.275]

15.6. HORMONES AND FEEDBACK CONTROLS — THE ADRENALS AND THYROID [pp.276–277]

15.7. FAST RESPONSES TO LOCAL CHANGES — PARATHYROIDS AND THE PANCREAS [pp.278–279]

15.8. SOME FINAL EXAMPLES OF INTEGRATION AND CONTROL [pp.280–281]

15.9. *Science Comes to Life:* GROWTH FACTORS [p.281]

Selected Words: negative feedback [p.276], positive feedback [p.276], *cortisol* [p.276], *gluconeogenesis* [p.276], "glucose sparing" [p.276], *hypoglycemia* [p.276], *aldosterone* [p.276], "water follows salt" [p.276], "fight–flight" response [p.277], *simple goiter* [p.277], *hypothyroidism* [p.277], *hyperthyroidism* [p.277], *Graves' disease* [p.277], *hyperparathyroidism* [p.278], *exocrine* cells [p.278], *endocrine* cells [p.278], *alpha cells* [p.278], *beta cells* [p.278], *delta cells* [p.278], *somatostatin* [p.278], *diabetes mellitus* [p.278], *metabolic acidosis* [p.278], "insulin-dependent" diabetes [p.279], "juvenile-onset diabetes" [p.279], "third eye" [p.280], *winter blues* [p.280], *puberty* [p.280], *atrial natriuretic peptide (ANP)* [p.280], *local signaling molecules* [p.280], *epidermal growth factor (EGF)* [p.281], *nerve growth factor (NGF)* [p.281]

Boldfaced, Page-Referenced Terms

[p.276] adrenal cortex _____

[p.276] glucocorticoids _____

[p.276] mineralocorticoids _____

[p.277] adrenal medulla _____

[p.277] thyroid gland _____

[p.278] parathyroid glands _____

[p.278] pancreatic islet _____

[p.278] glucagon _____

[p.278] insulin _____

[p.280] pineal gland _____

[p.280] biological clock _____

[p.280] thymus _____

[p.280] prostaglandins _____

[p.281] growth factors _____

Complete the Table

Complete the following table by matching the gland/organ and the hormone(s) produced by it to the descriptions of hormone action. Some glands/organs may be used more than once.

Gland/Organ

A. adrenal cortex [pp.275,276]
B. adrenal medulla [pp.275,277]
C. thyroid [pp.275,277]
D. parathyroids [pp.275,277]
E. testes [p.275]
F. ovaries [p.275]
G. pancreas (alpha cells) [p.278]
H. pancreas (beta cells) [p.278]
I. pancreas (delta cells) [p.278]
J. thymus [pp.275,280]
K. pineal [pp.275,280]

Hormone

a. thyroxine and triiodothyronine
b. glucagon
c. PTH
d. androgens (including testosterone)
e. somatostatin
f. thymosins
g. glucocorticoids (including cortisol)
h. estrogens
i. epinephrine
j. melatonin
k. insulin
l. progesterone
m. mineralocorticoids (including aldosterone)
n. calcitonin
o. norepinephrine

Gland/Organ	Hormone	Hormone Action
1.		Elevates calcium and phosphate levels in the bloodstream by stimulating bone cells to release these elements and the kidneys to convert them; also helps activate vitamin D
2.		Influences carbohydrate metabolism by control of food digestion; can block secretion of insulin and glucagon
3.		Required in egg maturation and release; prepares and maintains the uterine lining for pregnancy; influences growth and development
4.		Promote protein breakdown and conversion to glucose
5.		Lowers blood sugar level by stimulating glucose uptake by liver, muscle, and adipose cells; promotes protein and fat synthesis; inhibits protein conversion to glucose
6.		Required in sperm formation, genital development, and maintenance of sexual traits; influences growth and development
7.		Regulates metabolism; roles in growth and development
8.		Influences daily biorhythms; influences gonad development and reproductive cycles
9.		Raises blood level of sugar and fatty acids; increases heart rate and contraction force, the "fight–flight" response
10.		Plays roles in immune responses
11.		Raises blood sugar level by causing conversion of glycogen and amino acid to glucose in liver
12.		Prepares and maintains uterine lining for pregnancy; stimulates breast development
13.		Promotes sodium reabsorption; controls salt, water balance
14.		Promotes constriction or dilation of blood vessels
15.		Lowers calcium levels in blood

Dichotomous Choice

Circle one of two possibilities given between parentheses in each statement.

16. Insulin deficiency can lead to diabetes mellitus, a disorder in which the blood glucose level (rises/decreases) and glucose accumulates in the urine. [p.278]
17. In a person with diabetes mellitus, urination becomes (reduced/excessive), so that the body's water–solute balance becomes disrupted. [p.278]
18. Lacking a steady glucose supply, body cells of a person with diabetes mellitus begin breaking down fats and proteins for (energy/water). [p.278]
19. After a meal, blood glucose rises; pancreatic beta cells secrete (glucagon/insulin); targets use glucose or store it as glycogen. [p.278]
20. Blood glucose levels decrease between meals. (Glucagon/Insulin) is secreted by stimulated pancreas alpha cells; targets convert glycogen back to glucose, which then enters the blood. [p.278]
21. In ("type 1 diabetes"/"type 2 diabetes") the body mounts an immune response against its own insulin-secreting beta cells and destroys them. [p.279]
22. Juvenile-onset diabetes is also known as ("type 1 diabetes"/"type 2 diabetes"). [p.279]
23. In ("type 1 diabetes"/"type 2 diabetes"), insulin levels are close to or above normal, but target cells fail to respond to insulin. [p.279]
24. ("Type 1 diabetes"/"Type 2 diabetes") is usually manifested during middle age and is less dramatically dangerous than the other type. [p.279]
25. ("Type 1 diabetes"/"Type 2 diabetes") is appearing more commonly in young adults and is related to obesity. [p.279]

Matching

26. _____ glucocorticoids [p.276]

27. _____ cortisol [p.276]

28. _____ gluconeogenesis [p.276]

29. _____ hypoglycemia [p.276]

30. _____ mineralocorticoids [p.276]

31. _____ aldosterone [p.276]

32. _____ fight–flight response [p.277]

33. _____ adrenal medulla [p.277]

34. _____ adrenal cortex [p.276]

A. Secretions of adrenal medulla prepare the body for responding in times of emergency
B. Generally regulates the concentrations of mineral salts such as potassium and sodium present in extracellular fluid
C. Persistent low concentration of glucose in the blood; can develop from cortisol deficiency
D. Inner region of each adrenal gland; contains modified neurons that secrete epinephrine and norepinephrine
E. Generally influences metabolism in ways that help raise the level of glucose in the blood when it falls below a set point
F. Outer portion of each adrenal gland; secretes glucocorticoids and mineralocorticoids
G. The primary glucocorticoid; promotes protein breakdown and stimulates the liver to take up amino acids, dampens uptake of blood glucose, promotes fat breakdown
H. The most abundantly produced mineralocorticoid; stimulates reabsorption of sodium ions and excretion of potassium ions at the distal nephron tubules
I. Process in which liver cells synthesize glucose from amino acids

Fill-in-the-Blanks

Thyroxine (T4) and triiodothyronine (T3) are the main hormones secreted by the human (35) _____ [p.277] gland. They affect a person's overall (36) _____ [p.277] rate, growth, and development. The thyroid also makes (37) _____ [p.277], a hormone that lowers the level of calcium (and phosphate) in the blood. The synthesis of thyroid hormones requires (38) _____ [p.277], which is obtained from the diet. In the absence of iodine, blood levels of these hormones decrease. The anterior pituitary responds by secreting (39) _____ [p.277]. Excess TSH overstimulates the thyroid gland and causes it to enlarge. This tissue enlargement leads to an enlargement of the gland called simple (40) _____ [p.277]. Insufficient thyroid output is called (41) _____ [p.277]. Hypothyroid adults tend to be (42) _____ [p.277], sluggish, intolerant of cold, and sometimes feel confused and depressed. Simple goiter is no longer common in areas where people use (43) _____ _____ [p.277]. When blood levels of thyroid hormones become too high, (44) _____ [p.277] results. The most common disorder of this type is known as (45) _____ [p.271] disease, which appears to be an autoimmune disorder in which a thyroid-stimulating antibody binds to thyroid cells and causes overproduction of thyroid hormones.

The (46) _____ [p.278] glands are four glands located on the back of the human thyroid that respond homeostatically to chemical change in the immediate surroundings. They secrete (47) _____ [p.278] hormone (PTH). This hormone stimulates bone cells to release calcium and phosphate ions and the nephrons of the (48) _____ [p.278] to conserve them. PTH also helps to activate vitamin D. The activated form is a hormone that enhances (49) _____ [p.278] absorption from ingested food. In vitamin D deficiency, too little calcium and phosphorus are absorbed, so that bones develop improperly. This ailment is called (50) _____ [p.278].

Choice

For questions 51–64, choose from the following integration and control examples:

a. pineal gland [p.280]   b. thymus gland [p.280]   c. heart [p.280]   d. prostaglandins [p.280]
e. growth factors [p.281]   f. pheromones [p.281]   g. local signaling molecules [p.280]

51. _____ Produces ANP, a hormone with various effects that include regulating blood pressure

52. _____ Produced by many animals; released outside of the individual and then pass through air
or water to reach another individual

53. _____ An ancient photosensitive organ in the brain

54. _____ Actions are confined to the immediate vicinity of change

55. _____ Located behind the breastbone, between the lungs

56. _____ EGF influences many cell types

57. _____ Hormones known as *thymosins*

58. _____ Menstrual cramping is one effect

59. _____ Examples are prostaglandins and growth factors

60. _____ Secretes melatonin into the blood and cerebrospinal fluid

61. _____ EGF, IGF, and NGF

62. _____ T lymphocytes multiply, differentiate, and mature in this gland

63. _____ Evidence shows that human behavior may be affected by these

64. _____ Decreased melatonin secretion may trigger human puberty

---

Self-Quiz

Multiple Choice

_____ 1. The anterior lobe of the _____
governs the release of hormones from
other endocrine glands, while the posterior lobe stores hormones secreted by the
_____ . [p.272]
a. pituitary; hypothalamus
b. pancreas; hypothalamus
c. thyroid; parathyroid glands
d. hypothalamus; pituitary
e. pituitary; thalamus

_____ 2. ADH is sometimes called vasopressin
because it increases _____ .
[p.272]
a. heart rate
b. sweat production
c. water excretion
d. blood vessel diameter
e. blood pressure

_____ 3. The anterior lobe of the pituitary secretes
_____ different hormones,
while the posterior lobe secretes
_____ hormones. [p.272]
a. two, six
b. six, six
c. six, zero
d. six, two
e. two, two

Choice

For questions 4–6, choose from the following answers:

a. PTH   b. cortisol   c. aldosterone   d. calcitonin   e. melatonin

4. _____ lowers the level of calcium and phosphate in the blood [p.277]
5. _____ stimulates kidneys to reabsorb sodium ions and secrete potassium ions [p.276]
6. _____ affects sleep/wake cycles [p.280]

For questions 7–9, choose from the following answers:

a. adrenal medulla   b. adrenal cortex   c. thyroid   d. anterior pituitary   e. posterior pituitary

7. _____ The _____ produces glucocorticoids that help maintain the blood level of glucose and suppress inflammatory responses. [p.276]

8. _____ The gland that is most closely associated with emergency situations is the _____. [p.277]

9. _____ The _____ gland regulates the basic metabolic rate. [p.277]

Multiple Choice

_____ 10. If all sources of calcium were eliminated from your diet, your body would secrete more _____ in an effort to release calcium stored in your body and send it to the tissues that require it. [p.272]
    a. parathyroid hormone
    b. aldosterone
    c. calcitonin
    d. mineralocorticoids
    e. none of the above

Matching

Choose the most appropriate description for each term.

11. _____ ACTH [p.272]

12. _____ ADH [p.272]

13. _____ calcitonin [p.277]

14. _____ cortisol [p.276]

15. _____ epinephrine and norepinephrine [p.277]

16. _____ estrogen [p.275]

17. _____ glucagon [pp.278]

18. _____ insulin [p. 278]

19. _____ melatonin [p.280]

20. _____ oxytocin [p.272]

21. _____ parathyroid hormone [p.278]

22. _____ progesterone [p.275]

23. _____ GH (STH) [p.273]

24. _____ testosterone [p.275]

25. _____ thymosins [p.280]

26. _____ thyroxine [p.275]

27. _____ TSH [p.272]

A. Raises the glucose level in the blood
B. Influences daily biorhythms, gonad development, and reproductive cycles
C. Affects development of male sexual traits; required for sperm formation
D. Increase heart rate and control blood volume; the "emergency hormones"
E. Essential for egg maturation and maintenance of secondary sexual characteristics in the female
F. The water-conservation hormone; released from posterior pituitary
G. Lowers blood sugar by signaling cells to take in glucose; promotes synthesis of proteins and fats
H. Stimulates adrenal cortex to secrete steroid hormones
I. Causes increase of calcium in the blood
J. Influences overall metabolic rate, growth, and development
K. Roles in immunity
L. Triggers uterine contractions during labor and causes milk release during nursing
M. Prepares and maintains uterine lining for pregnancy; stimulates breast development
N. Inhibits uptake of blood glucose by muscle cells; primary glucocorticoid
O. Lowers calcium levels in blood
P. Secreted by anterior pituitary; stimulates release of thyroid hormones
Q. Secreted by anterior pituitary; enhances growth in young animals, especially cartilage and bone

Chapter Objectives/Review Questions

This section lists general and detailed chapter objectives that can be used as review questions. You can make maximum use of these items by writing answers on a separate sheet of paper. Fill in answers where blanks are provided. To check for accuracy, compare your answers with information given in the chapter or glossary.

1. _____ cells have receptors for a specific signaling molecule; this molecule may alter the behavior of the cells in response to it. [p.268]
2. Where is a hormone made, how does it travel to its target, and what happens when it arrives there? [p.268]
3. Collectively, sources of hormones are referred to as the _____ system. [p.268]
4. Name and define the three kinds of hormonal interaction. [p.268]
5. Locate and name the components of the human endocrine system on a diagram such as Figure 15.2 of the main text. [p.269]
6. Contrast the proposed mechanisms of hormonal action on target cell activities by (a) steroid hormones and (b) nonsteroid hormones. [pp.270–271]
7. State the relationship between both the anterior and posterior pituitary lobes and the hypothalamus. [p.272]

8. Identify the hormones released from the posterior lobe of the pituitary and state their target tissues. [p.272]
9. Identify the hormones produced by the anterior lobe of the pituitary and tell which target tissues or organs each acts on. [pp.272–273]
10. Most hypothalamic hormones acting in the anterior lobe are _____ that cause target cells to secrete hormones of their own. Some are _____ that slow down secretion from their targets. [p.273]
11. Pituitary dwarfism, gigantism, and acromegaly are all associated with abnormal secretion of _____ by the pituitary gland. [p.274]
12. Describe the major human hormone sources together with their secretions, main targets, and primary actions, as shown in Table 15.3 of the main text. [p.275]
13. Name the hormones secreted by alpha, beta, and delta pancreatic cells; identify the effect of each. [p.278]
14. Describe the symptoms of diabetes mellitus and distinguish between type 1 and type 2. [pp.278–279]
15. The adrenal _____ secretes glucocorticoids. [p.276]
16. Define *hypoglycemia*; describe the roles of the hypothalamus and the anterior pituitary in this condition. [p.276]
17. The most abundantly produced mineralocorticoid is _____ ; cite its function. [p.276]
18. The _____ _____ is the part of the adrenal gland that is involved in response to stress. [p.277]
19. List the features of the "fight–flight" response. [p.277]
20. Describe the characteristics of hypothyroidism and hyperthyroidism. [p.277]
21. Name the glands that secrete PTH and give the function of this hormone. [p.278]
22. Describe the ailment called *rickets* and cite its cause. [p.278]
23. Thymosins are secreted by the _____ gland; they appear to be related to the proper functioning of the _____ system. [p.280]
24. The pineal gland secretes the hormone _____ ; give two examples of the action of this hormone. [p.280]
25. Give two examples that illustrate the effects of local signaling molecules. [p.280]
26. More than sixteen different kinds of the fatty acids called _____ have been identified in tissues throughout the body; list their major effects. [p.280]
27. Define *pheromone*; cite a possible function in humans. [p.280]

Integrating and Applying Key Concepts

Suppose you suddenly quadruple your already high daily consumption of calcium. State which organs would be affected and tell how they would be affected. Name two hormones whose levels would most probably be affected and tell whether your body's production of them would increase or decrease. Suppose you continue this high rate of calcium consumption for 10 years. Can you predict which organs would be subject to the most stress as a result?

16

REPRODUCTIVE SYSTEMS

Interactive Exercises

Selected Words: epididymides (singular: epididymis) [p.286], *spermatogonia* [p.288], *mitosis* [p.288], *meiosis* [p.288], *gametes* [p.288], "haploid" number [p.288], "diploid" number [p.288], *primary spermatocytes* [p.288], *secondary spermatocytes* [p.288], *spermatids* [p.288], *spermatozoa* [p.288], *spermatogenesis*

Boldfaced, Page-Referenced Terms

[p.285] testes (singular: testis) _____

_____

[p.285] ovaries _____

_____

[p.285] secondary sexual traits _____

_____

[p.286] seminiferous tubules _____

_____

[p.286] vas deferentia (singular: vas deferens) _____

_____

[p.286] semen _____

_____

[p.286] seminal vesicles _____

_____

[p.287] prostate gland _____

_____

[p.287] bulbourethral glands _____

_____

[p.288] sperm _____

_____

[p.288] Sertoli cells _____

_____

[p.288] acrosome _____

_____

[p.288] Leydig cells _____

_____

[p.288] testosterone _____

_____

[p.289] LH (luteinizing hormone) _____

_____

[p.289] FSH (follicle-stimulating hormone) _____

_____

Fill-in-the-Blanks

The numbered items in the accompanying illustrations represent missing information; fill in the numbered blanks in the following narrative to supply the missing information on each illustration. Some illustrated structures are numbered more than once to aid identification.

Just inside the walls of seminiferous tubules are cells called (1) _____ [p.288]. These cells undergo continuous divisions, including a type of division called (2) _____ [p.288] and a type called (3) _____ [p.288]. This process results in the specialized haploid male reproductive cells called *sperm*. Spermatogonia develop into (4) _____ _____ [p.288], which, after a meiotic division known as (5) _____ _____ [p.288], are termed (6) _____ _____ [p.288]. A second division, known as (7) _____ _____ [p.288], results in immature sperm known as (8) _____ [p.288], which gradually develop into spermatozoa, or simply (9) _____ [p.288]— the male gametes. The (10) "_____" [p.288] of each sperm arises at the very end of the process, which takes 9 to 10 weeks. These developing cells receive nourishment and chemical signals from adjacent (11) _____ [p.288] cells. A mature sperm has a tail, a midpiece, and a(n) (12) _____ [p.288]. Within the head, a nucleus contains DNA organized into chromosomes. An enzyme-containing cap, the (13) _____ [p.288], covers most of the head. Its enzymes help sperm penetrate the extracellular material around an egg at fertilization. In the midpiece, (14) _____ [p.288] supply energy for the tail's whiplike movements.

Human sperm are not quite mature when they leave the (15) _____ [p.286]. First they enter a pair of long, coiled ducts, the (16) _____ [p.286]. When a male becomes sexually aroused, muscle contractions in the walls of reproductive organs propel sperm into and through a pair of thick-walled tubes, the (17) _____ _____ [p.286]. From there, contractions propel sperm through a pair of ejaculatory ducts and then through the (18) _____ [p.286] as it passes through the penis and to the outside. During the trip to the urethra, glandular secretions become mixed with sperm. The result is semen, a thick fluid that is eventually expelled from the penis. A pair of (19) _____ [p.286] vesicles secrete the sugar fructose, which nourishes the sperm cells. Seminal vesicles also secrete certain prostaglandins, signaling molecules able to induce muscle contraction. Secretions from the (20) _____ [p.287] gland probably help buffer the acidic vaginal environment. A pair of (21) _____ [p.287] glands secrete a mucus-rich fluid into the urethra that neutralizes urine traces when the male is sexually aroused. (22) _____ _____ [p.288], or interstitial cells, are located in tissue between the seminiferous tubules in testes.

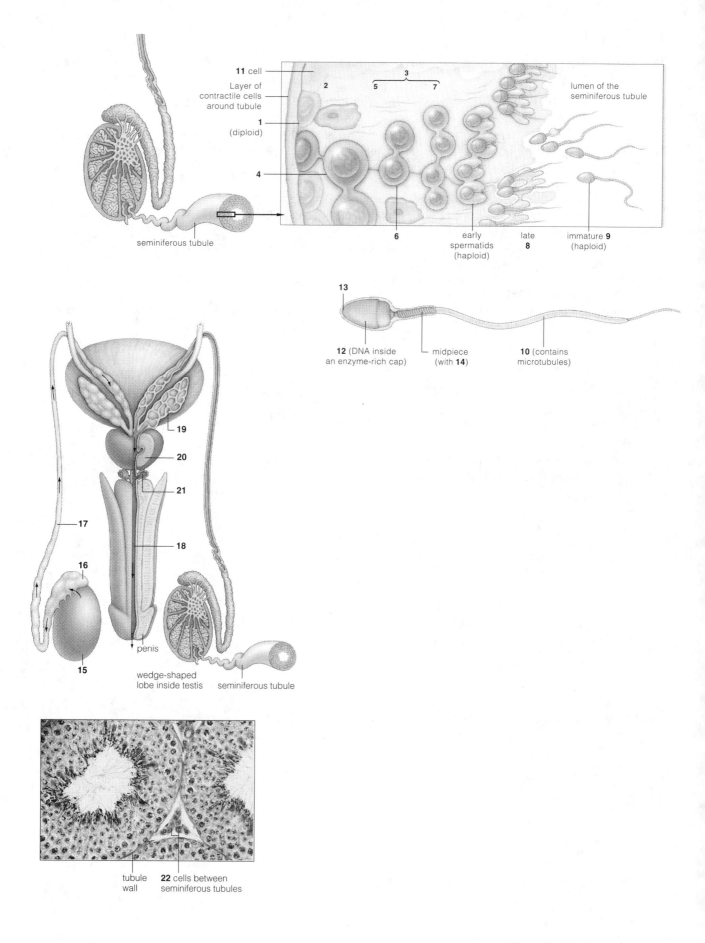

11 cell

Layer of
contractile cells
around tubule

1
(diploid)

4

seminiferous tubule

3

2

5   **7**

lumen of the
seminiferous tubule

6

early
spermatids
(haploid)

late
8

immature **9**
(haploid)

13

12 (DNA inside
an enzyme-rich cap)

midpiece
(with **14**)

10 (contains
microtubules)

19

20

21

17

18

16

15

penis

wedge-shaped
lobe inside testis

seminiferous tubule

tubule
wall

22 cells between
seminiferous tubules

Dichotomous Choice

Circle one of two possibilities given between parentheses in each statement.

23. Testosterone is secreted by (Leydig cells/the hypothalamus). [p.288]
24. (Testosterone/FSH) governs the growth, form, and functions of the male reproductive tract. [pp.288–289]
25. Sexual behavior and secondary sexual traits are associated with (LH/testosterone). [p.289]
26. LH and FSH are secreted by the (anterior/posterior) lobe of the pituitary gland. [p.289]
27. The (testes/hypothalamus) govern(s) sperm production by controlling interactions among testosterone, LH, and FSH. [p.289]
28. When blood levels of testosterone (increase/decrease) beyond a certain set point, the hypothalamus secretes GnRH, stimulating the anterior pituitary lobe to release LH and FSH, which the bloodstream distributes to the testes. [p.289]
29. Within the testes, (LH/FSH) acts on Leydig cells; they secrete testosterone, which stimulates diploid germ cells to become sperm. [p.289]
30. Sertoli cells have (LH/FSH) receptors; this compound is crucial to establishing spermatogenesis at the time of puberty. [p.289]
31. When blood testosterone levels (increase/decrease) past a set point, feedback loops to the hypothalamus slow down testosterone secretion and sperm formation. [p.289]

16.3. THE FEMALE REPRODUCTIVE SYSTEM [pp.290–291]

16.4. HOW OOCYTES DEVELOP [pp.292–293]

16.5. VISUAL SUMMARY OF THE MENSTRUAL CYCLE [p.294]

Selected Words: *fallopian tube* [p.290], *cervix* [p.290], *vagina* [p.290], *vulva* [p.290], *labia majora* [p.290], *labia minora* [p.290], *primary* oocyte [p.290], *secondary* oocyte [p.290], *menarche* [p.291], *menopause* [p.291], *endometriosis* [p.291], *first polar body* [p.292], *fimbriae* [p.293], *implantation* [p.293]

Boldfaced, Page-Referenced Terms

[p.290] oocytes _____

_____

[p.290] oviduct _____

_____

[p.290] uterus _____

_____

[p.290] endometrium _____

_____

[p.290] clitoris _____

_____

[p.290] menstrual cycle _____

_____

[p.290] menstruation _____

_____

[p.290] ovulation _____

_____

[p. 290] estrogens _____

_____

[p.290] progesterone _____

_____

[p.292] granulosa cells _____

_____

[p.292] follicle _____

_____

[p.292] zona pellucida _____

_____

[p.292] secondary oocyte _____

_____

[p.292] corpus luteum _____

_____

Fill-in-the-Blanks

The numbered items on the accompanying illustration represent missing information; fill in the numbered blanks in the following narrative to supply the missing information on the illustration.

An oocyte (immature egg) is released from a(n) (1) _____ [p.290]. When the oocyte is released from either ovary, it moves into a(n) (2) _____ [p.290] and is transported to the (3) _____ [p.290], a hollow, pear-shaped organ in which a baby can grow and develop. The wall of the uterus consists of a thick layer of smooth muscle, the (4) _____ [p.290], and an interior lining, the (5) _____ [p.290], which includes epithelial tissue, connective tissue, glands, and blood vessels. The lower portion of the uterus is the (6) _____ [p.286]. A muscular tube, the (7) _____ [p.290], extends from the cervix to the body surface and receives the penis and sperm and functions as part of the birth canal. The external female genitals are collectively called the vulva. Outermost is a pair of fat-padded skin folds, the (8) _____ _____ [p.290]. Those folds enclose a smaller pair of skin folds, the (9) _____ _____ [p.290], which are highly vascularized but have no fatty tissue. The smaller folds partly enclose the (10) _____ [p.290], a small organ sensitive to stimulation that is developmentally analogous to the penis. The opening of the (11) _____ [p.290] is about midway between the clitoris and the vaginal opening.

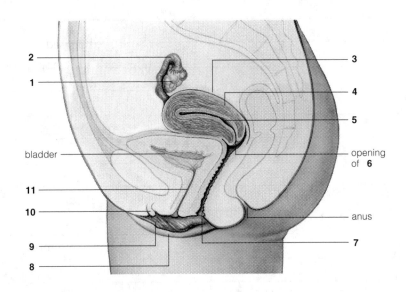

Outlining

Fill in the outline below covering the reproductive cycle of female humans.

   I. The (12) _____-phase menstrual cycle [pp.290–291]
      A. (13) _____ phase
         1. (14) _____ — marks the first day of a new cycle
         2. Endometrial disintegration and rebuilding
         3. (15) _____ maturation
      B. (16) _____ — release of an oocyte from an ovary
      C. (17) _____ phase

    1. (18) _____ _____ forms

    2. (19)_____ is primed for pregnancy

  D. Terms related to the menstrual cycle

    1. (20) _____ — first menstruation, between ages 10 and 16

    2. (21) _____ — menstrual cycles stop; occurs in late 40s or early 50s

    3. (22)_____ — disorder in which endometrial tissue spreads outside of uterus

II. The (23) _____ cycle — steps leading to ovulation [pp.292–293]

  A. Primary oocyte maturation

    1. About (24) _____ exist at age seven

    2. Found near the surface of an (25) _____

    3. Surrounded and nourished by the (26) _____ cells

    4. Follicle grows as a result of anterior pituitary secretions of (27) _____ and LH

    5. Noncellular (28) _____ _____ forms around oocyte

    6. FSH and LH stimulate cells outside zona pellucida to secrete (29) _____

    7. Primary oocyte completes meiosis I, giving rise to a (30) _____ oocyte and the first (31) _____ body

  B. Ovulation

    1. Surge of LH causes the (32) _____ to swell and rupture

    2. (33) _____ _____ and first polar body are released

    3. Secondary oocyte released into abdominal cavity and drawn into a(n) (34)_____ by the beating of the cilia or fimbriae

    4. (35)_____ typically occurs in the oviduct

    5. At fertilization, the oocyte completes meiosis II and is a mature (36) _____

III. Hormones prepare the uterus for pregnancy [p.293]

  A. Before ovulation

    1. Estrogens stimulate growth of the (37) _____ and its glands

    2. Cells of follicle wall secrete (38) _____ as well as estrogens

  B. At ovulation — estrogens cause the (39) _____ to secrete a thin mucus for sperm to swim through

  C. After ovulation

    1. LH surge leads to development of the yellowish glandular (40) _____ _____

    2. Corpus luteum secretes progesterone, which maintains the (41) _____ during a pregnancy

    3. While corpus luteum persists, a pituitary-ordered decrease in FSH prevents other (42) _____ from developing

    4. Corpus luteum breaks down after about 12 days if (43) _____ of a developing embryo into the endometrium does not occur

    5. When progesterone and estrogen levels drop, the (44) _____ breaks down and menstruation occurs

    6. The cycle begins again as rising levels of (45) _____ stimulate the repair and growth of the endometrium

Analyzing Diagrams

Study the diagram below to correctly complete the following dichotomous choice statements. [p.294]

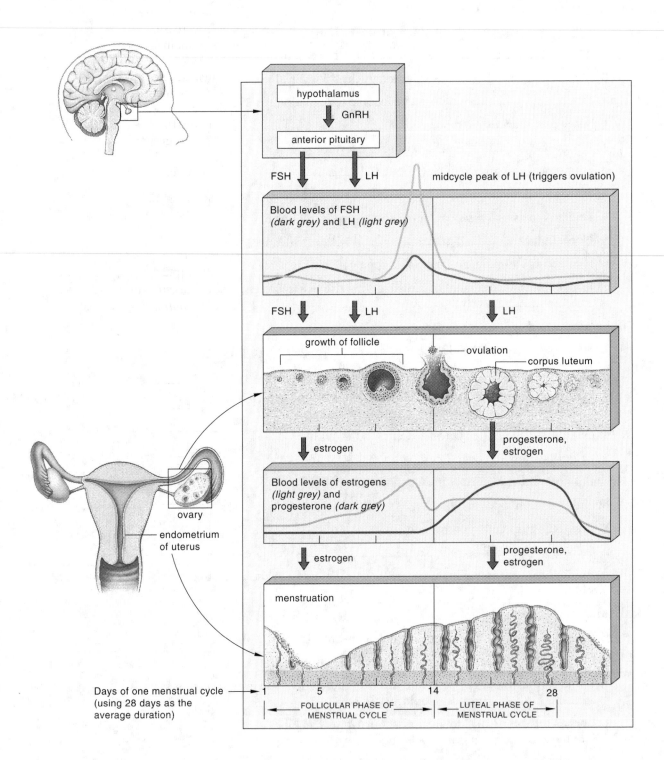

46. With increasing levels of FSH and LH secreted by the pituitary, a follicle (shrinks/grows).
47. Ovulation occurs when there is a sharp surge of (FSH/LH) levels.
48. FSH and LH levels (decrease/increase) after ovulation.
49. As a follicle develops prior to ovulation, estrogen levels (decrease/increase).
50. After ovulation, estrogen levels drop slightly, while progesterone levels (decrease/increase).
51. As long as the corpus luteum remains, estrogen and progesterone levels (remain stable/continue to increase).
52. As long as the corpus luteum remains, the endometrium is (very thin/fully developed).
53. The endometrium is fully developed during the (follicular/luteal) stage of the menstrual cycle.
54. The menstrual cycle begins with (ovulation/menstruation).
55. Progesterone appears to cause the endometrium to (deteriorate/be maintained).
56. Estrogen production continues throughout the existence of a follicle, while progesterone is only made after the formation of the (corpus luteum/endometrium).
57. Fertilization is possible (throughout/around the middle of) the menstrual cycle.

16.6. SEXUAL INTERCOURSE, ETC. [p.295]

16.7. CONTROLLING FERTILITY [p.296–297]

Selected Words: abstinence [p.296], *rhythm method* [p.296], *sympto-thermal method* [p.296], *withdrawal* [p.296], *douching* [p.296], *tubal ligation* [p.296], *diaphragm* [p.297], *cervical cap* [p.297], *contraceptive sponge* [p.297], *intrauterine device* (IUD) [p.297], *condoms* [p.297], "*female condom*" [p.297], *birth control pill* [p.297], *Norplant* [p.297], *morning-after pill* [p.297], "*sterilize*" [p.297]

Boldfaced, Page-Referenced Terms

[p.295] coitus _____

_____

[p.295] orgasm _____

_____

[p.295] fertilization _____

_____

Matching

Match each birth control option with its description (place capital letters in the short blanks). Complete the exercise by selecting the "effectiveness" category and placing that lowercase letter within the parentheses provided with each birth control option.

1. _____ ( ) abstinence [p.296]

2. _____ ( ) rhythm or sympto-thermal method [p.296]

3. _____ ( ) withdrawal [p.296]

4. _____ ( ) douching [p.296]

5. _____ ( ) spermicidal foams and jellies [pp.296–297]

6. _____ ( ) diaphragm plus spermicide [pp.296–297]

7. _____ ( ) cervical cap [pp.296–297]

8. _____ ( ) contraceptive sponge plus spermicide [pp.296–297]

9. _____ ( ) IUD [pp.296–297]

10. _____ ( ) condoms (good quality) [pp.296–297]

11. _____ ( ) female condom [pp.296–297]

12. _____ ( ) oral contraceptive (the "pill") [pp.296–297]

13. _____ ( ) Depo-Provera [pp.296–297]

14. _____ ( ) Norplant [pp.296–297]

15. _____ ( ) morning-after pills [pp.296–297]

16. _____ ( ) vasectomy [p.296]

17. _____ ( ) tubal ligation [p.296]

A. Each vas deferens is severed and tied off
B. Removal of the penis from the vagina before ejaculation
C. A small plastic or metal device that is placed in the uterus and interferes with implantation
D. Rinsing out the vagina with a chemical after intercourse
E. An implant inserted under the skin of a woman's upper arm; contains a hormone that prevents implantation
F. No sexual intercourse
G. The oviducts are cauterized or cut and tied off
H. Injection of progestin to inhibit ovulation
I. Interfere with hormones that control events between ovulation and implantation
J. A soft, disposable disk that contains a spermicide and covers the cervix; wetted and inserted up to 24 hours before intercourse
K. Most widely used method of fertility control; contains synthetic estrogens and progesterones; suppresses release of LH and FSH, normally required for eggs to mature
L. Avoiding intercourse during the woman's fertile period; uses daily temperature readings
M. Thin, tight-fitting sheaths of latex or animal skin worn over the penis during intercourse
N. Flexible, dome-shaped device that is inserted into the vagina and positioned over the cervix before intercourse
O. Toxic to sperm; packaged in an applicator and placed in the vagina just before intercourse; not reliable unless used with a diaphragm or condom
P. Variation on the diaphragm but smaller and can be left in place up to three days with a single dose of spermicide
Q. Latex pouch inserted into the vagina

a. Extremely effective

b. Highly effective

c. Effective

d. Moderately effective

e. Fairly effective

f. Unreliable

Short Answer

18. List some future options for fertility control. [p.297] _____

_____

_____

_____

16.8. COPING WITH INFERTILITY [p.298]

16.9. *Choices: Biology and Society*: DILEMMAS OF FERTILITY CONTROL [p.299]

Selected Words: infertile [p.298], "reproductive technology" [p.298], *artificial insemination by donor* (AID) [p.298], *in vitro fertilization* (IVF) [p.298], *zygotes* [p.298], *IVF with embryo transfer* [p.298], "surrogate mother" [p.298], GIFT [p.298], ZIFT [p.298], *Roe v. Wade* [p.299]

Choice

For questions 1–13, choose from the following:

a. *in vitro* fertilization   b. intrafallopian transfers   c. artificial insemination

1. _____ AID [p.298]

2. _____ Literally means "fertilization in glass" [p.298]

3. _____ ZIFT [p.298]

4. _____ A technique that can produce several viable embryos at one time; those not used in a given procedure can be frozen and stored for long periods of time [p.298]

5. _____ GIFT [p.298]

6. _____ The fate of unused embryos has prompted ethical debates and even bitter court battles between divorcing couples [p.298]

7. _____ An anonymous donor can sell his sperm to a "sperm bank," which then charges a fee for insemination [p.298]

8. _____ A couple's sperm and oocytes are collected and then placed into an oviduct (fallopian tube); fertilization rate is about 20 percent [p.298]

9. _____ Zygotes are transferred to a solution that will support further development [p.298]

10. _____ A fertile female "donor" is inseminated with sperm from a male whose female partner is infertile [p.298]

11. _____ A single sperm is injected into an egg using a tiny glass needle [p.298]

12. _____ Oocytes and sperm are brought together in a laboratory dish, where fertilization can give rise to a zygote; the zygote is then placed in one of the woman's oviducts [p.298]

13. _____ If the donor becomes pregnant, the developing embryo is transferred to the infertile woman's uterus [p.298]

Self-Quiz

Choice

For questions 1–2, choose from the following answers:

> a. cervix   b. oviduct   c. urethra   d. uterus   e. vulva

1. The _____ is the lower, narrowed portion of the uterus. [p.293]

2. The _____ is a pathway from the ovary to the uterus. [p.293]

For questions 3–6, choose from the following answers:

> a. Leydig (or interstitial) cells   b. seminiferous tubules   c. vas deferens
> d. epididymis   e. Sertoli cells

3. Sperm mature and are stored in the _____ . [p.286]

4. The _____ connects a structure on the surface of the testis with the ejaculatory duct. [p.286]

5. Testosterone is produced by the _____ . [p.288]

6. Male gametes are produced by meiosis in the _____ . [p.288]

Multiple Choice

_____ 7. Male reproductive functions are controlled by the hormones _____ .
[p.288]
   a. LH, FSH, and progesterone
   b. estrogen, FSH, and LH
   c. testosterone, LH, and FSH
   d. testosterone and FSH

Choice

For questions 8–12, choose from the following answers:

> a. corpus luteum   b. developing early embryo   c. follicle   d. hypothalamus   e. pituitary

8. _____ Provides a midcycle surge of LH to trigger ovulation [p.292]

9. _____ Secretes follicle-stimulating hormone (FSH) and luteinizing hormone (LH) [p.292]

10. _____ Cells outside the zona pellucida secrete estrogens; estrogen-containing fluid starts to increase [p.292]

11. _____ Secretes GnRH, which stimulates the pituitary to begin secreting LH and FSH [p.292]

12. _____ Secretes some estrogen and progesterone [p.293]

For questions 13–17, choose from the following answers:

a. contraceptive sponge   b. sympto-thermal or rhythm   c. diaphragm   d. IUD
e. spermicidal foam and spermicidal jelly

13. _____ A small plastic or metal device that is inserted into the uterus and interferes with implantation [p.297]

14. _____ A soft, disposable disk that contains a spermicide and covers the cervix [p.297]

15. _____ A flexible, dome-shaped device that is inserted into the vagina just before intercourse [p.297]

16. _____ Substances that are toxic to sperm, packaged in an applicator, and placed in the vagina just before intercourse [p.297]

17. _____ The avoidance of intercourse during the woman's fertile period [p.296]

For questions 18–20, choose from the following answers:

a. *in vitro* fertilization   b. intrafallopian transfers   c. artificial insemination

18. _____ The introduction of semen into the vagina or uterus by artificial means, usually a syringe, around the time of ovulation [p.298]

19. _____ GIFT and ZIFT [p.298]

20. _____ Conception occurs externally [p.298]

Chapter Objectives/Review Questions

This section lists general and detailed chapter objectives that can be used as review questions. You can make maximum use of these items by writing answers on a separate sheet of paper. Fill in answers where blanks are provided. To check for accuracy, compare your answers with information given in the chapter or glossary.

1. The primary reproductive organs are sperm-producing _____ [p.285] in males and egg-producing _____ [p.285] in females.
2. In humans the primary reproductive organs also produce _____ _____ [p.285], which influence reproductive functions and _____ [p.285] sexual traits.
3. Explain when and how gonads begin to differentiate in early human embryos. [p.285]
4. Follow the path of a mature sperm from the seminiferous tubules to the urethral exit. List every structure encountered along the path and state its contribution to the nurture of the sperm. [pp.286–287]
5. List, in order, the stages of spermatogenesis. [pp.288–289]
6. Name the three hormones that directly control sperm formation and form part of feedback loops between the hypothalamus, anterior pituitary, and testes. [pp.288–289]
7. Diagram the structure of a sperm, label its components, and state the function of each. [pp.288–289]
8. Ovulation is the release of a primary _____ from the ovary. [p.290]
9. Name the event that brings about ovulation as well as the other hormonal events that bring about the onset and finish of menstruation. [pp.292–294]
10. Describe the origin and functions of the corpus luteum. [pp.293]
11. The four hormones that control egg maturation and release as well as changes in the endometrium are _____ , _____ , _____ , and _____ ; they are part of feedback loops involving the hypothalamus, anterior pituitary, and ovaries. [pp.292,294]
12. In both males and females, _____ from the hypothalamus stimulates the anterior pituitary to release LH and FSH. [pp.289,292]

13. List the physiological events that bring about erection of the penis during sexual stimulation, and explain the process of ejaculation. [p.295]
14. Trace the path of a sperm from the urethral exit to the place where fertilization normally occurs. Mention, in correct sequence, all major structures of the female reproductive tract that are passed along the way, and state the principal function of each structure. [p.295]
15. The _____ is a method of birth control that also helps prevent the spread of sexually transmitted diseases. [p.297]
16. Two different types of surgical birth control are _____ _____ and _____ . [p.296]
17. Describe in vitro fertilization as a method of overcoming infertility. [p.298]
18. Distinguish between ZIFT and GIFT as methods of intrafallopian transfer. [p.298]
19. Generally define *artificial insemination* and *AID*. [p.298]

Integrating and Applying Key Concepts

What percentage of humans on the planet today do you think resulted from unplanned pregnancies? As technology allows increasingly better control of fertility, what effect do you think this might have on size of families or age of parents having families? What effect will it have on societies that currently do not have easily available birth control? Assuming that technology will continue to be available to some groups or populations before others, what effect do you think this will have on increases or decreases in the relative sizes of populations?

17

DEVELOPMENT AND AGING

CHAPTER INTRODUCTION

THE SIX STAGES OF DEVELOPMENT
Three primary tissues form
Organogenesis, growth, and tissue specialization

THE BEGINNINGS OF YOU — EARLY STEPS IN
DEVELOPMENT
Fertilization unites sperm and oocyte
Cleavage produces a multicellular embryo
Implantation

VITAL MEMBRANES OUTSIDE THE EMBRYO
The placenta: A pipeline for oxygen, nutrients, and
other substances

HOW THE EARLY EMBRYO TAKES SHAPE
Gastrulation establishes the body's basic plan
Internal body regions get their shape and structure
during morphogenesis
The folding of sheets of cells is an important part
of morphogenesis

THE FIRST EIGHT WEEKS — HUMAN FEATURES
EMERGE
Miscarriage

DEVELOPMENT OF THE FETUS
The second trimester: Movements begin
Organ systems mature during the third trimester
Special features of the blood and circulatory
system of a fetus

BIRTH AND BEYOND
Labor has three stages
Nourishing the newborn

Choices: Biology and Society: SHOULD EMBRYOS BE
CLONED?

MOTHER AS PROVIDER, PROTECTOR, AND
POTENTIAL THREAT
Good maternal nutrition is vital
Risk of infections
Harm from drugs and alcohol
Harm from cigarette smoke

Science Comes to Life: PRENATAL DIAGNOSIS:
DETECTING BIRTH DEFECTS

THE PATH FROM BIRTH TO ADULTHOOD
Transitions from birth to adulthood
Adulthood is also a time of bodily change

WHY DO WE AGE?
The "programmed life span" hypothesis
The "cumulative assaults" hypothesis

AGING SKIN, MUSCLE, BONE, AND TRANSPORT
SYSTEMS
Skin, muscles, and bones
Aging of the cardiovascular and respiratory
systems

AGE-RELATED CHANGES IN SOME OTHER BODY
SYSTEMS
The nervous system and senses
Reproductive systems and sexuality
Immunity, nutrition, and the urinary system

Interactive Exercises

CHAPTER INTRODUCTION [p.303]

17.1. THE SIX STAGES OF DEVELOPMENT [pp.304–305]

17.2. THE BEGINNINGS OF YOU — EARLY STEPS IN DEVELOPMENT [pp.306–307]

17.3. VITAL MEMBRANES OUTSIDE THE EMBRYO [pp.308–309]

Selected Words: *blastomere* [p.304], *capacitation* [p.306], *identical twins* [p.306], *fraternal twins* [p.306], *ectopic (tubal) pregnancy* [p.307], *chorionic villi* [p.308], *chorionic villus sampling* [p.309]

Boldfaced, Page-Referenced Terms

[p.304] gametes _____

_____

[p.304] fertilization _____

_____

[p.304] zygote _____

_____

[p.304] cleavage _____

_____

[p.304] morula _____

_____

[p.304] gastrulation _____

_____

[p.304] germ layers _____

_____

[p.304] endoderm _____

_____

[p.304] mesoderm _____

_____

[p.304] ectoderm _____

_____

[p.304] organogenesis _____

_____

[p.304] growth and tissue specialization _____

_____

[p.304] cell determination _____

_____

[p.305] cell differentiation _____

_____

[p.305] morphogenesis _____

_____

[p.306] ovum (plural: ova) _____

_____

[p.306] blastocyst _____

_____

[p.306] trophoblast _____

_____

[p.306] inner cell mass _____

_____

[p.307] implantation _____

_____

[p.308] embryonic disk _____

_____

[p.308] extraembryonic membranes _____

_____

[p.308] yolk sac _____

_____

[p.308] amnion _____

_____

[p.308] allantois _____

_____

[p.308] umbilical cord _____

_____

[p.308] chorion _____

_____

[p.308] placenta _____

_____

Sequence

Arrange the following events in correct chronological sequence. Write the letter of the first step next to 1, the letter of the second step next to 2, and so on.

1. _____

2. _____

3. _____

4. _____

5. _____

6. _____

A. Gastrulation [p.304]
B. Fertilization [p.304]
C. Cleavage [p.304]
D. Growth and tissue specialization [p.304]
E. Organogenesis (organ formation) [p.304]
F. Gamete formation [p.304]

Complete the Table

7. Complete the following table by entering the correct embryonic germ layer (ectoderm, mesoderm, or endoderm) that forms the tissues and organs listed. [p.304]

Germ Layer	Tissues/Organs
a.	Muscle
b.	Nervous tissue
c.	Epithelial lining of the GI tract and lungs
d.	Cardiovascular system (blood vessels, heart)
e.	Epidermis (skin)
f.	Reproductive and excretory organs
g.	Parts of tonsils, thyroid and parathyroid glands
h.	Most of the skeleton
i.	Connective tissues of the gut and integument

Fill-in-the-Blanks

When the three germ layers have formed, they separate into subgroups of cells. This signals the beginning of a phase called (8) _____ [p.304], or organ formation. During this phase, different sets of cells get their basic biological identities and give rise to different tissues and organs. The final stage of development is known as growth and tissue (9) _____ [p.304]. There are three key processes of development. The first is cell (10) _____ [p.304]. It establishes the eventual fate of the cell and its descendants. The second process is known as cell (11) _____ [p.305]. This is a gene-guided process by which cells in different locations in the embryo become specialized. The third process is known as (12) _____ [p.305]. This process produces the shape and structure of particular body regions. It also involves localized cell division and growth, as well as (13) _____ [p.305] of cells and entire tissues from one site to another. Controlled death and folding of sheetlike tissues can also occur in this process.

As an example, morphogenesis at the ends of limb buds first produced (14) _____ - [p.305] shaped hands at the ends of your arms; then skin cells between lobes in the paddles (15) _____ [p.306] on cue, leaving separate fingers. Keep in mind that as a developing organism comes to the end of each developmental stage, the embryo has become more (16) _____ [p.305] than it was in the previous stage.

Matching

Choose the most appropriate description for each term.

17. _____ capacitation [p.306]

18. _____ zona pellucida [p.306]

19. _____ ovum [p.306]

20. _____ identical twins [p.306]

21. _____ fraternal twins [p.306]

22. _____ morula [p.306]

23. _____ blastocyst [p.306]

24. _____ trophoblast [p.306]

25. _____ inner cell mass [p.306]

26. _____ implantation [p.307]

27. _____ ectopic pregnancy [p.307]

A. The solid ball of cells reaching the uterus three or four days after fertilization
B. Produced by separation of the two cells formed by the first cleavage of the zygote; two independent embryos develop
C. About one week after fertilization, epithelial cells of the blastocyst become embedded in the endometrium
D. Acrosome enzymes clear a path through this outer layer of the egg
E. Caused by implantation of a fertilized egg in an oviduct or in the abdominal wall
F. Produced when two different eggs are fertilized at roughly the same time by two different sperm
G. Chemical process that weakens the acrosome membrane of a sperm
H. Surface layer of epithelial cells on a blastocyst
I. The clump of cells located inside the blastocyst that will develop into an embryo
J. Produced along with a polar body by meiotic cell division of the secondary oocyte
K. A ball of cells that consists of a surface epithelium and an inner clump of cells to one side of the ball

Matching

Match each numbered structure on the accompanying illustrations with its correct letter. [pp.306–307]

28. _____
29. _____
30. _____
31. _____
32. _____
33. _____
34. _____
35. _____
36. _____
37. _____
38. _____
39. _____
40. _____
41. _____
42. _____

A. Blastocyst
B. Endometrium
C. Fertilization
D. Four-cell stage
E. Implantation
F. Inner cell mass
G. Morula
H. Ovary
I. Oviduct (fallopian tube)
J. Ovulation
K. Opening of cervix
L. Trophoblast
M. Two-cell stage
N. Uterus
O. Zygote

Find these numbered structures on the illustrations below and on the preceding page and arrange them in correct sequential order: 29, 30, 31, 33, 40. [pp.306–307]

43. _____ would occur first

44. _____ second

45. _____ third

46. _____ fourth

47. _____ last in the sequence

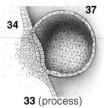

34

37

33 (process)

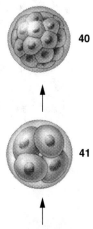

37

38

39

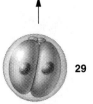

40

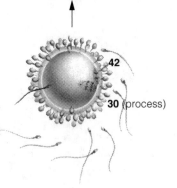

41

29

42

30 (process)

Choice

For questions 48–62, choose from the following:

a. yolk sac   b. amnion   c. allantois   d. umbilical cord   e. chorion   f. placenta

48. _____   An intimate association of the embryonic chorion and the superficial cells of the mother's endometrial lining. [p.308]

49. _____   A protective membrane around the embryo and other membranes. [p.308]

50. _____   Membrane that develops from the trophoblast and continues the secretion of HCG that began when the blastocyst implanted. [p.308]

51. _____   Links the embryo with the placenta. [p.308]

52. _____   Source of early blood cells and of germ cells that will become the gametes. [p.308]

53. _____   Innermost membrane that develops into a fluid-filled sac that surrounds the embryo (later, the fetus). [p.308]

54. _____   As this membrane develops, mesoderm on its surface gives rise to blood vessels that become housed within the umbilical cord. [p.308]

55. _____   Parts of this membrane give rise to the embryo's digestive tube. [p.308]

56. _____   Fluid within this membrane keeps the embryo from drying out, absorbs shocks, and acts as insulation. [p.308]

57. _____   Houses blood vessels that link the embryo and the placenta. [p.308]

58. _____   Through this tissue, the embryo receives nutrients and oxygen from the mother and sends out wastes to the mother's bloodstream in return. [pp.308–309]

59. _____   The "maternal side" of this structure consists of a layer of endometrial tissue containing arterioles and venules. [p.308]

60. _____   Maintains the uterine lining for the first three months of pregnancy. [p.308]

61. _____   Cells from the small projections of this membrane are used to screen for birth defects. [p.309]

62. _____   In addition to oxygen and nutrients, many other substances taken in by the mother — including alcohol, caffeine, drugs, pesticide residues, the AIDS virus, and toxins in cigarette smoke — can cross this structure. [p.309]

17.4. HOW THE EARLY EMBRYO TAKES SHAPE [pp.310–311]

17.5. THE FIRST EIGHT WEEKS — HUMAN FEATURES EMERGE [pp.312–313]

17.6. DEVELOPMENT OF THE FETUS [pp.314–315]

Selected Words: "first trimester" [p.310], primitive streak [p.310], *notochord* [p.310], *coelom* [p.311], *spina bifida* [p.311], "vertebrate" [p.312], *vernix caseosa* [p.314], *respiratory distress syndrome* [p.314], *foramen ovale* [p.315], *ductus arteriosus* [p.315]

Boldfaced, Page-Referenced Terms

[p.310] neural tube _____

_____

[p.310] somites _____

_____

[p.313] fetus _____

_____

Matching

Choose the most appropriate answer for each term.

1. _____ embryonic period [p.311]

2. _____ apoptosis (programmed cell death) [p.311]

3. _____ gastrulation [pp.310–311]

4. _____ neural tube [p.310]

5. _____ somites [pp.310–311]

6. _____ coelom [p.311]

7. _____ neurulation [p.311]

8. _____ spina bifida [p.311]

9. _____ gonad development [p.313]

10. _____ fetus [p.313]

11. _____ miscarriage [p.313]

A. Paired blocks of mesoderm that give rise to most bones and skeletal muscles of neck and trunk together with their dermal coverings

B. Begins to develop in both sexes by the second half of the first trimester

C. Birth defect that occurs when the neural tube fails to develop properly; a portion of the spine may be exposed

D. Designation for an embryo after eight weeks; organ systems have formed

E. Formed by spaces that open up in the mesoderm and then coalesce to form a larger cavity

F. Begins shortly after fertilization and lasts for eight weeks

G. Important developmental stage initiated by the time a woman has missed her first period

H. Spontaneous expulsion of the uterine contents; occurs in 20 percent of all conceptions

I. First stage in the development of the nervous system from ectodermal cells

J. Enzymatic destruction of cells to help sculpt body parts

K. Its forerunner is the primitive streak; gives rise to the brain and spinal cord

Fill-in-the-Blanks

When the fetus is three months old, soft, fuzzy hair, the (12) _____ [p.314], covers the fetal body. The skin is wrinkled and protected by a thick, cheesy coating called the (13) _____ _____ [p.314]. The (14) _____ [p.314] trimester of human development extends from the start of the fourth month to the end of the sixth. Facial muscles move, and near the end of the trimester, the mother feels arm and leg movements. Eyelids and eyelashes form.

The (15) _____ [p.314] trimester extends from the seventh month until birth. Not until the middle of the third trimester can the baby survive on its own. Babies born before seven months' gestation often suffer from respiratory (16) _____ [p.314] syndrome. The circulatory system takes a detour on its way to independence. Because the fetus exchanges gases and receives nutrients via the mother's bloodstream prior to birth, the fetal circulatory system develops temporary vessels that bypass the lungs and liver. At birth, normal circulatory routes begin functioning. Two (17) _____ [p.314] arteries within the umbilical cord transport deoxygenated blood and metabolic wastes from the fetus to the placenta. Oxygenated blood, enriched with nutrients, returns from the placenta to the fetus in the (18) _____ [p.315] vein.

Other temporary vessels divert blood past the (19) _____ [p.315] and (20) _____ [p.315]. These organs do not develop as rapidly as some others, because (by way of the placenta) the mother's body can perform their functions. Fetal lungs are (21) _____ [p.315] and do not become functional for gas exchange until the newborn takes its first breaths outside the womb. A little of the blood entering the heart's right (22) _____ [p.315] flows into the right ventricle and moves on to the lungs, but most of it travels through a gap in the interior heart wall called the (23) _____ _____ [p.315] or into an arterial duct, the ductus arteriosus, that entirely bypasses the nonfunctioning lungs.

Likewise, most blood bypasses the fetal liver because the mother's liver performs most liver functions until (24) _____ [p.315]. Nutrient-laden blood from the placenta travels through a venous duct past the liver and on to the (25) _____ [p.315], which pumps it to body tissues. At birth, blood pressure in the heart's left atrium (26) _____ [p.315]. Normally, this causes a valvelike flap of tissue to close off the (27) _____ _____ [p.315], which gradually seals. The closure separates the (28) _____ [p.315] and (29) _____ [p.315] circuits of blood flow, and the arterial duct collapses. The (30) _____ [p.315] duct gradually closes during the first few weeks after birth.

Labeling

Identify each indicated part of the following illustrations.

31. _____ _____ [p.310]
32. _____ _____ [p.310]
33. _____ _____ [p.310]
34. _____ _____ [p.310]
35. _____ _____ [p.310]
36. _____ _____ [p.310]
37. _____ _____ [p.310]

38. _____ [p.310]
39. _____ _____ [p.310]
40. _____ [p.312]
41. _____ _____ [p.312]
42. _____ [p.312]
43. _____ [p.312]
44. _____ [p.312]

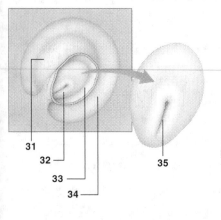
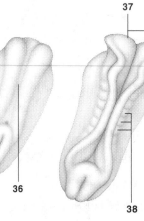
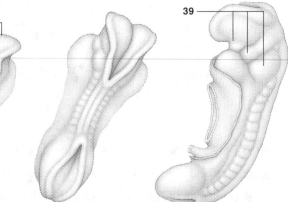

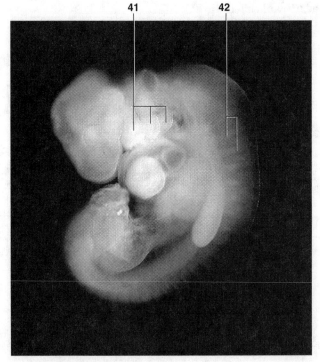

A human embryo at (**40**) weeks after conception.

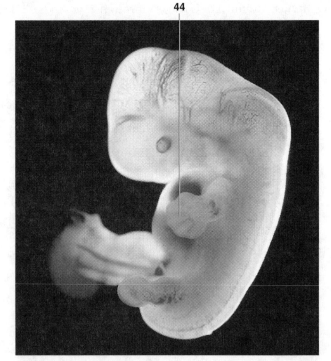

A human embryo at (**43**) weeks after conception.

Short Answer

45. Briefly describe the human embryo in the final week of the embryonic period. [p.313]

_____

_____

_____

17.7. BIRTH AND BEYOND [pp.316–317]

17.8. *Choices: Biology and Society:* SHOULD EMBRYOS BE CLONED? [p.317]

17.9. MOTHER AS PROVIDER, PROTECTOR, AND POTENTIAL THREAT [pp.318–319]

17.10. *Science Comes to Life:* PRENATAL DIAGNOSIS: DETECTING BIRTH DEFECTS [p.320]

17.11. THE PATH FROM BIRTH TO ADULTHOOD [p.321]

Selected Words: "due date" [p.316], "labor" [p.316], *breech* position [p.316], afterbirth [p.316], "donor" embryo [p.317], *fetal alcohol syndrome* (FAS) [p.319], *amniocentesis* [p.320], *chorionic villus sampling* (CVS) [p.320], *preimplantation diagnosis* [p.320], *fetoscopy* [p.320], *neonate* [p.321], "growth spurts" [p.321], *senescence* [p.321]

Boldfaced, Page-Referenced Terms

[p.316] parturition _____

_____

[p.317] lactation _____

_____

Fill-in-the-Blanks

(1) _____ [p.316], or birth, takes place about 39 weeks after fertilization. The birth process,

or (2) "_____" [p.316], begins when smooth muscle in the uterus starts to contract. This

process is divided into three stages. In the first, uterine contractions push the fetus against the mother's

(3) _____ [p.316], which gradually dilates to a diameter of about 10 centimeters, or 4 inches. The

(4) _____ [p.316] sac ruptures during the first contraction stage. The second stage is the actual

(5) _____ [p.316] of the fetus. This stage is usually brief — under (6) _____ [p.316]

hours. After the baby is expelled, usually headfirst, the third stage of labor begins. Complications can

develop if the baby begins to emerge in a "bottom first" or (7) _____ [p.316] position, and the

attending physician may use hands or forceps to aid the delivery. Uterine contractions of the third stage of

labor force fluid, blood, and the placenta or (8) _____ [p.316] from the mother's body. The

(9) _____ _____ [p.316] is now severed. Without the placenta to remove wastes,

(10) _____ _____ [p.316] builds up in the baby's blood. These and other factors,

including handling by medical personnel, stimulate control centers in the brain that respond by triggering

(11) _____ [p.316]— the newborn's crucial first breath. The reminder of this last stage is the scar we call the (12) _____ [p.316], the site where the umbilical cord was once attached.

Under the influence of estrogen and progesterone, mammary glands and ducts grew within the mother's breasts. Only the colorless fluid (13) _____ [p.317] is produced for a few days. This fluid is low in fat but rich in proteins, antibodies, minerals, and vitamin A. Then prolactin secreted by the pituitary gland stimulates milk production, or (14) _____ [p.317]. (15) _____ [p.317] released from the pituitary acts to force milk into mammary ducts.

Short Answer

16. Compose a few statements about the critical importance of maternal lifestyle during pregnancy. [pp.318–319] _____

_____

_____

_____

Dichotomous Choice

Circle one of two possible answers given between parentheses in each statement.

17. (Amniocentesis/CVS) uses tissue from the chorionic villi of the placenta for prenatal diagnosis of genetic defects. [p.320]
18. (Amniocentesis/CVS) samples fluid from within the amnion that contains some fetal cells and chemicals for prenatal diagnosis of genetic defects. [p.320]
19. Using methods of (CVS/preimplantation diagnosis), an embryo "conceived" by in vitro fertilization is analyzed for genetic defects using recombinant DNA technology. [p.320]
20. An individual from infancy to about age 12 or 13 years is in the (childhood/pubescent) period. [p.321]
21. During the first two weeks after birth, an individual is a(n) (newborn or neonate/infant). [p.321]
22. Human bone formation and growth is completed when an individual is an (adolescent/adult). [p.321]
23. An individual is (pubescent/adolescent) when secondary sexual traits develop. [p.321]
24. An individual from puberty until about three or four years later is (pubescent/adolescent). [p.321]
25. Progressive cellular and bodily deterioration is built into the life cycle of all organisms; the process is called (senescence/adulthood). [p.321]
26. In the prenatal period, a ball of cells with a surface layer and inner cell mass is the (morula/blastocyst). [p.317]

17.12. WHY DO WE AGE? [p.322]

17.13. AGING SKIN, MUSCLE, BONES, AND TRANSPORT SYSTEMS [p.323]

17.14. AGE-RELATED CHANGES IN SOME OTHER BODY SYSTEMS [pp.324–325]

Selected Words: "cross-links" [p.322], *telomeres* [p.322], *neurofibrillary tangles* [p.324], *beta amyloid* [p.324], *Alzheimer's disease* [p.324], *apolipoprotein E* [p.324], *hormone replacement therapy* (HRT) [p.325], *urinary incontinence* [p.325]

Matching

Choose the most appropriate description for each term.

1. _____ aging or senescence [p.321]
2. _____ cumulative damage to DNA [p.322]
3. _____ aging of the skin, muscles, and skeleton [p.323]
4. _____ aging in the cardiovascular and respiratory systems [p.323]
5. _____ aging of the nervous system and senses [pp.324–325]
6. _____ aging of the reproductive system and changes in sexuality [p.325]
7. _____ aging of the immune system [p.325]
8. _____ aging of the digestive system [p.325]
9. _____ aging of the urinary system [p.325]

A. Menopause, hot flashes, and HRT; after about age 50, men begin to take longer to achieve erection and may experience prostate enlargement
B. Muscles of bladder and urethra weaken; urinary incontinence; loss of nephrons
C. Number of T cells falls; B cells become less active; ability to recognize self markers on body cells declines; more prone to autoimmune diseases
D. Free radical damage of proteins, DNA and other biological molecules; cells lose the capacity for DNA self-repair
E. Walls of alveoli break down; enlarged heart; smaller and weaker heart muscles; plaques; high blood pressure
F. Number of fibroblasts in dermis decreases; elastic fibers are replaced with more-rigid collagen; skin becomes thinner and less elastic; bones become weaker, more porous, and brittle; joint breakdown and osteoarthritis
G. Begins around age 40; physical and physiological changes continue until we die
H. Neurofibrillary tangles, beta amyloid protein fragments, and Alzheimer's disease (AD)
I. Glands in mucous membranes of stomach and large intestine gradually deteriorate; fewer digestive enzymes are secreted

Self-Quiz

Multiple Choice

_____ 1. Shortly after fertilization, the zygote is subdivided into a multicellular embryo during a process known as _____ . [p.304]
   a. meiosis
   b. parthenogenesis
   c. embryonic induction
   d. cleavage
   e. invagination

_____ 2. The morula is transformed into a(n) _____ . [p.304]
   a. zygote
   b. blastocyst
   c. gastrula
   d. third germ layer
   e. organ

_____ 3. Muscles differentiate from _____ tissue. [p.304]
   a. ectoderm
   b. mesoderm
   c. endoderm
   d. parthenogenetic
   e. yolk

_____ 4. The nervous system differentiates from _____ tissue. [p.304]
   a. ectoderm
   b. mesoderm
   c. endoderm
   d. parthenogenetic
   e. yolk

Choice

For questions 5–6, choose from the following answers:

a. blastocyst   b. allantois   c. yolk sac   d. oviduct   e. amnion

5. _____ The _____ is a pathway from the ovary to the uterus. [p.306]

6. _____ The _____ is the innermost membrane of the extraembryonic membranes. [p.308]

Matching

7. _____ HCG [p.307]

8. _____ second trimester [p.314]

9. _____ umbilical cord [pp.308,309,314]

10. _____ first trimester [pp.310–311]

11. _____ mature egg [p.306]

12. _____ fetus [p.313]

13. _____ amnion [p.308]

14. _____ placenta [p.308]

15. _____ chorion [p.308]

16. _____ lactation [p.317]

17. _____ third trimester [p.314]

18. _____ implantation [p.307]

19. _____ FAS [p.319]

20. _____ birth [p.316]

A. Outermost extraembryonic membrane; secretes HCG to maintain the uterine lining for three months

B. Period of development extending from the start of the fourth month to the end of the sixth

C. Blastocyst adheres to the uterine lining; cells invade maternal tissues

D. Fluid-filled sac immediately surrounding the embryo

E. Period of development extending from the seventh month until birth

F. Hormone secreted by the blastocyst; stimulates corpus luteum to continue estrogen and progesterone secretion to maintain uterine lining

G. Results from alcohol use during pregnancy; third most common cause of mental retardation in the U.S.

H. Sperm penetration of the oocyte's cytoplasm stimulates its formation in meiosis II

I. Period of development from fertilization to the end of the third month

J. Blood vessels of this structure are used to transport oxygen and nutrients for the embryo

K. Occurs about 39 weeks after fertilization; involves uterine contractions, cervical dilation, amnion rupture, fetal expulsion, and severed umbilical cord

L. Period beginning after completion of organ system formation; begins at about the ninth week

M. By way of this tissue, the embryo receives nutrients and oxygen from the mother and sends out wastes to her bloodstream

N. Period during which hormone-primed glands produce milk

Chapter Objectives/Review Questions

This section lists general and detailed chapter objectives that can be used as review questions. You can make maximum use of these items by writing answers on a separate sheet of paper. Fill in answers where blanks are provided. To check for accuracy, compare your answers with information given in the chapter or glossary.

1. Name each of the three germ layers, and generally list the organs formed from each. [p.308]
2. Describe early human embryonic development and distinguish among the following: zygote, fertilization, cleavage, morula, gastrulation, morphogenesis, and organogenesis. [pp.304–305]
3. What process begins during gastrulation that did not happen during cleavage? [pp.304–305]
4. When the three germ layers have formed and they split into subpopulations of cells, it marks the onset of _____ . [p.304]
5. The gene-guided process by which cells in different locations in the embryo becomes specialized is called _____ _____ . [p.304]
6. Describe the processes and products of morphogenesis. [p.304]
7. _____ is the process in which the region of cell membrane covering the sperm cell's acrosome becomes structurally unstable. [p.306]
8. Describe the process of implantation. [p.307]
9. A(n) _____ consists of a surface layer of cells called the *trophoblast* and a clump of cells to one side called the *inner cell mass.* [p.306]
10. How much time passes between fertilization and implantation? [p.307]
11. Name the four membranes that form around the early embryo, and give the major functions of each. [p.308]
12. The primitive streak is the forerunner of a _____ _____ , which will give rise to the brain and spinal cord. [p.310]
13. The embryo and mother exchange substances through the _____ . [p.308]
14. At what point in the development process does the embryo begin to be referred to as a fetus? [p.313]
15. Generally describe fetal development until birth. [pp.314–315]
16. Describe events occurring during the process of human birth. [pp.316–317]
17. Explain why the mother must be particularly careful of her diet, health habits, and lifestyle during the first trimester after fertilization. [pp.318–319]
18. List and characterize the stages of human development, beginning with a zygote and completing the list with old age. [p.321]
19. List factors affecting various body systems that may contribute to aging or _____ . [pp.321–325]

Integrating and Applying Key Concepts

If cell differentiation did not occur, how would the human body appear? If controlled cell death did not take place in a human embryo, how would its hands appear? If development did not occur, what would a thirty-year-old human look like?

18

LIFE AT RISK: INFECTIOUS DISEASE

Interactive Exercises

CHAPTER INTRODUCTION [p.329]

18.1. VIRUSES AND INFECTIOUS PROTEINS [pp.330–331]

18.2. BACTERIA — THE UNSEEN MULTITUDES [pp.332–333]

Selected Words: hepatitis C [p.329], *gastroenteritis* [p.329], *West Nile virus* [p.329], *Lyme disease* [p.329], *hemorrhagic fevers* [p.329], type 1 Herpes simplex [p.331], type 2 Herpes simplex [p.331], *infectious mononucleosis* [p.331], *reverse transcriptase* [p.331], *provirus* [p.331], *Creutzfeldt-Jakob disease* (CJD) [p.331], *bovine spongiform encephalitis* [p.331], *cell wall* [p.332], *spirochetes* [p.332], *bacterial flagellum* [p.332], *pili* (singular: pilus) [p.332], *prokaryotic fission* [p.332], "fertility" plasmid [p.332], <u>Salmonella</u> [p.332], <u>Streptococcus</u> [p.332], *strep throat* [p.332], <u>Borrelia</u> <u>burgdorferi</u> [p.332], *antiviral* drugs [p.333], "super bugs" [p.333], <u>Staphylococcus</u> <u>aureus</u> [p.333]

Boldfaced, Page-Referenced Terms

[p.329] sexually transmitted diseases _____

_____

[p.329] emerging diseases _____

_____

[p.329] reemerging diseases _____

_____

[p.330] virus _____

_____

[p.330] latency _____

_____

[p.331] retrovirus _____

_____

[p.331] prions _____

_____

[p.332] plasmids _____

_____

[p.333] antibiotic _____

_____

Choice

a. zoonosis   b. sexually transmitted   c. transmitted by contaminated food and/or water
d. emerging disease   e. reemerging disease

1. _____ tuberculosis [p.329]
2. _____ genital herpes [p.329]
3. _____ *E. coli* O157:H7 [p.329]
4. _____ Lyme disease [p.329]
5. _____ West Nile virus [p.329]
6. _____ dengue fever [p.329]
7. _____ tapeworm infection [p.329]
8. _____ *Salmonella* [p.329]
9. _____ viral influenza [p.329]
10. _____ chlamydial infections [p.329]
11. _____ hemorrhagic fever [p.329]
12. _____ anthrax [p.329]

Short Answer

13. List three reasons for the increasing number of emerging and reemerging diseases. [p.329] _____

_____

_____

14. A virus has been described as "bad news wrapped in protein." What is the "bad news" component? [p.329]

_____

_____

_____

True/False

If the statement is true, write a "T" in the blank. If the statement is false, make it correct by writing the word(s) in the blank that should take the place of the underlined word(s).

_____ 15. A virus <u>is</u> a cell. [p.330]

_____ 16. A virus invades a host cell in order to <u>feed</u>. [p.330]

_____ 17. Viruses that infect bacteria are <u>bacteriophages</u>. [p.330]

_____ 18. The protein coat immediately surrounding the nucleic acid of a virus is called the <u>envelope</u>. [p.330]

_____ 19. The genetic material of a virus is DNA <u>or</u> RNA. [p.330]

_____ 20. Proteins extending from the viral coat bind to receptors on <u>other viruses</u>. [p.330]

Sequence

Show the correct order of the events listed by placing their letters in the blanks beside the numbers. [p.330]

21. _____   A. Assembly of viral nucleic acids and proteins into new virus particles

22. _____   B. Entry of the virus or its genetic material into the host cell's cytoplasm

23. _____   C. Synthesis of enzymes and other proteins, using host cell machinery

24. _____   D. Release of new virus particles from the cell

25. _____   E. Attachment to the host cell

26. _____   F. Replication of viral nucleic acid

Dichotomous Choice

27. A given virus can often attack only one type of cell because of differences in (receptors/lipids) at the cell surface. [p.330]
28. Usually, a virus (kills its host quickly/enters a period of latency). [pp.330–331]
29. Latent viruses are (inactivated/reactivated) by stressors such as sunburn, illness, and emotional upsets. [p.331]
30. Infectious mononucleosis is caused by the (type 2 Herpes simplex/Epstein-Barr) virus. [p.331]
31. Viruses that use reverse transcriptase to copy DNA from viral RNA are called (retroviruses/latent viruses). [p.331]
32. Creutzfeldt-Jakob disease and mad cow disease (bovine spongiform encephalitis) are caused by (retroviruses/prions). [p.331]
33. A prion is a misfolded (lipid/protein). [p.331]

Fill-in-the-Blanks

Bacteria are (34) _____ [p.333] with no nucleus or other membrane-bound organelles. In most, there is a cell wall made of (35) _____ [p.333]. A ball-shaped bacterium is a(n) (36) _____ [p.333], a rod-shaped bacterium is a(n) (37) _____ [p.333], and a twisted bacterium is a (38) _____ [p.333]. Some bacteria have a protein filament called a(n) (39) _____ [p.333] for locomotion. (40) _____ [p.333] are filaments used to stick to surfaces. Bacteria reproduce by (41) _____ _____ [p.333]. Bacterial cells have (42) _____ [p.333] circular chromosome, making cell division simple. Some bacteria can divide as often as every (43) _____ [p.333] minutes. Bacterial cells may have small circles of DNA containing just a few genes. These are called (44) _____ [p.333]. One type, a(n) (45) "_____" [p.333] plasmid, carries genes that allow the bacterium to transfer plasmid DNA to other bacterial cells.

18.3. INFECTIOUS PROTOZOA AND WORMS [p.334]

18.4. *Science Comes to Life:* MALARIA — HERE TODAY . . . AND TOMORROW AND TOMORROW [p.335]

18.5. THE SPREAD OF DISEASE AND PATTERNS OF OCCURRENCE [pp.336–337]

18.6. *Focus on Your Health:* TB, CONTINUED [p.337]

Selected Words: <u>Entamoeba</u> <u>histolytica</u> [p.334], *amoebic dysentery* [p.334], <u>Giardia</u> <u>lamblia</u> [p.334], *giardiasis* [p.334], <u>Trypanosoma</u> <u>brucei</u> [p.334], *African trypanosomiasis* [p.334], *sleeping sickness* [p.334], *cryptosporidosis* [p.334], <u>Cryptosporidium</u> <u>parvum</u> [p.334], <u>Ascaris</u> [p.334], *malaria* [p.335], <u>Plasmodium</u> [p.335], <u>Anopheles</u> [p.335], *sickle-cell anemia* [p.335], "contagious" [p.336], <u>Salmonella</u> [p.332], <u>Shigella</u> [p.332], *intermediate host* [p.336], *ringworm* [p.336], *gonorrhea* [p.337], *tuberculosis* (TB) [p.337], *consumption* [p.337], <u>Mycobacterium</u> <u>tuberculosis</u> [p.337]

Boldfaced, Page-Referenced Terms

[p.334] parasites _____

_____

[p.336] disease vector _____

_____

[p.336] nosocomial infection _____

_____

[p.336] sporadic disease _____

_____

[p.336] endemic disease _____

_____

[p.337] epidemic _____

_____

[p.337] pandemic _____

_____

[p.337] virulence _____

_____

Matching

1. _____ giardiasis
2. _____ pinworm
3. _____ African sleeping sickness
4. _____ amoebic dysentery
5. _____ cryptosporidosis

A. *Entamoeba histolytica;* common in areas without sewage treatment [p.334]
B. *Giardia lamblia;* widespread cause of explosive diarrhea and "rotten egg" belches [p.334]
C. *Trypanosoma brucei;* spread by tsetse fly; invades central nervous system and is fatal if untreated [p.334]
D. Most common worm infection in developed countries; infection route is oral–fecal [p.334]
E. *Cryptosporidium parvum;* watery diarrhea, due to fecal contamination of food or water [p.334]

Fill-in-the-Blanks

Malaria kills about (6) _____ _____ [p.335] people each year. The cause of malaria is the protozoan (7) _____ [p.335]. This parasite is transmitted from one human to another by the female (8) _____ [p.335] mosquito. The classic symptoms include shaking, chills, a high (9) _____ [p.335], and drenching sweats. After the initial illness, the symptoms may subside, but in time a malaria sufferer may develop (10) _____ [p.335] and a grossly enlarged liver. Malaria's main stronghold has been tropical (11) _____ [p.335], although it is now found in many other areas. There are some strains of *Plasmodium* that are (12) _____ [p.335] to antimalarial drugs. There are also insecticide-resistant (13) _____ [p.335].

14. _____ virulence

15. _____ sporadic disease

16. _____ contact with a vector

17. _____ nosocomial infection

18. _____ endemic disease

19. _____ inhaling pathogens

20. _____ pandemic

21. _____ direct contact

22. _____ epidemic

23. _____ indirect contact

A. Insects carry a pathogen from an infected organism to a new host [p.336]
B. Acquired in a hospital [p.336]
C. Ability of a pathogen to cause serious disease — depends on rate of invasion, severity of damage, and targeted tissues [p.337]
D. Epidemic breaks out in several countries in a given time span; AIDS is an example [p.337]
E. Transfer of pathogen by a handshake, kiss, or other contact [p.336]
F. Transfer of a pathogen by a cough or sneeze, as with influenza [p.336]
G. Breaks out irregularly and affects relatively few people, as with whooping cough [p.336]
H. Disease rate increases to a level above what was predicted, as with cholera in Peru in 1991 [p.337]
I. Transfer of a pathogen by touching objects previously in contact with an infected person, or contaminated food or water [p.336]
J. Disease that occurs more or less continuously, such as the common cold or ringworm [p.336]

Short Answer

24. What fraction of the world population today is infected with the tuberculosis pathogen? What must be done to insure the effectiveness of antibiotic treatment for TB? [p.337] _____

_____

_____

_____

18.7. THE HUMAN IMMUNODEFICIENCY VIRUS AND AIDS [pp.338–339]

18.8. TREATING AND PREVENTING HIV INFECTION AND AIDS [p.340]

Selected Words: "indicator diseases" [p.338], *transcription* [p.339], *CD4 cells* [p.339], drug "cocktail" [p.340], "entry inhibitors" [p.340], *antigenic forms* [p.340]

Boldfaced, Page-Referenced Terms

[p.338] human immunodeficiency virus (HIV) _____

_____

True/False

If the statement is true, write a "T" in the blank. If the statement is false, make it correct by writing the word(s) in the blank that should take the place of the underlined word(s).

_____ 1. AIDS is <u>one</u> disease caused by infection with HIV. [p.338]

_____ 2. HIV destroys T cells, crippling the <u>cardiovascular</u> system. [p.338]

_____ 3. Certain types of pneumonia, cancer, yeast infections, and drug-resistant tuberculosis are "<u>indicator diseases</u>" of AIDS. [p.338]

_____ 4. The HIV virus can enter the body from infected body fluids that come into contact with <u>cuts or abrasions</u>. [p.338]

_____ 5. Infected mothers <u>cannot</u> transfer HIV to their babies during pregnancy, birth, and breast-feeding. [p.338]

_____ 6. In recent years, more young adults in the United States have died from <u>hepatitis C</u> than from any other cause. [p.338]

_____ 7. HIV is a retrovirus with <u>DNA</u> as its genetic material. [p.338]

Sequence

Show the correct order of the events listed by placing their letters in the blanks beside the numbers. [p.338]

8. _____   A. Copies of HIV circulate in bloodstream (produces flulike symptoms)

9. _____   B. HIV enters the cell, and more viruses are produced and bud out of cell

10. _____   C. Huge reservoirs of infected T cells accumulate in lymph nodes

11. _____   D. Body mounts an immune response, but infection rate overwhelms it

12. _____   E. HIV docks with a CD4 receptor on helper T cell

13. _____   F. Number of helper T cells drops and symptoms appear such as weight loss, fatigue, nausea, night sweats, enlarged lymph nodes, etc., and eventually an "indicator disease"

True/False

If the statement is true, write a "T" in the blank. If the statement is false, make it correct by writing the word(s) in the blank that should take the place of the underlined word(s).

_____ 14. There is <u>no</u> known cure for AIDS. [p.340]

_____ 15. Integrase and protease inhibitors target steps in the DNA <u>replication</u> process. [p.340]

_____ 16. Some HIV genes become incorporated into the <u>RNA</u> of the patient. [p.340]

_____ 17. Present treatments for AIDS consist of an <u>antibiotic</u> and two anti-HIV drugs. [p.340]

_____ 18. Drugs called "entry inhibitors" prevent HIV from entering the <u>bloodstream</u>. [p.340]

_____ 19. The biggest problem in developing a vaccine against HIV is that it mutates extremely <u>slowly</u>. [p.340]

_____ 20. The current downward trend in HIV-related deaths in the United States is due largely to improved <u>drug treatments</u>. [p.340]

18.9. *Focus on Your Health:* PROTECTING YOURSELF — AND OTHERS — FROM SEXUAL DISEASE [p.341]

18.10. A TRIO OF COMMON STDS [pp.342–343]

18.11. A ROGUE'S GALLERY OF VIRAL STDS AND OTHERS [pp.344–345]

Selected Words: *scientific* advice [p.341], *chlamydial infections* [p.342], <u>Chlamydia</u> <u>trachomitis</u> [p.342], *pelvic inflammatory disease* (PID) [p.342], <u>Neisseria</u> <u>gonorrhea</u> [p.342], *syphilis* [p.343], <u>Treponema</u> <u>pallidium</u> [p.343], *primary stage* of syphilis [p.343], *secondary stage* of syphilis [p.343], *tertiary stage* of syphilis [p.343], *genital warts* [p.344], *Pap smear* [p.344], <u>Haemophilus</u> <u>ducreyi</u> [p.345], *pubic lice* [p.345], "crabs" [p.345], "nits" [p.345], *scabies* [p.345], *vaginitis* [p.345], <u>Candida</u> <u>albicans</u> [p.345], *candidiasis* [p.345], <u>Trichomonas</u> <u>vaginalis</u> [p.345], *trichomoniasis* [p.345]

Boldfaced, Page-Referenced Terms

[p.344] genital herpes _____

_____

[p.344] human papillomavirus (HPV) _____

_____

[p.344] hepatitis B virus (HBV) _____

_____

[p.345] chancroid _____

_____

Choice

For questions 1–11, choose from the following:

    a. caused by a bacterium   b. caused by a virus   c. caused by an arthropod
                    d. caused by a protozoan or fungus

1. _____   genital herpes infection [p.344]
2. _____   syphilis [p.343]
3. _____   AIDS [p.348]
4. _____   hepatitis [p.344]
5. _____   gonorrhea [p.342]
6. _____   pubic lice [p.345]
7. _____   genital warts [p.344]
8. _____   chlamydial infection [p.342]
9. _____   vaginitis [p.345]
10. _____   scabies [p.345]
11. _____   chancroid [p.345]

For questions 12–22, choose from the following:

a. no cure exists   b. treated with antibiotics
c. treated by surgery or by freezing or burning the affected area   d. the only treatment is rest
e. treated by applying antiparasitic drugs to the infested area(s)

12. _____ gonorrhea [pp.342–343]

13. _____ AIDS [p.340]

14. _____ genital warts [p.344]

15. _____ pubic lice [p.345]

16. _____ genital herpes [p.344]

17. _____ hepatitis B [p.345]

18. _____ pelvic inflammatory disease [p.342]

19. _____ chlamydial infection [p.342]

20. _____ syphilis [p.343]

21. _____ scabies [p.345]

22. _____ chancroid [p.345]

Self-Quiz

For questions 1–14, choose the letter that applies from the following:

a. AIDS   b. gonorrhea   c. syphilis   d. chlamydial infection   e. pelvic inflammatory disease
f. genital herpes   g. genital warts   h. vaginitis

1. _____ Benign, painless clustered growths on the penis, cervix, or around the anus; probable cause of cervical cancer. [p.344]

2. _____ The virus slowly cripples the immune system and opens the door to "opportunistic" infections. [p.338]

3. _____ Caused by a motile, corkscrew-shaped bacterium, *Treponema pallidum.* [p.343]

4. _____ Serious complication of some STDs; severe abdominal pain, scarred oviducts leading to abnormal pregnancies and sterility. [p.342]

5. _____ A bacterium, *Neisseria gonorrhoeae,* enters mucous membranes and causes more noticeable symptoms in males; prompt treatment quickly cures this disease. [p.342]

6. _____ Requires direct contact with viruses or sores that contain them; small painful blisters occur; the virus is reactivated sporadically; about 45 million people in the United States are infected with it. [p.344]

7. _____ Caused by a parasitic bacterium that invades cells of the genital and urinary tracts; results in swollen lymph nodes and often leads to PID. [p.342]

8. _____ The pathogen causing this disease enters a host through contact between infected body fluids and a cut or abrasion. [p.338]

9. _____ Being cured of infection does not confer on the person immunity to the bacterium that causes the disease; the use of condoms can prevent infection. [p.343]

10. _____ Produces a chancre (localized ulcer) one to eight weeks following infection. [p.343]

11. _____ Can lead to lesions in the eyes that cause blindness in babies born to mothers with the disease. [p.344]

12. _____ The tertiary stage of this disease begins when pathogens enter the brain, and the ultimate result is insanity and death. [p.343]

13. _____ Yeast infections and trichomoniasis are examples of this type of infection found in females. [p.345]

14. _____ Between flare-ups, the virus remains latent within the nervous system. [p.344]

Chapter Objectives/Review Questions

This section lists general and detailed chapter objectives that can be used as review questions. You can make maximum use of these items by writing answers on a separate sheet of paper. To check for accuracy, compare your answers with information given in the chapter or glossary.

1. Describe the principal body forms of bacteria and viruses (inside and outside). [pp.330–332]
2. Explain how, with no nucleus or membrane-bound organelles, bacteria and viruses reproduce themselves and obtain energy to carry on metabolism. [pp.330–333]
3. Describe four ways in which a person can be infected with HIV. [p.338]
4. Describe in detail which tissues HIV attacks, the viral behaviors that allow HIV to remain undetected in the body, the general schedule of attack, and the symptoms ultimately presented. [pp.334–335]
5. List the names and symptoms of 10 diseases that can be sexually transmitted by humans to their partners. [pp.338–339]
6. For each of 10 STDs, state how you can avoid being infected; or, if in spite of all your precautions, you did become infected, what would be the likely course of treatment. [pp.338–345]

Integrating and Applying Key Concepts

STDs can be at the least annoying and at worst deadly. Adult humans can have a happy and fulfilling sex life without ever contracting any of the pathogens discussed in this chapter. Make a written outline of the approach you would like to use with a child of your own to make certain he or she understands not only the pleasures that can be achieved, but also the possible dangers of sexual activity.

19

CELL REPRODUCTION

Interactive Exercises

CHAPTER INTRODUCTION [p.349]

19.1. DIVIDING CELLS: THE BRIDGE BETWEEN GENERATIONS [pp.350–351]

Selected Words: chromatin [p.350], *condensed* chromosome [p.350], "double helix" [p.350], *karyotype* [p.351], *homologues* [p.351]

Boldfaced, Page-Referenced Terms

[p.350] reproduction _____

_____

[p.350] life cycle _____

_____

[p.350] mitosis _____

_____

[p.350] meiosis _____

_____

[p.350] germ cells _____

_____

[p.350] chromosome _____

_____

[p.350] sister chromatids _____

_____

[p.350] centromere _____

_____

[p.351] somatic cells _____

_____

[p.351] diploid cell (2*n*) _____

_____

[p.351] chromosome number _____

_____

[p.351] homologous chromosomes _____

_____

Matching

Choose the most appropriate description for each term.

1. _____ centromere [p.350]
2. _____ diploid cell [p.351]
3. _____ chromatin [p.350]
4. _____ homologous chromosomes [p.351]
5. _____ chromosome [p.350]
6. _____ germ cells [p.350]
7. _____ condensed chromosome [p.350]
8. _____ life cycle [p.350]
9. _____ somatic cells [p.351]
10. _____ sister chromatids [p.350]
11. _____ mitosis and meiosis [p.350]
12. _____ chromosome number [p.351]

A. Mass of uncondensed chromosomes
B. Cells in which meiosis occurs; oogonia and spermatogonia
C. Indicates the number of each type of chromosome normally present in a cell
D. Tightly coiled, rod-shaped condition of a chromosome during cell division
E. Division processes in humans and other eukaryotes that sort and divide DNA between daughter cells
F. A pair of physically similar chromosomes (one from each parent) with genetic instructions for the same trait
G. A DNA molecule with attached proteins
H. Body cells that are diploid
I. A recurring series of events in which individuals grow, develop, maintain themselves, and reproduce according to instructions encoded in DNA
J. Constricted area of a chromosome; attachment point for microtubules
K. Two attached threads of a duplicated chromosome
L. Any cell possessing two of each type of chromosome

19.2. THE CELL CYCLE [p.352]

19.3. *Science Comes to Life:* IMMORTAL CELLS — THE GIFT OF HENRIETTA LACKS [p.353]

Selected Words: "HeLa" cells [p.353]

Boldfaced, Page-Referenced Terms

[p.352] cell cycle _____

_____

[p.352] interphase _____

_____

Labeling

Identify the stage in the cell cycle indicated by each number. [p.352]

1. _____
2. _____
3. _____
4. _____
5. _____
6. _____
7. _____
8. _____
9. _____
10. _____

Matching

Link each of the following time spans with the most appropriate number in the preceding labeling section. Answers may be used more than once.

11. _____ Period after duplication of DNA, during which the cell prepares for division [p.352]

12. _____ Period of nucleus and cytoplasm division [p.352]

13. _____ DNA replication occurs [p.352]

14. _____ Period of cell growth before DNA duplication [p.352]

15. _____ Longest phase of the cell cycle [p.352]

16. _____ Period of cytoplasmic division [p.352]

17. _____ Period that includes G_1, S, and G_2 [p.352]

Short Answer

18. Define and describe the value of *HeLa cells* to biologists. [p.353] _____

_____

_____

_____

19.4. A TOUR OF THE STAGES OF MITOSIS [pp.354–355]

19.5. HOW THE CYTOPLASM DIVIDES [p.356]

19.6. *Focus on Our Environment:* CONCERNS AND CONTROVERSIES OVER IRRADIATION [p.357]

19.7. A CLOSER LOOK AT THE CELL CYCLE [pp.358–359]

Selected Words: "prometaphase" [p.355], "shelf life" [p.355], microtubule "poison" [p.359]

Boldfaced, Page-Referenced Terms

[p.354] prophase _____

_____

[p.354] metaphase _____

_____

[p.354] anaphase _____

_____

[p.354] telophase _____

_____

[p.354] spindle apparatus _____

_____

[p.356] cytokinesis _____

_____

[p.356] cleavage furrow _____

_____

[p.358] nucleosome _____

_____

[p.358] kinetochores _____

_____

Labeling-Matching

Identify each of the following mitotic stages by entering the correct stage in the blank beneath the sketch. Select from *late prophase, transition to metaphase (prometaphase), interphase — parent cell, metaphase, early prophase, telophase, interphase — daughter cells,* and *anaphase.* [pp.354–355] Complete the exercise by matching and entering the letter of the correct phase description (choose from list below) in the parentheses after each label.

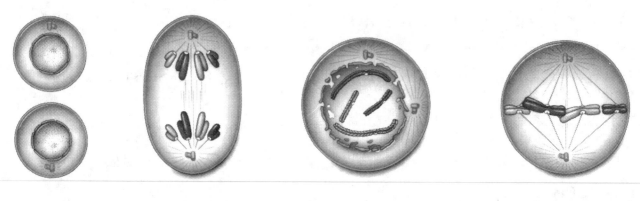

1. _____ ( )   2. _____ ( )   3. _____ ( )   4. _____ ( )

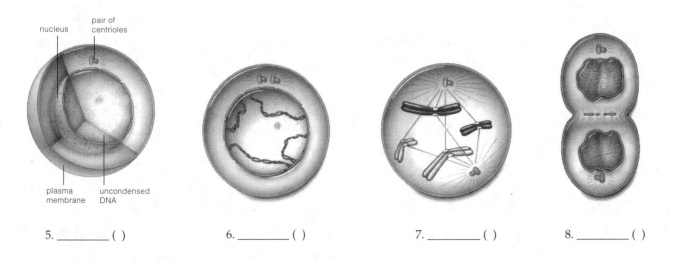

5. _____ ( )   6. _____ ( )   7. _____ ( )   8. _____ ( )

A. Attachment between two sister chromatids of each chromosome breaks; the two are now chromosomes in their own right; they move to opposite spindles. [p.355]
B. Microtubules that form the spindle apparatus enter the nuclear region; microtubules become attached to the sister chromatids of each chromosome. [p.355]
C. The DNA and its associated proteins start to condense into the threadlike chromosome form. [p.354]
D. Chromosomes are now fully condensed and lined up at the equator of the spindle. [p.355]
E. DNA is duplicated and the cell prepares for division. [p.354]
F. Two daughter cells have formed, each diploid and with two of each type of chromosome. [p.355]
G. Chromosomes continue to condense; new microtubules are assembled, and they move one of two centrioles toward the opposite end of the cell: the nuclear envelope begins to break up. [p.354]
H. New patches of membrane join to form nuclear envelopes around the decondensing chromosomes. [p.355]

Fill-in-the-Blanks

(9) _____ [p.354], the first stage of mitosis, is evident when chromosomes become visible in the light microscope as threadlike forms. Each was duplicated earlier, during (10) _____ [p.354]. Each already consists of two (11) _____ _____ [p.354] joined together at the (12) _____ [p.354]. Early in prophase, both (13) _____ [p.354] twist and fold into a more compact form. By late prophase, all the (14) _____ [p.354] will be condensed into thicker, rod-shaped forms. Cytoplasmic microtubules break apart into their tubulin subunits. The subunits reassemble near the nucleus, as new (15) _____ [p.354]. Many microtubules will extend from one spindle (16) _____ [p.354] or the other to the (17) _____ [p.354] of a chromosome. Others do not make contact with chromosomes but extend from the poles and overlap each other. Many cells have two barrel-shaped (18) _____ [p.354] that were duplicated during interphase. By the time (19) _____ [p.354] is underway, the cell has two pairs of them. Microtubules begin moving one pair to the opposite pole of the newly forming spindle. (20) _____ [p.355] is a time during which the nuclear envelope breaks up completely, into numerous tiny, flattened vesicles. The (21) _____ [p.355] begin interacting with the microtubules. When each chromosome is harnessed by microtubules from both (22) _____ [p.355], a two-way pull orients the two sister (23) _____ [p.355] of a chromosome toward opposite poles. When all the duplicated chromosomes are aligned midway between the poles of a completed spindle, it is called (24) _____ [p.355]. During (25) _____ [p.355], the two sister chromatids of each chromosome separate and move to opposite poles by two mechanisms: microtubules attached to the centromere regions shorten, pulling the chromosomes to the poles, and the spindle elongates when overlapping microtubules ratchet past each other and push the spindle poles farther apart. Once separated from its sister, each (26) _____ [p.355] now becomes a chromosome in its own right. (27) _____ [p.355] begins when the two clusters of chromosomes arrive at opposite spindle poles. Chromosomes are no longer harnessed to microtubules, and they return to a more threadlike form. Vesicles of the old (28) _____ _____ [p.355] fuse to form, patch by patch, a new nuclear envelope that separates each cluster of chromosomes from the cytoplasm. With mitosis, each new nucleus has the same chromosome (29) _____ [p.355] as the parent nucleus. Once the two nuclei form, (30) _____ [p.355] is over — and so is (31) _____ [p.355].

Matching

Match the descriptions of cytokinesis with the four diagrams. [p.356]

32. _____

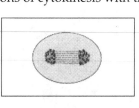

A. Continuing contractions of the microfilament rings
B. Mitosis is complete and the spindle is disassembling
C. The parent cell has divided into two daughter cells
D. Just beneath the plasma membrane, microfilament rings at the former spindle equator begin to contract

33. _____

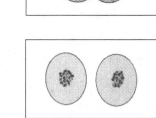

34. _____

35. _____

Short Answer

36. List some common natural sources of ionizing radiation. [p.357] _____

_____

_____

_____

37. Summarize the effects of ionizing radiation on cells. [p.357] _____

_____

_____

_____

38. Name some uses of irradiation in medicine and the food industry. [p.357] _____

_____

_____

_____

Matching

Choose the most appropriate description for each term.

39. _____ nucleosome [p.358]

40. _____ colchicine [p.359]

41. _____ kinetochore [p.358]

42. _____ histone [p.358]

A. A unit of DNA wound onto a histone "spool"
B. Type of protein found tightly bound to DNA
C. Poison that stops mitosis by interfering with spindle microtubules
D. Structures at centromeres where spindle microtubules dock to chromosomes

19.8. MEIOSIS — THE BEGINNINGS OF EGGS AND SPERM [pp.360–361]

19.9. A VISUAL TOUR OF THE STAGES OF MEIOSIS [pp.362–363]

Selected Words: *spermatogonia* [p.360], *oogonia* [p.360]

Boldfaced, Page-Referenced Terms

[p.360] reductional division _____

_____

[p.360] haploid number (*n*) _____

_____

[p.360] spermatogenesis _____

_____

[p.361] oogenesis _____

_____

True/False

If the statement is true, write a "T" in the blank. If the statement is false, make it correct by changing the underlined word(s) and writing the correct word(s) in the answer blank.

_____ 1. Meiosis occurs in the nucleus of dividing germ cells in ovaries or testes. [p.360]

_____ 2. Sperm and eggs are haploid cells referred to as germ cells. [p.360]

_____ 3. Meiosis involves a reductional division that halves the diploid chromosome number. [p.360]

_____ 4. Haploid cells possess pairs of homologous chromosomes. [p.360]

_____ 5. Gametes formed by meiosis have one member of each pair of homologous chromosomes possessed by the species. [p.360]

Matching

To review the major stages of meiosis, match the following written descriptions with the appropriate sketch. To simplify the process, assume that the cell in this model initially has only one pair of homologous chromosomes (one from a paternal source and one from a maternal source) and that crossing over does not occur. Complete the exercise by indicating diploid ($2n = 2$) or haploid ($n = 1$) chromosome number of the cell chosen in the parentheses following each blank.

6. _____ ( ) A pair of homologous chromosomes prior to S of interphase in a diploid germ cell [p.362]

7. _____ ( ) While the germ cell is in S of interphase, chromosomes are duplicated through DNA replication; the two sister chromatids are attached at the centromere [p.362]

8. _____ ( ) During meiosis I, each duplicated chromosome lines up with its partner, homologue to homologue [p.362]

9. _____ ( ) Also during meiosis I, the chromosome partners separate from each other in anaphase I; cytokinesis occurs, and each chromosome goes to a different cell [p.362]

10. _____ ( ) During meiosis II (in two cells), the sister chromatids of each chromosome are separated from each other; four haploid nuclei form; cytokinesis results in four cells (potential gametes) [p.363]

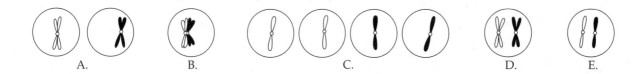

A.   B.   C.   D.   E.

Labeling-Matching

Identify each of the following meiotic stages by entering the correct stage of either meiosis I or meiosis II in the blank beneath the sketch. Choose from *prophase I, metaphase I, anaphase I, telophase I, prophase II, metaphase II, anaphase II,* and *telophase II.* [pp.362–363] Complete the exercise by matching and entering the letter of the correct phase description in the parentheses following each label.

A. The spindle is now fully formed; pairs of duplicated chromosomes are positioned midway between the poles of one cell. [p.362]

B. Centrioles have moved to opposite poles in each of two cells; a spindle forms, and microtubules attach the duplicated chromosomes to the spindle and begin moving them. [p.363]

C. Four daughter nuclei form; when the cytoplasm divides, each new cell has a haploid number of chromosomes, all in the unduplicated state; one or all cells may develop into gametes. [p.363]

D. In one cell, each duplicated chromosome is pulled away from its homologue; the two are moved to opposite spindle poles. [p.362]

E. Each chromosome is drawn up close to its homologue; crossing over and swapping of segments (genetic recombination) occurs. [p.362]

F. All the chromosomes are now aligned at the spindle's equator; this is occurring in two haploid cells. [p.363]

G. Two haploid cells form, but chromosomes are still in the duplicated state. [p.362]

H. Chromatids of each chromosome separate; former "sister chromatids" are now chromosomes in their own right and are moved to opposite poles. [p.363]

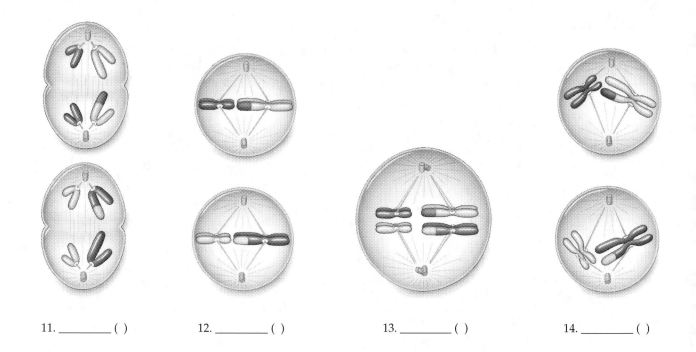

11. _____ ( )   12. _____ ( )   13. _____ ( )   14. _____ ( )

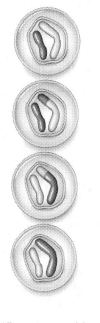

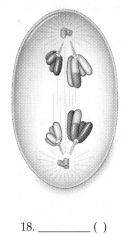

15. _____ ( )   16. _____ ( )   17. _____ ( )   18. _____ ( )

Sequence

Arrange the following entities in correct order of development, entering a "1" next to the stage that appears first and a "5" next to the stage that completes the process of spermatogenesis. Complete the exercise by indicating if each cell is *n* or *2n* in the parentheses following each blank. Refer to Figure 19.13 in the text. [pp.360–361]

19. _____ ( )   primary spermatocyte

20. _____ ( )   sperm

21. _____ ( )   spermatid

22. _____ ( )   spermatogonium

23. _____ ( )   secondary spermatocyte

Short Answer

24. List the major differences between oogenesis and spermatogenesis. [pp.360–361] _____

_____

_____

_____

19.10. THE SECOND STAGE OF MEIOSIS — NEW COMBINATIONS OF PARENTS' TRAITS [pp.364–365]

19.11. MEIOSIS AND MITOSIS COMPARED [pp.366–367]

Selected Words: haploid number [p.364], *nonsister* chromatids [p.365], *maternal* chromosomes [p.365], *paternal* chromosomes [p.365], *8,388,608 combinations* [p.365]

Boldfaced, Page-Referenced Terms

[p.364] crossing over _____

_____

[p.365] genetic recombination _____

_____

[p.365] alleles _____

_____

[p.365] disjunction _____

_____

Matching

Choose the most appropriate description for each term.

1. _____ crossing over [pp.364–365]
2. _____ nonsister chromatids [p.365]
3. _____ chiasma [p.365]
4. _____ alleles [p.365]
5. _____ genetic recombination [p.365]
6. _____ maternal chromosomes [p.365]
7. _____ paternal chromosomes [p.365]
8. _____ disjunction [p.365]

A. X-shaped configuration indicating a crossover
B. Chromosomes inherited from mother
C. Interaction between two nonsister chromatids of a pair of homologues during prophase I
D. The separation of each homologue from its partner during anaphase I
E. One chromatid from each of a homologous pair of chromosomes
F. Alternative forms of genes
G. The result of chromosomes exchanging pieces between nonsister chromatids during prophase I
H. Chromosomes inherited from father

Complete the Table

9. Complete the following table by entering the word *mitosis* or *meiosis* in the blank adjacent to the statement describing one of the two processes. [pp.366–367]

Mitosis/Meiosis	Description
a.	Involves one division
b.	Functions in growth, repair, and maintenance
c.	Daughter cells are haploid
d.	Occurs in germ cells
e.	Involves two divisions
f.	Daughter cells have one chromosome from each homologous pair
g.	Daughter cells have the diploid chromosome number
h.	Completed when four daughter cells are formed
i.	Functions to form gametes

Matching

The cell model used in this exercise has two pairs of homologous chromosomes, one long pair and one short pair. Match the descriptions to the numbers of chromosomes shown in the following sketches. [pp.366–367]

10. _____ One cell at the beginning of meiosis II

11. _____ A daughter cell at the end of meiosis II

12. _____ Metaphase I of meiosis

13. _____ Metaphase of mitosis

14. _____ G_1 in a daughter cell following mitosis

15. _____ Prophase of mitosis

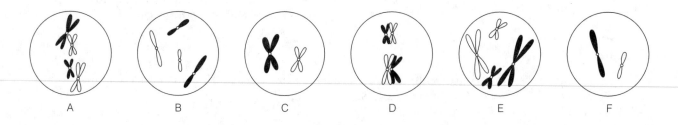

| A | B | C | D | E | F |

The following questions refer to the preceding sketches. (Write the number in the blank.) [pp.366–367]

16. _____ How many chromosomes are present in cell E?

17. _____ How many chromatids are present in cell E?

18. _____ How many chromatids are present in cell C?

19. _____ How many chromatids are present in cell D?

20. _____ How many chromosomes are present in cell F?

Self-Quiz

_____ 1. The DNA copying process occurs _____ . [p.352]
   a. between the growth phases of interphase
   b. immediately before prophase of mitosis
   c. during prophase of mitosis
   d. during G_1 growth

_____ 2. In eukaryotic cells, which of the following can occur during mitosis? [pp.354–355]
   a. Two cytoplasmic divisions to maintain the parental chromosome number
   b. The replication of DNA
   c. A long growth period
   d. The dissipation of the nuclear envelope and nucleolus

_____ 3. Being "diploid" means having _____ . [p.351]
   a. two chromosomes of each type in somatic cells
   b. twice the parental chromosome number
   c. half the parental chromosome number
   d. one chromosome of each type in somatic cells

_____ 4. Somatic cells are _____ cells; germ cells are _____ cells. [pp.350–351]
   a. meiotic; body
   b. body; body
   c. meiotic; meiotic
   d. body; meiotic

_____ 5. If a parent cell has 16 chromosomes and undergoes mitosis, the resulting cells will have _____ chromosomes. [p.355]
a. 64
b. 32
c. 16
d. 8
e. 4

_____ 6. The correct order of the stages of mitosis is _____ . [pp.354–355]
a. prophase, metaphase, telophase, anaphase
b. telophase, anaphase, metaphase, prophase
c. telophase, prophase, metaphase, anaphase
d. anaphase, prophase, telophase, metaphase
e. prophase, metaphase, anaphase, telophase

_____ 7. Which of the following does _not_ occur in prophase I of meiosis? [pp.362,364]
a. a cytoplasmic division
b. a cluster of four chromatids
c. homologues pairing tightly
d. crossing over

_____ 8. Crossing over is one of the most important events in meiosis because _____ . [pp.364–365]
a. it produces new combinations of alleles on chromosomes
b. homologous chromosomes must be separated into different daughter cells
c. the number of chromosomes allotted to each daughter cell must be halved
d. homologous chromatids must be separated into different daughter cells

_____ 9. The appearance of chromosome ends lapped over each other in meiotic prophase I provides evidence of _____ . [pp.364–365]
a. meiosis
b. crossing over
c. chromosomal aberration
d. fertilization
e. spindle fiber formation

_____ 10. Which of the following does _not_ increase genetic variation? [pp.365–366]
a. crossing over
b. random fertilization
c. prophase of mitosis
d. random homologue alignments at metaphase I

_____ 11. Which of the following is the most correct sequence of events in human life cycles? [p.350]
a. meiosis → fertilization → gametes → diploid organism
b. diploid organism → meiosis → gametes → fertilization
c. fertilization → gametes → diploid organism → meiosis
d. diploid organism → fertilization → meiosis → gametes

_____ 12. In sexually reproducing organisms, the zygote is _____ . [p.361]
a. an exact genetic copy of the female parent
b. an exact genetic copy of the male parent
c. a genetic mixture of male parent and female parent
d. unlike either parent genetically

_____ 13. The cell in the diagram is diploid with three pairs of chromosomes. From the number and pattern of chromosomes, the cell could be in _____ . [p.361]
a. the first division of meiosis
b. the second division of meiosis
c. mitosis
d. neither mitosis nor meiosis, because this stage is not possible in a cell with three pairs of chromosomes

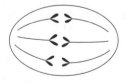

_____ 14. You are looking at a cell from the same organism as in the previous question. Now the cell is in _____ . [p.355]
  a. the first division of meiosis
  b. the second division of meiosis
  c. mitosis
  d. neither mitosis nor meiosis, because this stage is not possible in this organism

Chapter Objectives/Review Questions

This section lists general and detailed chapter objectives that can be used as review questions. You can make maximum use of these by writing answers on a separate sheet of paper. Fill in answers where blanks are provided. To check for accuracy, compare your answers with information given in the chapter or glossary.

1. What substance contains the instructions for making proteins? [p.350]
2. *Mitosis* and *meiosis* refer to the sorting and packaging of the cell's _____ . [p.350]
3. Distinguish between somatic cells and germ cells as to their location and function. [pp.350–351]
4. The eukaryotic chromosome is composed of _____ and _____ . [p.350]
5. The two attached threads of a duplicated chromosome are known as sister _____ . [p.348]
6. Describe the function of the portion of a chromosome known as a *centromere*. [p.350]
7. Any cell having two of each type of chromosome is said to be a _____ cell. [p.351]
8. Define the term *karyotype*. [p.351]
9. State the criteria that two chromosomes must meet to be called "homologous chromosomes." [p.351]
10. List, in order, the stages and the various activities occurring during the eukaryotic cell cycle. [p.352]
11. Briefly explain reductional division using the concepts of homologous chromosomes, diploid, and haploid. [p.360]
12. While diploid germ cells are in interphase, each chromosome is _____ ; each chromosome is then composed of two sister _____ . [pp.360,362]
13. What are the two different divisions of meiosis called? [p.360]
14. In prophase of meiosis I, homologous chromosomes exchange corresponding genes in a process known as _____ ; each homologue consists of _____ chromatids. [pp.362,364]
15. If the diploid chromosome number for humans is 46, the haploid number is _____ . [p.360]
16. Describe spermatogenesis in human males. [pp.360–361]
17. Describe oogenesis in human females. [p.361]
18. In comparing mitosis and meiosis, explain why it may be said that meiosis I is the reductional division and that meiosis II is exactly like a mitotic division. [pp.366–367]

Integrating and Applying Key Concepts

Runaway cell division is characteristic of cancer. Imagine the various points of the mitotic process that might be sabotaged in cancerous cells in order to halt their multiplication. Then try to imagine how one might discriminate between cancerous and normal cells in order to guide those methods of sabotage most effective in combating cancer.

20

OBSERVABLE PATTERNS OF INHERITANCE

Interactive Exercises

CHAPTER INTRODUCTION [p.371]

20.1. PATTERNS OF INHERITANCE [p.372]

20.2. SEGREGATION: ONE CHROMOSOME, ONE COPY OF A GENE [p.373]

20.3. DOING GENETIC CROSSES AND FIGURING PROBABILITIES [pp.374–375]

Selected Words: pair of homologous chromosomes [p.372], gene locus (pl. loci) [p.372], parental, *first* (and second) *filial* generation [p.374]

Boldfaced, Page-Referenced Terms

[p.372] genes _____

_____

[p.372] allele _____

_____

[p.372] homozygous condition _____

_____

[p.372] heterozygous condition _____

_____

[p.372] dominant _____

_____

[p.372] recessive _____

_____

[p.372] genotype _____

_____

[p.372] phenotype _____

_____

[p.373] monohybrid cross _____

_____

[p.373] segregation _____

_____

[p.374] P (generation) _____

_____

[p.374] F_1 (generation) _____

_____

[p.374] F_2 (generation) _____

_____

[p.374] Punnett square _____

_____

[p.374] probability _____

_____

Matching

Choose the most appropriate description for each term.

1. _____ genotype [p.372]
2. _____ alleles [p.372]
3. _____ heterozygous [p.372]
4. _____ dominant allele [p.372]
5. _____ phenotype [p.372]
6. _____ genes [p.372]
7. _____ recessive allele [p.372]
8. _____ homozygous [p.372]
9. _____ diploid cell [p.372]
10. _____ gene locus [p.372]

A. The different versions of a gene
B. Particular location of a gene on a chromosome
C. Describes an individual having a pair of different alleles
D. Gene whose effect is masked by its partner
E. Refers to an individual's observable traits
F. Refers to the alleles present in an individual organism
G. Gene whose effect masks the effect of its partner
H. Describes an individual with two of the same allele
I. Units of information about specific traits; passed from parents to offspring
J. Has a pair of genes for each trait, one on each of two homologous chromosomes

Fill-in-the-Blanks

To study the inheritance of a single trait, a(n) (11) _____ [p.373] cross may be done. To predict the outcome of a cross, the possible gametes that each parent can make must be determined. Each gamete receives (12) _____ [p.373] allele for each trait because each gamete contains only one copy of each (13) _____ [p.373]. The separation of the two genes of each pair during meiosis II is known as the principle of (14) _____ [p.373]. Construction of a(n) (15) _____ [p.374] square is a convenient tool for determining the probable outcome of genetic crosses. Because fertilization is a chance event, the rules of (16) _____ [p.374] apply to genetic crosses. Probability is a number between 0 and 1 that expresses the (17) _____ [p.374] of a particular event.

Problems

18. In humans, chin fissure (represented by C) is dominant over smooth chin (represented by c). A woman with the genotype Cc marries a man with the genotype cc. [pp.373–375]
    a. What two types of gametes would the woman make?
    b. What one type of gamete would the man make?
    c. Place these on the appropriate sides of the Punnett square below.
    d. Fill in the four squares of the Punnett square. What ratio of genotypes would you predict in the offspring?
    e. What ratio of phenotypes would you predict for the offspring?
    f. If the couple has one child with a smooth chin, what is the probability of their second child having a chin fissure?

19. The problem above can also be solved by using simple probability equations. The likelihood of the woman producing either C or c gametes is 1/2. The probability of the man producing c gametes is 1. [p.375]
    a. Show the equation and solution for predicting the probability of this couple having a Cc child.
    b. Show the equation and solution for predicting the probability of this couple having a cc child.
    c. What fraction of the offspring is predicted to have a chin fissure? What fraction is predicted to have a smooth chin?

20.4. THE TESTCROSS: A TOOL FOR DISCOVERING GENOTYPES [p.376]

20.5. HOW GENES FOR DIFFERENT TRAITS ARE SORTED INTO GAMETES [pp.376–377]

20.6. A CLOSER LOOK AT GENE-SORTING POSSIBILITIES [p.378]

Selected Words: "simple dominance" [p.376]

Boldfaced, Page-Referenced Terms

[p.376] testcross _____

_____

[p.376] independent assortment _____

_____

[p.376] dihybrid cross _____

_____

Problems

1. In humans, the gene for the ability to taste phenyl-thio-carbamine (PTC), *A*, is dominant; its allele for the inability to taste PTC is *a*. Two parents, one a taster and one a nontaster, have two children. One child is a taster and one child is a nontaster. Determine the genotype of the parent who is a taster. State the reasons for your answer. [p.376]
2. In humans, a dominant gene *C* results in chin fissure, and its allele *c* codes for smooth chin. A man who lacks chin fissure had a mother who exhibited chin fissure and a father who lacked chin fissure. The man is married to a woman with chin fissure whose mother lacked chin fissure but whose father had chin fissure. The man and woman are expecting their first child. State the probability that the child will have a smooth chin or a chin fissure. [p.376]
3. Albinos cannot form the pigments that normally produce skin, hair, and eye color, so albinos have white hair and pink eyes and skin (because the blood shows through). To be an albino, one must be homozygous recessive (*aa*) for the pair of genes that codes for the key enzyme in pigment production. Suppose a woman of normal pigmentation (*A*) with an albino mother marries an albino man. State the kinds of pigmentation possible for this couple's children, and specify the ratio of each kind of child the couple is likely to have. Show the genotype(s) and state the phenotype(s). [p.376]

Problem

When working genetics problems dealing with two gene pairs, one can visualize the independent assortment of gene pairs located on nonhomologous chromosomes into gametes by use of a fork-line device. Assume that in humans, pigmented eyes (*B*) are dominant (an eye color other than blue) over blue (*b*), and right-handedness (*R*) is dominant over left-handedness (*r*). To learn to solve a problem, cross the parents *BbRr* × *BbRr*. A 16-block Punnett square is required, with gametes from each parent arrayed on two sides of the Punnett square (see Figure 20.10 in the text). The gametes receive genes through independent assortment using a fork-line method, as shown below. [pp.376–378]

4. Array the gametes at the right on two sides of the Punnett square; combine these haploid gametes to form diploid zygotes within the squares. In the following blank spaces, enter the probability ratios derived within the Punnett square for the phenotypes listed. [pp.376–378]

   a. _____ pigmented eyes, right-handed
   b. _____ pigmented eyes, left-handed
   c. _____ blue-eyed, right-handed
   d. _____ blue-eyed, left-handed

   B b R r   ×   *B b R r*

   BR, Br, bR, br   *BR, Br, bR, br*

20.7. SINGLE GENES, MULTIPLE EFFECTS [p.379]

20.8. HOW CAN WE EXPLAIN LESS PREDICTABLE VARIATIONS? [pp.380–381]

20.9. *Choices: Biology and Society:* DESIGNER GENES? [p.381]

Selected Words: "expressed" [p.379], *osteogenesis imperfecta* [p.379], *sickle-cell anemia* [p.379], *sickle-cell trait* [p.379], "dormant" genes [p.379], *polydactyly* [p.380], *campodactyly* [p.380], "variable expressivity" [p.380], *eugenic engineering* [p.380]

Boldfaced, Page-Referenced Terms

[p.379] pleiotropy _____

_____

[p.379] codominance _____

_____

[p.379] multiple allele system _____

_____

[p.380] penetrance _____

_____

[p.381] polygenic traits _____

_____

[p.381] continuous variation _____

_____

Complete the Table

1. Complete the following table by supplying the principle of inheritance illustrated by each example. Choose from *pleiotropy, polygenic inheritance,* and *multiple alleles.*

Type of Inheritance	Example
a.	The *osteogenesis imperfecta* disorder exhibits a phenotype of several seemingly unrelated effects [p.379]
b.	A gene system with three or more alleles, such as the alleles I^A, I^B, and i for blood type [p.379]
c.	Human eye and skin color [p.381]
d.	People with sickle-cell trait (Hb^AHb^S) have red blood cells containing both normal and abnormal hemoglobin [p.379]

Choice

For questions 2–11, choose from the following concepts concerned with variable gene expression:

a. codominance   b. pleiotropy   c. penetrance   d. incomplete penetrance
e. variable expressivity   f. polygenic traits   g. eugenic engineering

2. _____ Continuous variation is seen when there is a range of differences for a trait in a population. [p.381]

3. _____ Some people who inherit the dominant allele for polydactyly have the normal number of digits, while others have more. [p.380]

4. _____ This form of inheritance is an explanation for traits such as human eye color and skin color. [p.381]

5. _____ Selection of desirable traits for babies by manipulating genes could result from the mapping of the human genome. [p.381]

6. _____ One gene has several seemingly unrelated effects. [p.379]

7. _____ Both of a pair of contrasting alleles that specify different phenotypes are expressed in a heterozygous individual. [p.379]

8. _____ All persons who are homozygous recessive for the cystic fibrosis gene have cystic fibrosis disease. [p.380]

9. _____ Ethicists have concerns about which forms of a trait will seen as desirable. [p.381]

10. _____ Some persons with campodactyly have immobile, bent fingers on both hands. In others, the trait shows up on one hand only. [p.380]

11. _____ $I^A I^B$ [p.379]

Self-Quiz

_____ 1. The best statement of Mendel's principle of independent assortment is that _____ . [p.374]
a. one allele is always dominant over another
b. hereditary units from the male and female parents are blended in the offspring
c. the two hereditary units that influence a certain trait separate during gamete formation
d. each hereditary unit is inherited separately from other hereditary units

_____ 2. One of two or more alternative forms of a gene for a single trait is a(n) _____ . [p.372]
a. chiasma
b. allele
c. autosome
d. locus

_____ 3. In the F_2 generation of a monohybrid cross involving complete dominance, the expected phenotypic ratio is _____ . [p.374]
a. 3:1
b. 1:1:1:1
c. 1:2:1
d. 1:1

_____ 4. Ethical concerns about "designer babies" stem from possible future technology called _____ . [p.381]
a. human genome project
b. gene therapy
c. selective breeding
d. eugenic engineering

_____ 5. In a testcross, individuals with a dominant phenotype are crossed to an individual known to be _____ for the trait. [p.376]
   a. heterozygous
   b. homozygous dominant
   c. homozygous
   d. homozygous recessive

_____ 6. A man with type A blood could be the father of any of the following except _____ . [p.379]
   a. a child with type A blood
   b. a child with type B blood
   c. a child with type O blood
   d. a child with type AB blood

_____ 7. A single gene that affects several seemingly unrelated aspects of an individual's phenotype is said to be _____ . [p.379]
   a. pleiotropic
   b. epistatic
   c. mosaic
   d. continuous

_____ 8. Suppose two individuals, each heterozygous for the same characteristic, are crossed. The characteristic involves complete dominance. The expected genotypic ratio of their progeny is _____ . [p.374]
   a. 1:2:1
   b. 1:1
   c. 100 percent of one genotype
   d. 3:1

_____ 9. If a homozygous dominant individual is mated to a homozygous recessive individual, the predicted genotypic ratio of the offspring will be _____ . [p.374]
   a. 1:1
   b. 1:2:1
   c. 100 percent of one genotype
   d. 3:1

_____ 10. The trait of skin color in humans exhibits _____ . [p.381]
   a. pleiotropy
   b. epistasis
   c. mosaic
   d. continuous variation

Chapter Objectives / Review Questions

This section lists general and detailed chapter objectives that can be used as review questions. You can make maximum use of these items by writing answers on a separate sheet of paper. Fill in answers where blanks are provided. To check for accuracy, compare your answers with information given in the chapter or glossary.

1. _____ _____ was the university-educated monk who experimented with garden peas and first discovered basic laws of heredity. [p.372]
2. _____ are units of information about specific traits; they are passed from parents to offspring. [p.372]
3. Define _allele_; how many alleles are present in the genotypes _Tt? tt? TT?_ [p.372]
4. Having two of the same allele for a trait is a _____ condition; if the two alleles are different, this is a _____ condition. [p.372]
5. Distinguish a dominant allele from a recessive allele. [p.372]
6. _____ refers to the genes present in an individual; _____ refers to an individual's observable traits. [p.372]
7. In a _____ cross, the inheritance of a single trait is followed. [p.373]
8. Mendel's theory of _____ states that during meiosis, the two genes of each pair separate from each other and end up in different gametes. [p.373]
9. Explain why probability is useful to genetics. [p.374]
10. Define the _testcross_ and cite an example. [p.376]
11. Mendel's theory of _____ _____ states that alleles for different traits are sorted into gametes independent of each other. [p.376]
12. Explain _codominance_. [p.379]
13. Explain why sickle-cell trait and type AB blood are good examples of codominance. [p.379]

14. The disorder known as *osteogenesis imperfecta* is a good example of _____ . [p.379]
15. Why is polydactyly said to be an incompletely penetrant genotype? [p.380]
16. Campodactyly is said to show _____ expressivity. [p.380]
17. State the criterion that qualifies a trait to be polygenic. [p.381]
18. Define *eugenic engineering* and list possible good and bad results from genetically manipulating human offspring. [p.381]

Integrating and Applying Key Concepts

Solve the following genetics problem: In humans, hypotrichosis (sparse body hair) is caused by gene *h*, which is recessive to normal, *H*. Pituitary dwarfism is caused by a deficiency of a growth hormone controlled by an recessive allele, *p*; normal growth is caused by dominant gene *P*. A hypotrichotic male who is homozygous for normal stature marries a woman who is homozygous for normal body hair but is a pituitary dwarf. Predict the appearance of their children. If a mating were to occur between two genotypes like those of their children, what prediction would you make concerning the appearance and the proportions of the characteristics those children might have?

21

CHROMOSOMES AND HUMAN GENETICS

Interactive Exercises

CHAPTER INTRODUCTION [p.385]

20.1. CHROMOSOMES AND INHERITANCE [p.386]

21.2. *Science Comes to Life:* **PICTURING CHROMOSOMES WITH KARYOTYPES** [p.387]

Selected Words: CFTR [p.385]

Boldfaced, Page-Referenced Terms

[p.386] genes _____

_____

[p.386] homologous chromosomes _____

_____

[p.386] alleles _____

_____

[p.386] crossing over _____

_____

[p.386] genetic recombination _____

_____

[p.386] X chromosome _____

_____

[p.386] Y chromosome _____

_____

[p.386] X-linked genes _____

_____

[p.386] Y-linked genes _____

_____

[p.386] sex chromosomes _____

_____

[p.386] autosomes _____

_____

[p.387] karyotype _____

_____

[p.384] X-linked gene _____

_____

[p.384] Y-linked gene _____

_____

[p.385] mutation _____

_____

Fill-in-the-Blanks

Heritable traits arise from units of information located on chromosomes called (1) _____ [p.386].
A pair of chromosomes that have the same length, shape, and gene sequence and interact during meiosis are
known as (2) _____ [p.386] chromosomes. (3) _____ [p.386] are slightly different
forms of the same gene that arose through mutation. An event known as (4) _____

_____ [p.386] occurring early in meiosis functions to exchange corresponding gene segments
between the members of a homologous pair and provide new combinations of genes. This results in
(5) _____ _____ [p.386], meaning new combinations of genes in offspring not present
in the parents. Human X and Y chromosomes that determine gender are examples of (6) _____
[p.386] chromosomes. Human females have two sex chromosomes, both designated (7) "_____"

[p.386]. Males have one X chromosome and another, smaller chromosome designated (8) "_____"
[p.386]. All other chromosomes are designated (9) _____ [p.386]. These chromosomes carry the
vast majority of the estimated tens of thousands of human genes.

The (10) _____ [p.387] of an individual is a cut-up, rearranged photograph of the chromosomes of a single somatic (body) cell. To culture cells for the preparation of such a display, a chemical
compound, colchicine, is placed in a cell medium to arrest all nuclear divisions at the (11) _____
[p.387] stage of mitosis. In the display, homologous pairs of chromosomes are matched up and arranged
from largest to smallest, with the (12) _____ _____ [p.387] placed last.

21.3. HOW SEX IS DETERMINED [pp.388–389]

21.4. LINKED GENES [p.389]

21.5. HUMAN GENETIC ANALYSIS [pp.390–391]

21.6. *Choices: Biology and Society:* PROMISE — AND PROBLEMS — OF GENETIC SCREENING [p.391]

Selected Words: SRY [p.388], "switched off" [p.389], Barr body [p.389], *anhidrotic ectodermal dysplasia*
[p.389], "linked" [p.389], *carrier* [p.390], *abnormal* [p.391], *genetic disorder* [p.391], *genetic screening* [p.391]

Boldfaced, Page-Referenced Terms

[p.388] X inactivation _____

_____

[p.390] pedigrees _____

_____

Complete the Table

1. Complete the Punnett square table at lower right,
   which brings Y-bearing and X-bearing sperm
   together randomly in fertilization. [p.388]

Dichotomous Choice

Answer the following questions related to the
completed Punnett square.

2. Human males transmit their Y chromosome only
   to their (sons/daughters). [p.388]
3. Human males receive their X chromosome only
   from their (mothers/fathers). [p.388]
4. Human mothers and fathers each provide an X
   chromosome for their (sons/daughters). [p.388]
5. While the Y chromosome is responsible for
   masculinizing an embryo, most of the genes on
   the X chromosome deal with (sexual/nonsexual)
   traits. [p.388]

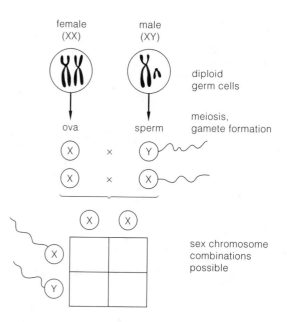

Matching

Choose the most appropriate description for each term.

6. _____ X inactivation [pp.388–389]

7. _____ Barr body [p.389]

8. _____ linkage [p.389]

9. _____ chromosome mapping [p.389]

10. _____ pedigree [p.390]

11. _____ carrier [p.390]

12. _____ abnormal [p.391]

13. _____ genetic disorder [p.391]

14. _____ genetic screening [p.391]

15. _____ SRY gene [p.388]

A. Designates a person who is heterozygous for a recessive trait that causes a genetic disease

B. An inherited condition causing mild to severe medical problems

C. Refers to genes located on the same chromosome

D. Early development mechanism that randomly switches off all or most of the genes on one of a female's X chromosomes

E. Plotting the positions of genes on a chromosome based on frequency of crossing over

F. A chart that is constructed to show the genetic connections among individuals

G. Y chromosome master gene that causes a new individual to develop testes

H. Means deviation from the average, such as having six toes on each foot rather than five

I. A condensed, inactivated X chromosome that may be seen under a light microscope

J. Procedures that can detect harmful alleles in carriers or in offspring

Problems

Early investigators realized that the frequency of crossing over (as detected in offspring with genetic recombinations of parental traits) could be used as a tool to map genes on chromosomes. Genetic maps are graphic representations of the relative positions and distances between genes on each chromosome (linkage group). Geneticists equate 1 percent crossing over (recombination) with one genetic map unit. In the following questions, a line is used to represent a chromosome with its linked genes.

16. Which of the following represents a chromosome that has undergone crossover and recombination? It is assumed that the organism involved is heterozygous with the genotype AaBbCc, with the alleles linked as shown at the left below. [p.389]

A	a		**a.** A	**b.** a	**c.** C	**d.** a
B	b		B	b	B	B
C	c		C	c	A	C

17. If genes A and C are twice as far apart on a chromosome as genes A and B, we would expect that crossing over occurs about ____ as between genes A and C as between genes A and B. [p.389]

    a. half as often   b. the same amount   c. twice as often   d. three times as often

21.7. INHERITANCE OF GENES ON AUTOSOMES [pp.392–393]

21.8. INHERITANCE OF GENES ON THE X CHROMOSOME [pp.394–395]

21.9. SEX-INFLUENCED INHERITANCE [p.396]

Selected Words: *phenylketonuria* (PKU) [p.392], *cystic fibrosis* (CF) [p.392], *Tay-Sachs disease* [p.392], *Huntington disorder* [p.393], *achondroplasia* [p.393], *familial hypercholesterolemia* [p.393], *hemophilia* [p.394], *"muscular dystrophy"* [p.395], *Duchenne muscular dystrophy* (DMD) [p.395], *dystrophin* [p.395], *red-green color blindness* [p.395], *faulty enamel trait* [p.395], *testicular feminizing syndrome* [p.395], *"androgen insensitivity"* [p.395], *sex-influenced* traits [p.396]

Choice

For questions 1–19, choose from the following patterns of inheritance.

a. autosomal recessive   b. autosomal dominant   c. X-linked recessive   d. X-linked dominant
e. sex-influenced inheritance

1. _____ The trait is expressed in heterozygotes (of either sex). [p.392]

2. _____ A genetic disease cannot be detected in a heterozygote (of either sex). [p.392]

3. _____ The recessive phenotype shows up far more often in males than in females. [p.394]

4. _____ Both parents of a person with a genetic disease may be heterozygous normal. [p.392]

5. _____ If one parent is heterozygous for a genetic disease and the other is homozygous recessive, there is a 50 percent chance that any child of theirs will inherit the disease. [p.392]

6. _____ The allele is expressed in homozygotes or heterozygotes. [p.392]

7. _____ Heterozygous normal parents can expect that one-fourth of their children will be affected by a disorder inherited in this manner. [p.392]

8. _____ These traits appear more frequently in one sex, or the phenotype differs depending on whether the individual is male or female. [p.396]

9. _____ A son cannot inherit the recessive allele from his father, but a daughter can. [p.394]

10. _____ Females can mask this gene; males cannot. [p.394]

11. _____ Individuals displaying this type of disorder will always be homozygous for the trait. [p.392]

12. _____ The allele behaves as a dominant in one sex and a recessive in the opposite sex; genes for these traits are carried on autosomes. [p.396]

13. _____ Heterozygous females will transmit the recessive allele to half their sons and half their daughters. Males can only transmit such alleles to their daughters. [p.394]

14. _____ Heterozygous females display the trait. Males are not homozygous or heterozygous for the trait. [p.395]

15. _____ All of the daughters of a man with a genetic disease will have the disease, but he cannot pass the allele to any of his sons. [p.395]

Problems

16. The autosomal allele that causes albinism, *a*, is recessive to the allele for normal pigmentation, *A*. A normally pigmented woman whose father is an albino marries an albino man whose parents are normal. They have three children, two normal and one albino. Give the genotypes for each person listed. [p.392] _____

_____

_____

17. Huntington disorder is a rare form of autosomal dominant inheritance, *H*; the normal gene is *h*. The disease causes progressive degeneration of the nervous system, with onset exhibited near middle age. An apparently normal man in his early twenties learns that his father has recently been diagnosed as having Huntington disorder. What are the chances that the son will develop Huntington disorder? [p.393] _____

_____

_____

18. A color-blind man and a woman with normal vision whose father was color blind have a son. Color blindness, in this case, is caused by an X-linked recessive gene. If only the male offspring are considered, what is the probability that this couple's son is color blind? [p.395] _____

_____

_____

19. Hemophilia A is caused by an X-linked recessive gene. A woman who is seemingly normal but whose father was a hemophiliac marries a normal man. What proportion of their sons will have hemophilia? What proportion of their daughters will have hemophilia? What proportion of their daughters will be carriers? [p.394] _____

_____

_____

20. The following pedigree shows the pattern of inheritance of color blindness in a family (people with the trait are indicated by blackened circles). What is the chance that the third-generation female indicated by the arrow (below) will have a color-blind son if she marries a normal male? A color-blind male? [p.395] _____

_____

_____

_____

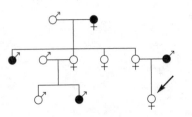

21. A nonbald man marries a nonbald woman whose mother was bald. What is the probability that their first child will be bald? [p.396] _____

_____

_____

_____

Complete the Table

22. Complete the following table by indicating whether the genetic disorder listed is caused by inheritance that is autosomal recessive, autosomal dominant, X-linked recessive, X-linked dominant, or sex-influenced.

Genetic Disorder	Inheritance Pattern
a. Achondroplasia [p.393]	
b. Hemophilia [p.394]	
c. Pattern baldness [p.396]	
d. Huntington disorder [p.393]	
e. Duchenne muscular dystrophy [p.395]	
f. Faulty enamel trait [p.395]	
g. Red-green color blindness [p.395]	
h. Phenylketonuria [p.392]	
i. Cystic fibrosis [p.392]	
j. Tay-Sachs disease [p.392]	
k. Familial hypercholesterolemia [p.393]	

21.10. HOW A CHROMOSOME'S STRUCTURE CAN CHANGE [pp.396–397]

21.11. CHANGES IN CHROMOSOME NUMBER [pp.398–399]

Selected Words: *deletion* [p.396], *cri-du-chat* [p.396], *duplications* [p.397], *translocation* [p.397], *aneuploidy* [p.398], *polyploidy* [p.398], "syndrome" [p.398], Down syndrome [p.398], *Turner syndrome* [p.399], XXX females [p.399], Klinefelter syndrome [p.399], XYY condition [p.399]

Boldfaced, Page-Referenced Terms

[p.397] mutation _____

_____

[p.398] nondisjunction _____

_____

[p.398] trisomic _____

_____

[p.398] monosomic _____

_____

Labeling-Matching

On rare occasions, chromosome structure becomes abnormally rearranged. Such changes may have profound effects on the phenotype of an organism. Label the following diagrams of abnormal chromosome structure as a deletion, a duplication, an inversion, or a translocation. Complete the exercise by matching and entering the letter of the proper description in the parentheses following each label. [pp.396–397]

A. The loss of a chromosome segment; an example is cri-du-chat disorder.
B. A gene sequence repeated several to many times on a chromosome.
C. The transfer of part of one chromosome to a nonhomologous chromosome; an example is when chromosome 14 ends up with a segment of chromosome 8.

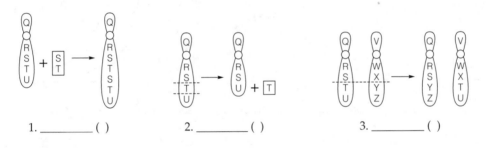

1. _____ ( )   2. _____ ( )   3. _____ ( )

Short Answer

4. Define the term *mutation*. Then list the three possible consequences of a mutation. _____

_____

_____

_____

Complete the Table

5. Complete the following table of important terms associated with chromosome number change in organisms. Choose from *monosomy, nondisjunction, aneuploidy, trisomy,* and *polyploidy.*

Category of Change	Description
a.	New individuals do not have an exact multiple of the normal haploid set of 23 chromosomes [p.398]
b.	New individuals have three or more of each type of chromosome; a lethal condition in humans [p.398]
c.	One or more pairs of chromosomes fail to separate during mitosis or meiosis [p.398]
d.	A gamete with an extra chromosome ($n + 1$) unites with a normal gamete at fertilization; the new individual will have three of one type of chromosome ($2n + 1$) [p.398]
e.	If a gamete is missing a chromosome, then the new individual resulting from fertilization will have only one of one type of chromosome ($2n - 1$) [p.398]

Short Answer

6. How does aneuploidy (an abnormal number of chromosomes) occur? [pp.398–399] _____

_____

_____

_____

Choice

For questions 7–16, choose from the following:

    a. Down syndrome   b. Turner syndrome   c. Klinefelter syndrome   d. XYY condition

7. _____ XXY male [p.399]

8. _____ Ovaries nonfunctional and secondary sexual traits fail to develop at puberty [p.399]

9. _____ Testes smaller than normal, sparse body hair, and some breast enlargement [p.399]

10. _____ Could only be caused by a nondisjunction in males [p.399]

11. _____ Moderate to severe mental retardation; about 40 percent of those affected develop heart defects [p.398]

12. _____ XO female; often miscarried early in pregnancy; distorted female phenotype [p.399]

13. _____ Males that tend to be taller than average but most are phenotypically normal [p.399]

14. _____ Injections of testosterone reverse feminized traits but not the mental retardation or sterility [p.399]

15. _____ Trisomy 21; abnormal skeleton development and muscles weaker than normal [p.398]

Self-Quiz

_____ 1. The evidence that human females have only one functional X chromosome is provided by _____ . [p.389]
   a. the Y chromosome
   b. the SRY gene
   c. sex determination
   d. the Barr body

_____ 2. Chromosomes other than those involved in sex determination are known as _____ . [p.386]
   a. nucleosomes
   b. heterosomes
   c. alleles
   d. autosomes

_____ 3. The farther apart two genes are on a chromosome, _____ . [p.389]
   a. the less likely that crossing over and recombination will occur between them
   b. the greater will be the frequency of crossing over and recombination between them
   c. the more likely they are to be in two different linkage groups
   d. the more likely they are to be deleted or duplicated

_____ 4. Karyotype analysis is _____ . [p.387]
   a. a means of detecting and reducing mutagenic agents
   b. a surgical technique that separates chromosomes that have failed to segregate properly during meiosis II
   c. used to detect chromosomal mutations
   d. a process that substitutes normal alleles for defective ones

_____ 5. Which of the following is not the result of an autosomal dominant gene? [p.390]
   a. Huntington disease
   b. Cystic fibrosis
   c. Achondroplasia
   d. Familial hypercholesterolemia

_____ 6. Red-green color blindness is an X-linked recessive trait in humans. A color-blind woman and a man with normal vision have a son. What are the chances that the son is color blind? If the parents ever have a daughter, what is the probability that the daughter will be color blind? (Consider only female offspring.) [p.395]
   a. 100 percent; 0 percent
   b. 50 percent; 0 percent
   c. 100 percent; 100 percent
   d. 50 percent; 100 percent

_____ 7. Suppose that a hemophilic male (X-linked recessive allele) and a female carrier for the hemophilic trait have a nonhemophilic daughter with Turner syndrome. Nondisjunction could have occurred in _____ . [pp.394, 398,399]
   a. both parents
   b. neither parent
   c. the father only
   d. the mother only

_____ 8. Nondisjunction involving the X chromosome occurs during oogenesis and produces two kinds of eggs, XX and O (no X chromosome). If normal Y sperm fertilize the two types, which genotypes are possible? [p.398]
   a. XX and XY
   b. XXY and YO
   c. XYY and XO
   d. XYY and YO

_____ 9. A person with $2n + 1$ chromosomes is _____ . [p.398]
   a. monosomic
   b. aneuploid
   c. polyploid
   d. tetraploid

_____ 10. Through a nondisjunction, a female has a juvenile phenotype, 45 chromosomes, nonfunctional ovaries, and a webbed neck. This fits the description of _____ . [p.399]
   a. Klinefelter syndrome
   b. XYY condition
   c. Turner syndrome
   d. XXX condition

Chapter Objectives/Review Questions

This section lists general and detailed chapter objectives that can be used as review questions. You can make maximum use of these items by writing answers on a separate sheet of paper. Fill in answers where blanks are provided. To check for accuracy, compare your answers with information given in the chapter or glossary.

1. The units of information about heritable traits are known as _____ . [p.386]
2. Diploid (2*n*) cells have pairs of _____ chromosomes. [p.386]
3. _____ are different forms of the same gene that arise through mutation. [p.386]
4. State the circumstances required for crossing over and describe the significance of the results. [p.389]
5. Name and describe the sex chromosomes in human males and females. [p.388]
6. Distinguish between chromosomes and autosomes. [p.386]
7. Distinguish between an X-linked gene and a Y-linked gene. [p.386]
8. Define *karyotype*; briefly describe its preparation and value. [p.387]
9. How are the sex chromosomes involved in sex determination? [p.388]
10. Explain meiotic segregation of sex chromosomes to gametes and the subsequent random fertilization that determines sex in many organisms. [p.388]
11. A newly identified region of the Y chromosome called _____ appears to be the master gene for sex determination. [p.388]
12. Explain the process of X inactivation and the Barr body. [pp.388–389]
13. The term _____ refers to genes being located on the same chromosome. [p.389]
14. State the relationship between crossover frequency and the location of genes on a chromosome. [p.389]
15. A(n) _____ chart or diagram is used to study genetic connections between individuals. [p.390]
16. In respect to genetic traits, _____ simply means deviation from the average, and a genetic _____ is an inherited condition that causes mild to severe medical problems. [p.391]
17. A(n) _____-_____ trait appears more frequently in one sex than the other; pattern baldness is an example. [p.396]
18. When gametes or cells of an affected individual end up with one more or one less than the correct number of chromosomes (not an exact multiple of the normal haploid set), it is known as _____ ; relate this to monosomy and trisomy. [p.398]
19. Having three or more sets of chromosomes is called _____ . [p.398]
20. _____ is the failure of the chromosome pairs to separate during either mitosis or meiosis (most significant during gamete formation). [p.398]
21. Trisomy 21 is known as _____ syndrome; Turner syndrome has the chromosome constitution _____ ; most _____ females develop normally; XXY chromosome constitution is _____ syndrome; taller-than-average males with normal male phenotypes have the _____ condition. [p.399]
22. Describe the cause and characteristics of Turner syndrome, XXX condition, Klinefelter syndrome, and the XYY condition. [p.399]

Integrating and Applying Key Concepts

1. The parents of a young boy bring him to their doctor. They explain that the boy does not seem to be going through the same vocal developmental stages as his older brother. The doctor orders a common cytogenetics test to be done, and it reveals that the young boy's cells contain two X chromosomes and one Y chromosome. Describe the test that the doctor ordered, and explain how and when such a genetic result, XXY, most logically occurred. What treatment would the doctor most likely prescribe?
2. Solve the following genetics problem. Show rationale, genotypes, and phenotypes. A husband sues his wife for divorce, arguing that she has been unfaithful. His wife gave birth to a girl with a fissure in the iris of her eye, an X-linked recessive trait. Both parents have normal eye structure. Can the genetic facts be used to argue for the husband's suit? Explain your answer.

22

DNA, GENES, AND BIOTECHNOLOGY

Interactive Exercises

CHAPTER INTRODUCTION [p.403]

22.1. DNA: A DOUBLE HELIX [pp.404–405]

22.2. PASSING ON GENETIC INSTRUCTIONS [pp.406–407]

Selected Words: thymine dimers [p.406], "point mutations" [p.406], *expansion mutations* [p.407], *fragile X syndrome* [p.407], *Huntington disorder* [p.407], *neurofibromatosis* [p.407]

Boldfaced, Page-Referenced Terms

[p.404] nucleotides _____

[p.404] base pairs _____

_____

[p.405] gene _____

_____

[p.406] DNA replication _____

_____

[p.406] semiconservative replication _____

_____

[p.406] DNA polymerases _____

_____

[p.406] gene mutations _____

_____

[p.407] base-pair substitution _____

_____

Short Answer

1. List the three parts of a nucleotide. [p.404] _____

_____

_____

_____

Complete the Table

2. Complete the table below showing important characteristics of the four nucleotides found in DNA. [p.404]

Name of Nucleotide	Abbreviation	Single- or Double-Ring	Base It Pairs With
a.	A	e.	i.
b.	G	f.	j.
c.	T	g.	k.
d.	C	h.	l.

True/False

If the statement is true, write a "T" in the blank. If the statement is false, make it correct by changing the underlined word(s) and writing the correct word(s) in the answer blank.

_____ 3. DNA is composed of <u>two</u> different types of nucleotides. [p.404]

_____ 4. In the DNA of every species, the amount of <u>adenine</u> present always equals the amount of thymine, and the amount of <u>cytosine</u> always equals the amount of guanine (A = T and C = G). [p.404]

_____ 5. In a nucleotide, the phosphate group is attached to the <u>nitrogen-containing base</u>, which is attached to the <u>five-carbon sugar</u>. [p.404]

_____ 6. Watson and Crick built their model of DNA in the early <u>1950s</u>. [p.403]

_____ 7. Hydrogen bonding pairs guanine with <u>cytosine</u> and adenine with <u>thymine</u>. [p.404]

_____ 8. The shape of the DNA molecule is often described as a <u>spiral ladder</u>. [p.404]

_____ 9. The two DNA strands run in <u>opposite</u> directions. [p.404]

_____ 10. The two strands of DNA are held together by <u>covalent</u> bonds between pairs of bases. [p.404]

Labeling

Identify each indicated part of the following DNA illustration. Choose from these answers: *phosphate group, double-ring nitrogen base, single-ring nitrogen base, nucleotide, deoxyribose, hydrogen bond, and covalent bond.*

11. _____ [p.405]

12. _____ [p.405]

13. _____ -ring nitrogen base [p.405]

14. _____ -ring nitrogen base [p.405]

15. _____ -ring nitrogen base [p.405]

16. _____ -ring nitrogen base [p.405]

17. A complete _____ [p.405]

18. _____ bond [p.405]

19. _____ bond [p.405]

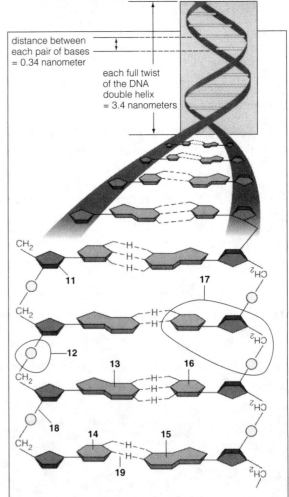

Labeling

20. The term *semiconservative replication* refers to the fact that each new DNA molecule resulting from the replication process is "half old, half new." In the accompanying illustration, complete the replication required in the middle of the molecule by adding the required letters representing the missing nucleotide bases. Recall that ATP energy and the appropriate enzymes are actually required in order to complete this process. [p.404]

T- _____ _____ -A

G- _____ _____ -C

A- _____ _____ -T

C- _____ _____ -G

C- _____ _____ -G

C- _____ _____ -G

old new   new   old

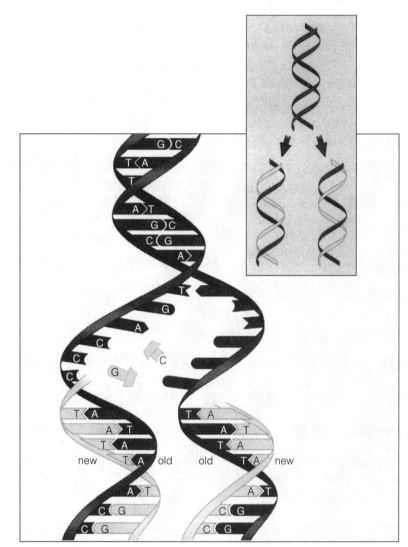

Short Answer

21. Define the word *gene*. [p.405] _____

_____

_____

_____

Fill-in-the-Blanks

Sister chromatids are the result of (22) _____ _____ [p.406]. DNA replication occurs as the double helix unwinds and hydrogen bonds between strands are broken, leaving the (23) _____ [p.406] exposed. Each parent strand remains intact as a new companion strand is assembled on it, one (24) _____ [p.406] at a time. DNA replication is said to be (25) _____ [p.406] because one strand of each completed DNA molecule is "new" while the other is from the starting molecule.

Individual nucleotides are assembled into the new portion of the double helix by enzymes called (26) _____ _____ [p.406]. These, along with DNA ligases and other enzymes, also function in DNA (27) _____ [p.406]. It has been estimated that a human cell must repair breaks in a single strand of DNA up to (28) _____ _____ [p.406] times every hour. When errors that occur during DNA replication are not detected and corrected, the result is a(n) (29) _____ [p.406].

DNA is also vulnerable to damage from chemicals, ionizing radiation, and (30) _____ _____ [p.406]. UV light can cause a type of damage known as a (31) _____ _____ [p.406], in which adjacent thymine bases become linked.

Small-scale changes in the nucleotide sequence are called "point mutations" or (32) _____ _____ [p.406]. In a(n) (33) _____-_____ _____ [p.407], one base is wrongly paired with another. Sickle-cell anemia is caused by this type of mutation. Other mutations involve an extra base (34) _____ [p.407] into a gene, or a base (35) _____ [p.407] from a gene. When a particular nucleotide sequence is repeated over and over, this is called a(n) (36) _____ [p.407] mutation, such as the one that causes fragile X syndrome. Bits of DNA that move from one location to another are called (37) _____ _____ [p.407]. They cause mutations, such as the one responsible for neurofibromatosis.

Mutations often affect the structure and functioning of (38) _____ [p.407]. They are only inherited when they take place in the cells of (39) _____ [p.407] that produce gametes. Mutations may be harmful, neutral, or (40) _____ [p.407].

22.3. DNA INTO RNA — STEP ONE IN MAKING PROTEINS [pp.408–409]
22.4. READING THE GENETIC CODE [pp.410–411]
22.5. TRANSLATING THE GENETIC CODE INTO PROTEIN [pp.412–413]

Selected Words: nucleotide "cap" [p.408], "tail" [p.409], pre-mRNA [p.409], "triplets" [p.410], "three-bases-at-a-time" [p.410], "wobble effect" [p.411], *initiation* [p.412], *elongation* [p.412], *termination* [p.412], *polysomes* [p.413]

Boldfaced, Page-Referenced Terms

[p.408] RNA _____

_____

[p.408] uracil _____

_____

[p.408] transcription _____

_____

[p.408] translation _____

_____

[p.408] ribosomal RNA (rRNA) _____

_____

[p.408] messenger RNA (mRNA) _____

_____

[p.408] transfer RNA (tRNA) _____

_____

[p.408] RNA polymerases _____

_____

[p.408] promoter _____

_____

[p.409] introns _____

_____

[p.409] exons _____

_____

[p.409] regulatory proteins _____

_____

[p.410] codons _____

_____

[p.410] genetic code _____

_____

[p.410] anticodon _____

_____

Complete the Table

1. Complete the following table to summarize the molecular differences between DNA and RNA. [p.406]

DNA Sugar	DNA Bases	RNA Sugar	RNA Bases
a.	b.	c.	d.

Fill-in-the-Blanks

The two steps from genes to proteins are called (2) _____ [p.408] and (3) _____ [p.408]. In (4) _____ [p.408], single-stranded molecules of RNA are assembled on DNA templates in the nucleus. In (5) _____ [p.408], the RNA molecules are shipped from the nucleus into the cytoplasm, where they are used as templates for assembling (6) _____ [p.408] chains. Following translation, one or more of these chains are folded into (7) _____ [p.408] molecules.

Complete the Table

8. Three types of RNA are translated from DNA in the nucleus (from genes that code only for RNA). Complete the following table, which summarizes information about these RNA molecules. [p.406]

RNA Molecule	Abbreviation	Description/Function
Ribosomal RNA	a.	
Messenger RNA	b.	
Transfer RNA	c.	

Short Answer

9. Cite the key differences between DNA replication and transcription (both processes involve DNA and occur in the nucleus). [p.408] _____

_____

_____

_____

Sequence

Arrange the steps of transcription in hierarchical order, with the earliest step first and the latest step last.

10. _____ A. A termination base sequence serves as a signal to release the RNA transcript. [p.409]

11. _____ B. Proteins help position an RNA polymerase on the DNA so that it binds with promoter. [p.408]

12. _____ C. Newly formed pre-RNA is unfinished; introns are removed and exons are spliced together; the mRNA then leaves the nucleus. [p.409]

13. _____ D. A promoter is a base sequence that signals the start of a gene being transcribed. [p.408]

14. _____ E. The enzyme moves along the DNA, joining the RNA nucleotides together. [p.409]

Completion

15. Suppose the following line represents the DNA strand that will act as a template for the production of mRNA through the process of transcription. Complete the blanks below the DNA strand with the sequence of complementary bases that will represent the message carried by mRNA from DNA to the ribosome in the cytoplasm. [pp.406–407]

TAC — AAG — ATA — ACA — TTA — TTT — CCT — ACC — GTC — ATC

_____ — _____ — _____ — _____ — _____ — _____ — _____ — _____ — _____ — _____

(transcribed single strand of mRNA)

Labeling-Matching

Newly transcribed mRNA contains more genetic information than is necessary to code for a chain of amino acids. Before the mRNA leaves the nucleus for its ribosome destination, an editing process occurs as certain portions of nonessential information are snipped out. Identify each indicated part of the following illustration; use abbreviations for the nucleic acids. Choose from *tail, intron, DNA, mature RNA transcript, cap,* and *exon*. Complete the exercise by matching and entering the letter of the description in the parentheses after each label. [p.409]

16. _____ ( )

17. _____ ( )

18. _____ ( )

19. _____ ( )

20. _____ ( )

21. _____ _____ _____ ( )

A. Regions that are translated into proteins [p.409]
B. Base sequences that do not get translated into an amino acid sequence [p.409]
C. Presence of cap and tail; introns snipped out and exons spliced together [p.409]
D. Protective region on the terminal end of a pre-mRNA transcript [p.409]
E. The region of the DNA template strand to be copied [p.408]
F. A nucleotide added to mRNA as transcription starts [p.408]

transcription into mRNA

(snipped out) (snipped out)

Matching

22. _____ codon [p.410]

23. _____ sixty-four [p.410]

24. _____ genetic code [p.410]

25. _____ ribosome [p.411]

26. _____ wobble effect [p.411]

27. _____ anticodon [p.410]

28. _____ the "stop" codons [p.410]

29. _____ three-at-a-time [p.410]

A. Composed of two subunits made in the nucleus and sent to the cytoplasm
B. Reading frame of the nucleotide bases in mRNA
C. UAA, UAG, UGA (three of sixty-four codons)
D. A sequence of three nucleotide bases on tRNA that can pair with a specific mRNA codon
E. Flexibility in codon–anticodon pairing at the third base
F. Name for each base triplet in mRNA
G. The number of codons in the genetic code
H. Provides cells with basic instructions for synthesizing proteins

Complete the Table

30. Complete the following table, which distinguishes the stages of translation. [p.412]

Translation Stage	Description
a.	An initiator tRNA binds with the small ribosomal subunit and attaches to one end of the mRNA; this unit moves along the mRNA until it encounters the start codon (AUG) of the mRNA transcript; after this, a large ribosomal unit binds with a small one.
b.	A polypeptide chain forms as the mRNA strand passes between the ribosomal subunits; some proteins in the ribosome function as enzymes, joining amino acids together in the sequence dictated by mRNA codons; they catalyze the formation of a peptide bond between adjacent amino acids.
c.	A stop codon is reached and there is no corresponding anticodon; now the ribosome interacts with certain release factor proteins; this causes the ribosome as well as the polypeptide chain to detach from the mRNA; the detached chain may join the cytoplasmic pool of free proteins or enter the cytomembrane system.

31. Find your answer to question 15 and enter it on the line provided. Deduce the composition of the tRNA anticodons that would pair with the specific mRNA codons as these tRNAs deliver the amino acids that are identified here to the binding sites of the small ribosomal subunit. [p.410]

mRNA _____-_____-_____-_____-_____-_____-_____-_____-_____-_____

tRNA _____-_____-_____-_____-_____-_____-_____-_____-_____-_____

32. From the mRNA transcript in exercise 31, use Figure 22.11 in the text to identify the composition of the amino acids in the polypeptide sequence.

_____-_____-_____-_____-_____-_____-

_____-_____-_____ [p.410]

(amino acids)

Labeling-Matching

A summary of the flow of genetic information in protein synthesis is useful as an overview. Identify the indicated parts of the following illustration by filling in the blanks with the names of the appropriate *structures* or *functions*. Choose from the following: *mRNA, tRNA, polypeptide, ribosome subunits, ribosome, translation,* and *transcription*. Complete the exercise by matching and entering the letter of the description (A–G) in the parentheses after each label. [pp.408–409,411–412]

33. _____ ( )

34. _____ _____ ( )

35. _____ _____ ( )

36. _____ _____ ( )

37. _____ ( )

38. _____ ( )

39. _____ ( )

A. Carries the genetic code from the nucleus to the cytoplasm, where it will be translated into a protein product [p.410]

B. Includes three stages: initiation, chain elongation, and chain termination [p.412]

C. May serve as an enzyme, a member receptor or channel, or any number of functions; made of amino acids [p.408]

D. Join together at initiation of translation [p.411]

E. RNA molecules are produced on DNA templates in the nucleus [p.408]

F. Place where translation occurs [p.411]

G. Will pick up specific amino acid for delivery to ribosome for translation [p.410]

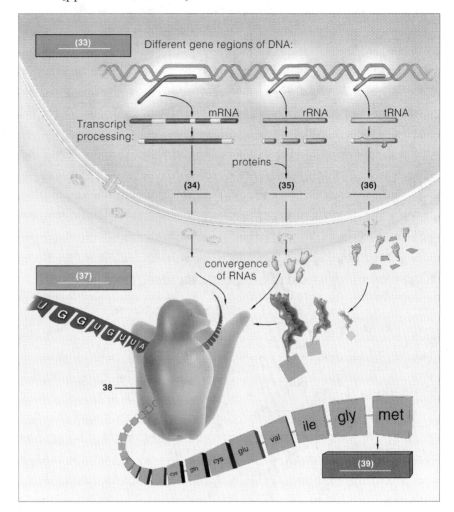

22.6. TOOLS FOR "ENGINEERING" GENES [pp.414–415]

22.7. "SEQUENCING" DNA [p.416]

Selected Words: "sticky" [p.414], plasmids [p.414], "cloned" [p.414], "cloning vector" [p.415], *probe* [p.416]

Boldfaced, Page-Referenced Terms

[p.414] recombinant DNA technology _____

_____

[p.414] genetic engineering _____

_____

[p.414] restriction enzyme _____

_____

[p.414] genome _____

_____

[p.415] polymerase chain reaction (PCR) _____

_____

[p.416] DNA sequencing _____

_____

[p.416] gene library _____

_____

Short Answer

1. Describe and distinguish between the bacterial chromosome and plasmids present in a bacterial cell. [p.414] _____

_____

_____

_____

Fill-in-the-Blanks

Genetic experiments have been occurring in nature for billions of years as a result of (2) _____ [p.414], crossing over and recombination, and other events. Humans now are causing genetic change by using (3) _____ _____ [p.414] technology, in which researchers cut and splice together gene regions from different (4) _____ [p.414]. The modified molecules are then inserted into bacteria or other cells that can rapidly (5) _____ [p.414] genetic material and then divide. This new technology also is the basis for (6) _____ _____ [p.414], in which genes are isolated, modified, and inserted back into the same organism or a different one.

Bacterial (7) _____ [p.414] are circular DNA molecules containing only a few genes. A(n) (8) _____ _____ [p.414] can cut apart specific sequences of DNA. Fragments of

DNA with the same (9) "_____" [p.414] ends will combine and form a recombinant DNA molecule. Foreign DNA inserted into a plasmid is called a DNA (10) _____ [p.414] because a bacterium will replicate the engineered plasmid, producing many "clones." As bacteria with engineered plasmids reproduce, they make much more of, or (11) _____ [p.419], the foreign DNA.

Choice

a. DNA clones   b. plasmids   c. cloning vectors   d. recombinant DNA technology
e. restriction enzymes   f. DNA amplification   g. genome   h. polymerase chain reaction

12. _____ Specialized knowledge that forms the basis for genetic engineering [p.414]
13. _____ All the DNA in a haploid set of a species' chromosomes
14. _____ Rapid way to copy DNA fragments [p.415]
15. _____ Cut(s) DNA molecules [p.414]
16. _____ Rapid division of DNA clones [p.415]
17. _____ Small, circular DNA molecules that carry only a few genes [p.414]
18. _____ Multiple, identical copies of DNA fragments that have been inserted into bacterial plasmids [p.415]
19. _____ Plasmids that serve to deliver foreign DNA into a host cell [p.415]

Short Answer

20. How does PCR differ from the use of cloning vectors for DNA amplification? [p.415] _____

_____

_____

_____

21. What is the function of primers used in PCR? [p.415] _____

_____

_____

_____

Fill-in-the-Blanks

_____ 22. Determining the order of nucleotides in a gene or other piece of DNA [p.416]

_____ 23. A type of radioactively labeled DNA fragment that can locate a specific nucleotide sequence [p.416]

_____ 24. A mixture of DNA fragments containing genes of interest [p.416]

22.8. THE HUMAN GENOME PROJECT [pp.416–417]

22.9. SOME APPLICATIONS OF BIOTECHNOLOGY [pp.418–419]

22.10. *Choices: Biology and Society:* ISSUES FOR A BIOTECHNOLOGICAL SOCIETY [p.420]

22.11. ENGINEERING BACTERIA, ANIMALS, AND PLANTS [p.421]

22.12. *Science Comes to Life:* MR. JEFFERSON'S GENES [p.422]

Selected Words: "Book of Life" [p.416], Expressed Sequence Tags (ESTs) [p.416], "shotgun sequencing" [p.416], "junk DNA" [p.416], SNPs [p.417], *amylotrophic lateral sclerosis* [p.417], EST [p.417], *transfection* [p.418], "suicide tags" [p.419], "lipoplexes" [p.419], *restriction fragment length polymorphisms* (RFLPs) [p.419], *transgenic* [p.421], *bioremediation* [p.421], "micro-injected" [p.421], "super mouse" [p.421]

Boldfaced, Page-Referenced Terms

[p.418] gene therapy _____

_____

[p.419] DNA fingerprint _____

_____

True/False

If the statement is true, write a "T" in the blank. If the statement is false, make it correct by changing the underlined word(s) and writing the correct word(s) in the answer blank.

_____ 1. The human genome consists of about 2.9 billion nucleotide pairs with perhaps <u>100,000</u> genes. [p.416]

_____ 2. The Human Genome Project began in 1990 and was predicted to take <u>five</u> years to complete. [p.416]

_____ 3. Craig Venter developed a system called <u>"shotgun sequencing"</u> which sped up the analysis of the human genome. [p.416]

_____ 4. Over 90 percent of the human genome had been sequenced by <u>2002</u>. [p.416]

_____ 5. The coding portions of human DNA make up about <u>80</u> percent of our DNA. [p.416]

_____ 6. There are over 1.4 million SNPs in the human genome, including many that result in different <u>alleles</u>. [p.416]

_____ 7. The findings of the Human Genome Project may allow therapeutic drugs to be <u>customized</u> for individuals and particular situations. [p.416]

_____ 8. The Human Genome Project seeks to create maps of where specific genes are on chromosomes — this <u>has not</u> been successful. [p.416]

Fill-in-the-Blanks

In (9) _____ _____ [p.418], one or more normal genes are inserted into a person's body to replace mutated genes and correct a genetic defect. Smaller genes can be carried into a host cell by (10) _____ [p.418], while larger ones must enter in some other way. In (11) _____ [p.418], DNA is integrated into exposed cells in a laboratory culture, but it has not been very successful. Many gene therapy trials involve inserting a small gene into a(n) (12) _____ [p.418] that then infects a host cell and carries the correct gene in. Research with SCID and cystic fibrosis has been only marginally successful. Continuing efforts are focused on the development of more effective (13) _____ [p.418].

Gene therapy has been most successful at treating (14) _____ [p.419]. Genes for a(n) (15) _____ [p.419] have been introduced into cultured tumor cells and the cells returned to the body. This may stimulate T cells to recognize cancerous cells and (16) _____ [p.419] them. (17) _____ [p.419] carrying plasmids that encode a protein marker result in the immune system attacking cancer cells.

Dichotomous Choice

Circle one of two possible answers given between parentheses in each statement.

18. The unique set of (DNA fragments/coding genes) in each individual is called a DNA fingerprint. [p.419]
19. The short DNA segments that differ greatly from person to person are called (tandem repeats/riff-lips). [p.419]
20. In DNA fingerprinting, DNA fragments are separated by a procedure called (column chromatography/gel electrophoresis). [p.419]
21. RFLPs can be detected by electrophoresis after being cut out of DNA by (polymerase enzymes/restriction enzymes). [p.419]

Short Answer

22. List five hotly debated concerns about human manipulation of DNA. [p.420] _____

_____

_____

_____

23. Name some benefits of human manipulation of DNA. [p.420] _____

_____

_____

_____

Fill-in-the-Blanks

Any genetically engineered organism that carries foreign genes is (24) _____ [p.421].

(25) _____ [p.421] in bioengineered bacteria carry many human genes so that useful human

(26) _____ [p.421] can be made easily. Growth hormone, (27) _____ [p.421], and

interferon are examples of such products. Animal cells can be (28) _____-_____

[p.421] with foreign DNA. This may allow sheep, pigs, and cows to become sources of important

(29) _____ _____ [p.421]. Plants may be engineered to improve traits such as

(30) _____ [p.421] to pests or herbicides and to improve crop yields.

Self-Quiz

_____ 1. A DNA molecule is built from four kinds of _____ . [p.404]
  a. base pairs
  b. nitrogen-containing bases
  c. phosphates
  d. nucleotides

_____ 2. In DNA, base pairing occurs between _____ . [p.404]
  a. cytosine and uracil
  b. adenine and guanine
  c. adenine and uracil
  d. adenine and thymine

_____ 3. A single strand of DNA with the base-pairing sequence C-G-A-T-T-G is compatible only with the sequence _____ . [p.404]
  a. C-G-A-T-T-G
  b. G-C-T-A-A-G
  c. T-A-G-C-C-T
  d. G-C-T-A-A-C

_____ 4. During DNA replication, individual nucleotides are assembled onto a parent DNA strand by _____ . [p.406]
  a. thymine dimers
  b. DNA polymerases
  c. ribosomes
  d. codons

_____ 5. Transcription _____ . [p.408]
  a. occurs on the surface of the ribosome
  b. is the final process in the assembly of a protein
  c. occurs during the synthesis of RNA by use of a DNA template
  d. is catalyzed by DNA polymerase

_____ 6. _____ carry(ies) amino acids to ribosomes, where they are linked into the primary structure of a polypeptide. [p.408]
  a. mRNA
  b. tRNA
  c. Introns
  d. rRNA

_____ 7. Transfer RNA differs from other types of RNA because it _____ . [p.408]
  a. transfers genetic instructions from cell nucleus to cytoplasm
  b. specifies the amino acid sequence of a particular protein
  c. carries an amino acid at one end
  d. contains codons

_____ 8. _____ is an enzyme that dominates the process of transcription. [p.408]
  a. RNA polymerase
  b. DNA polymerase
  c. Phenylketonuriase
  d. Transfer RNA

_____ 9. _____ and _____ are found in RNA but not in DNA. [p.408]
  a. Deoxyribose; thymine
  b. Deoxyribose; uracil
  c. Uracil; ribose
  d. Thymine; ribose

_____ 10. Each "word" in the DNA and mRNA language consists of _____ "letters." [p.408]
  a. three
  b. four
  c. five
  d. more than five

_____ 11. The genetic code is composed of _____ different kinds of codons. [p.410]
  a. three
  b. twenty
  c. sixteen
  d. sixty-four

_____ 12. Initiation, elongation, and termination are all stages of _____ . [p.412]
  a. replication
  b. translation
  c. transcription
  d. mutagenesis

13. Small, circular molecules of DNA in bacteria, often used as cloning vectors, are called _____ . [p.414]
    a. plasmids
    b. DNA probes
    c. RFLPs
    d. cDNA

14. _____ enzymes are used to cut genes in recombinant DNA research. [p.414]
    a. Ligase
    b. Restriction
    c. Transcriptase
    d. DNA polymerase

15. Genetically engineered bacteria may become major weapons in cleaning up environmental pollution, a process called _____ . [p.421]
    a. recombinant DNA technology
    b. bioremediation
    c. DNA library construction
    d. DNA sequencing

16. A DNA probe is a known DNA sequence that is _____ . [p.416]
    a. resistant to antibiotics
    b. used to make restriction enzymes
    c. radioactively labeled
    d. found in human noncoding DNA

17. Amplification results in _____ . [p.415]
    a. plasmid integration
    b. bacterial conjugation
    c. cloned DNA
    d. production of DNA ligase

18. A commonly used method of DNA amplification is _____ . [p.415]
    a. polymerase chain reaction
    b. gene expression
    c. genome mapping
    d. RFLPs

19. Restriction fragment length polymorphisms are valuable because they _____ . [p.421]
    a. reduce the risks of genetic engineering
    b. provide an easy way to sequence the human genome
    c. allow fragmenting DNA without enzymes
    d. identify individuals by unique DNA sequences

20. Transgenic species are those that carry _____ . [p.421]
    a. micro-injected DNA
    b. one or more foreign genes
    c. human genes
    d. transfected cells

Chapter Objectives/Review Questions

This section lists general and detailed chapter objectives that can be used as review questions. You can make maximum use of these items by writing answers on a separate sheet of paper. Fill in answers where blanks are provided. To check for accuracy, compare your answers with information given in the chapter or glossary.

1. Aided by clues in X-ray images, _____ and _____ painstakingly devised a correct model of the DNA molecule. [p.403]
2. Explain what is meant by pairing of nitrogen-containing bases (base-pairing) and explain the mechanism that causes bases of one DNA strand to join with bases of the other strand. [p.404]
3. Assume that the two parent strands of DNA have been separated and that the base sequence on one parent strand is A-T-T-C-G-C; the base sequence that will complement that parent strand during replication is _____ . [p.404]
4. Explain what is meant by "Each parent strand is conserved in each new DNA molecule." [p.406]
5. Briefly describe the spontaneous DNA mutations known as *base-pair substitutions, frameshift mutations,* and *expansion mutations.* [pp.412–413]
6. Cite an example of a change in one DNA base pair that has profound effects on the human phenotype. [p.407]
7. State how RNA differs from DNA in structure and function and indicate which features RNA has in common with DNA. [p.408]

8. _____ RNA combines with certain proteins to form the ribosome; _____ RNA carries genetic information for protein construction from the nucleus to the cytoplasm; _____ RNA picks up specific amino acids and moves them to the area of mRNA and ribosome. [p.408]
9. Describe the process of transcription and indicate three ways in which it differs from replication. [pp.408–409]
10. What mRNA code would be formed from the following DNA code: TAC-CAT-GAG-ACC-GCC-ACT? [p.408]
11. Transcription starts at a(n) _____ , a specific sequence of bases on one of the two DNA strands that signals the start of a gene. [p.408]
12. Actual coding portions of a newly transcribed mRNA are called _____ ; _____ are the noncoding portions. [p.409]
13. What is the function of regulatory proteins in the process of transcription? [p.409]
14. Explain the triplet nature of the genetic code. [p.410]
15. Describe the relationships between DNA and mRNA and between mRNA and tRNA that insure the synthesis of the correct protein. [p.410]
16. Name the three stages of translation and tell what happens in each. [p.414]
17. List the means by which natural genetic experiments occur. [p.418]
18. Define recombinant DNA technology. [p.414]
19. _____ are small, circular, self-replicating molecules of DNA or RNA within a bacterial cell. [p.414]
20. Some bacteria produce _____ enzymes that cut apart DNA molecules at specific base sequences; such DNA fragments may have "_____ ends" capable of base-pairing with other DNA molecules. [p.414]
21. Describe PCR as a major method of DNA amplification. [p.415]
22. Explain what a DNA probe is and what a gene library is. [p.416]
23. Discuss the history and findings of the Human Genome Project. [p.416]
24. Define "gene therapy" and describe techniques used in gene therapy trials. [p.418]
25. Why is a "DNA fingerprint" given that name? [p.419]
26. DNA fragments from different people show unique restriction _____ length _____ . [p.419]
27. What are transgenic organisms and how are they "created" by humans? [p.421]

Integrating and Applying Key Concepts

Genes code for specific polypeptide sequences. Not every substance in living cells is a polypeptide. Explain how genes might be involved in the production of a storage carbohydrate (such as glycogen) that is constructed from simple sugars.

23

GENES AND DISEASE: CANCER

Interactive Exercises

CHAPTER INTRODUCTION [p.427]

23.1. CANCER: CELL CONTROLS GO AWRY [pp.428–429]

23.2. THE GENETIC TRIGGERS FOR CANCER [pp.430–431]

23.3. *Focus on Our Environment:* **ASSESSING THE CANCER RISK FROM ENVIRONMENTAL CHEMICALS** [p.432]

Selected Words: neoplasm [p.428], *benign* tumor [p.428], *poorly differentiated* [p.428], contact inhibition [p.429], malignant [p.429], *angiogenin* [p.429], *retinoblastoma* [p.430], "precarcinogens" [p.431], cancer "promoters" [p.431]

Boldfaced, Page-Referenced Terms

[p.428] tumor _____

_____

[p.428] dysplasia _____

_____

[p.429] metastasis _____

_____

[p.430] carcinogenesis _____

_____

[p.430] proto-oncogenes _____

[p.430] oncogene _____

_____

[p.430] tumor suppressor gene _____

_____

[p.430] carcinogens _____

_____

Matching

Choose the most appropriate description for each term.

1. _____ dysplasia [p.428]

2. _____ neoplasm [p.428]

3. _____ malignant [p.429]

4. _____ poorly differentiated cell [p.428]

5. _____ tumor [p.428]

6. _____ metastasis [p.429]

7. _____ angiogenin [p.429]

8. _____ benign [p.428]

A. Growth factor secreted by cancer cells; promotes blood vessel growth around cancer cells
B. Cancer cells lacking clear structural specializations; display abnormally large nuclei, less cytoplasm, and disorganized cytoskeletons
C. Cancer cells breaking away from a tumor and establishing new cancer sites
D. Refers to a noncancerous, slow-growing, and well-differentiated tumor; often enclosed by a capsule
E. A defined mass of overgrown tissue
F. "Bad form"; abnormal changes in tissue cells
G. Means "new growth"; a defined mass of tissue overgrowth
H. A cancer with the ability to metastasize.

Dichotomous Choice

Circle one of two possible answers given between parentheses in each statement.

9. The transformation of a normal cell into a cancerous one is called (metastasis/carcinogenesis). [p.430]
10. (Oncogenes/Proto-oncogenes) are normal genes regulating cell growth and development. [p.430]
11. A DNA segment capable of inducing cancer in a normal cell is a(n) (oncogene/proto-oncogene). [p.430]
12. An oncogene (does/does not) respond to controls over cell division. [p.430]
13. A(n) (oncogene/proto-oncogene) acting alone is not wholly responsible for the onset of malignant cancer. [p.430]
14. Onset of cancer requires mutations in several genes, including mutation of at least one (tumor suppressor gene/proto-oncogene). [p.430]
15. Retinoblastoma, or childhood eye cancer, develops when a child inherits (one/two) normal copies of a tumor suppressor gene. [p.430]
16. p53 appears to help (cause/prevent) cancerous changes in many types of cancer. [p.430]
17. When p53 mutates, the resulting faulty protein seems to (inhibit/promote) the development of cancer, possibly by activating an oncogene. [p.430]
18. The mutation of an oncogene or a related suppressor gene or a change in chromosome structure that moves an oncogene into a new position can cause the oncogene to be (activated/inactivated). [p.430]

Labeling

Identify each numbered part of the following illustration that traces the steps of carcinogenesis. [p.437]

19. _____
20. _____ _____
21. _____-_____
22. _____ _____ _____
23. _____
24. _____
25. _____
26. _____
27. _____
28. _____
29. _____

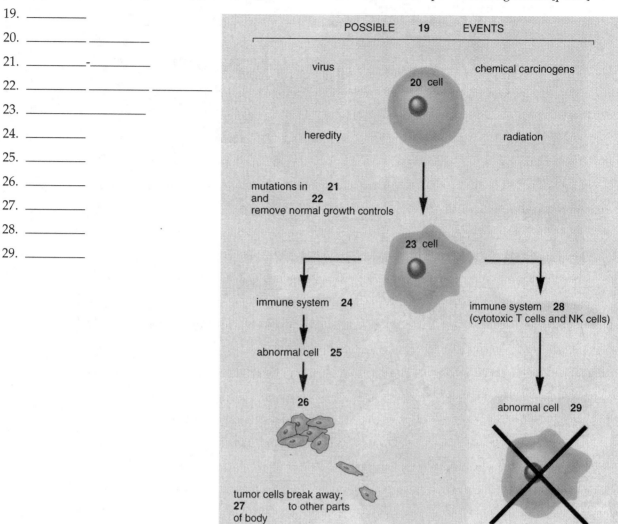

POSSIBLE   **19**   EVENTS

virus   chemical carcinogens

20 cell

heredity   radiation

mutations in   **21**
and   **22**
remove normal growth controls

23 cell

immune system   **24**   immune system   **28**
(cytotoxic T cells and NK cells)

abnormal cell   **25**

26

abnormal cell   **29**

tumor cells break away;
27   to other parts
of body

Complete the Table

30. Complete the following table by providing names of specific carcinogens or mechanism failures acting as carcinogenesis triggers for the general categories provided.

Carcinogen	Description
a. Inherited cancer [p.430]	
b. Viruses [p.430]	
c. Chemical carcinogens [pp.430–431, 432]	
d. Radiation [p.431]	
e. Immunity breakdowns [p.431]	

23.4. DIAGNOSING CANCER [p.433]

23.5. TREATING AND PREVENTING CANCER [p.434]

Selected Words: *DNA probe* [p.433], *adjuvant therapy* [p.434], *chronic myelogenous leukemia* [p.434], cancer "vaccines" [p.440]

Boldfaced, Page-Referenced Terms

[p.433] tumor markers _____

_____

[p.433] biopsy _____

_____

[p.434] chemotherapy _____

_____

[p.434] immunotherapy _____

_____

Completion

Complete the following statements of the seven warning signs of cancer. [p.433]

1. Change in _____ or _____ habits and function.
2. A _____ that does not heal.
3. Unusual _____ or _____ discharge.
4. Thickening or _____ .
5. _____ or difficulty swallowing.
6. Obvious change in a _____ or _____ .
7. Nagging _____ or _____ .

Matching

Choose the most appropriate description for each term.

8. _____ interleukins [p.434]

9. _____ medical imaging [p.433]

10. _____ monoclonal antibodies [pp.433,434]

11. _____ tumor markers [p.433]

12. _____ adjuvant therapy [p.434]

13. _____ immunotherapy [p.434]

14. _____ cancer screening [p.433]

15. _____ interferon [p.434]

16. _____ biopsy [p.433]

17. _____ chemotherapy [p.434]

18. _____ cancer "vaccines" [p.434]

19. _____ DNA probe [p.433]

A. Treatment with substances that trigger a strong immune response against cancer cells
B. The use of drugs to kill cancer cells
C. Definitive cancer detection tool; a small piece of tissue is removed from the body and analyzed
D. Substances produced by cancer cells or normal cells in response to cancer; detected in blood tests
E. Substances that stimulate cytotoxic T cells to recognize and destroy body cells with abnormal surface proteins
F. Diagnosis using MRI, X rays, ultrasound, and CT
G. A treatment combining surgery and a less toxic dose of chemotherapy
H. Examples are mammograms, Pap test, testicle self-examination, and breast self-examination
I. "Magic bullets" that deliver lethal doses of radiation or anticancer drugs to tumor cells
J. Signaling molecules produced by lymphocytes; cytotoxic T cells grown in IL-2 become activated
K. Activates cytotoxic T cells and natural killer cells; they recognize and kill various types of cancer
L. A snippet of radioactively labeled DNA used to locate gene mutations or alleles associated with some types of inherited cancers

23.6. SOME MAJOR TYPES OF CANCER [p.435]

23.7. CANCERS OF THE BREAST AND REPRODUCTIVE SYSTEM [pp.436–437]

23.8. A SURVEY OF OTHER COMMON CANCERS [pp.438–439]

Selected Words: *sarcomas* [p.435], *carcinomas* [p.435], *adenocarcinoma* [p.435], *lymphomas* [p.435], *leukemias* [p.435], *modified radical mastectomy* [p.435], *lumpectomy* [p.435], *Pap smear* [p.437], *squamous cell carcinomas* [p.438], *adenocarcinomas* (or *large-cell carcinomas*) [p.438], *small-cell carcinoma* [p.438], *Wilms tumor* [p.438], *non-Hodgkin lymphoma* [p.438], *Hodgkin's disease* [p.438], *Burkitt lymphoma* [p.438]

Fill-in-the-Blanks

Cancers of connective tissues (muscle and bone) are (1) _____ [p.435]. Cancers that arise

from epithelium, including cells of the skin and epithelial linings of internal organs, are known as

(2) _____ [p.435]. (3) _____ [p.435] begin in the body or ducts of a gland.

(4) _____ [p.435] are cancers of lymphoid tissues in organs such as lymph nodes. Cancers

of blood-forming regions such as stem cells in bone marrow are (5) _____ [p.435].

Choice

For questions 6–37, choose from the following major types of cancer:

a. breast   b. uterine and ovarian   c. testis and prostate   d. oral and lung

e. digestive system and related organs   f. urinary system   g. blood and lymphatic system   h. skin

6. _____ Risk factors are fair skin and exposure to UV radiation in sunlight [p.439]

7. _____ Hodgkin's disease [p.438]

8. _____ Warning signs include a change in bowel habits, rectal bleeding, and blood in the feces [p.438]

9. _____ Most common among smokers and people who use smokeless tobacco, especially if they are also heavy alcohol drinkers [p.438]

10. _____ About one woman in nine develops this type of cancer [p.436]

11. _____ Kills more people than any other type of cancer [p.438]

12. _____ Non-Hodgkin lymphoma [p.438]

13. _____ These cancers are usually adenocarcinomas of duct cells and often not detected until they have spread to other organs [p.438]

14. _____ Develop in lymphoid tissues [p.438]

15. _____ Cancers in which stem cells in bone marrow overproduce white blood cells [p.439]

16. _____ Malignant melanoma is the most dangerous kind and is a cancer of melanin-producing cells [p.439]

17. _____ Cancers associated with heavy alcohol consumption and a diet rich in smoked, pickled, and salted foods [p.438]

18. _____ Burkitt lymphoma [p.438]

19. _____ Nonsmokers, particularly children and spouses inhaling secondhand tobacco smoke, have a significant risk for this cancer [p.438]

20. _____ The second leading cause of cancer deaths in men after lung cancer [p.437]

21. _____ Self-examination should occur every month, about a week after the menstrual period [p.436]

22. _____ This cancer is often lethal because symptoms, mainly an enlarged abdomen, do not appear until the cancer is advanced and has metastasized [p.437]

23. _____ A blood test called the PSA test screens for tumor marker associated with this type of cancer [p.437]

24. _____ This cancer has now surpassed breast cancer as the leading cancer killer of women [p.438]

25. _____ Squamous cell carcinomas, adenocarcinomas, large-cell carcinomas, and small-cell carcinomas [p.438]

26. _____ Wilm's tumor, one of the most common of all childhood cancers [p.438]

27. _____ Basal cell carcinomas and squamous cell carcinomas usually are easily treated by minor surgery [p.439]

28. _____ The part of this system affected by cigarette smoking is the pancreas [p.438]

29. _____ Lumpectomy and modified radical mastectomy are treatments [p.436]

30. _____ Risk seems to increase with infection — such as HIV — that impairs immune system function [p.438]

31. _____ Drugs like tamoxifen are increasingly used to shrink these tumors [p.436]

32. _____ Pap smear tests cervical cells to detect precancerous phases [p.437]

33. _____ Chemotherapy using two compounds — vincristine and vinblastine — derived from species of periwinkle plants is effective [p.439]

34. _____ Warning signals include a nagging cough, shortness of breath, chest pain, bloody sputum, unexplained weight loss, and frequent respiratory infections or pneumonia [p.438]

35. _____ Risk factors include Down syndrome, exposure to chemicals such as benzene, and radiation exposure [p.439]

36. _____ Some types are the most common childhood cancers [p.439]

Self-Quiz

Multiple Choice

_____ 1. A defined mass of tissue known as a *tumor* can also be called a _____ . [p.428]
   a. benign tumor
   b. neoplasm
   c. malignant tumor
   d. capsule

_____ 2. Cancer cells that break away from a primary tumor and migrate to establish new cancer sites are said to be undergoing _____ . [p.429]
   a. activation
   b. dysplasia
   c. carcinogenesis
   d. metastasis

_____ 3. Of the following, which one is *not* characteristic of cancer cells? [pp.428–429]
   a. When a cancer cell divides, its daughter cells are normal
   b. They exhibit uncontrolled growth
   c. Metastasis
   d. They lack strong cell-to-cell adhesion

_____ 4. The p53 gene _____ . [p.430]
   a. initiates metastasis
   b. mutates and becomes a malignant cell
   c. is a suppressor gene effective in many tissue types
   d. is a type of oncogene

_____ 5. A DNA segment capable of inducing cancer in a normal cell is a(n) _____ . [p.430]
   a. tumor suppressor gene
   b. oncogene
   c. p53 gene
   d. proto-oncogene

_____ 6. Carcinogenesis is _____ . [p.430]
   a. the transformation of a normal cell into a cancerous one
   b. the production of carcinogens
   c. the dying process of tumor suppressor genes
   d. a breakdown in immunity that leads to cancer

_____ 7. Of the following, which may serve as a trigger for carcinogenesis? [pp.430–431]
   a. Viruses
   b. Chemicals
   c. Radiation
   d. Heredity
   e. All the above can trigger carcinogenesis

_____ 8. _____ is a cancer-screening technique. [p.439]
   a. Chemotherapy
   b. MRI
   c. Immune therapy
   d. Digital rectal examination
   e. Stimulation of cytotoxic T cells to recognize and destroy cancer cells

_____ 9. Adenocarcinomas are cancers of _____ . [p.435]
   a. lymph nodes
   b. connective tissues
   c. blood-forming regions
   d. a gland or its ducts
   e. epithelium

Choice

For questions 10–15, select the correct type of cancer from the following list:

a. oral and lung cancer   b. cancer of the skin and epithelial linings of internal organs
c. cancer of the digestive system and related organs   d. cancer of the blood and lymphatic system
e. female breast cancer; cancer of male and female reproductive system

10. PSA is a test for _____ . [p.437]
11. A malignant melanoma is a(n) _____ . [p.439]
12. Low-dose mammography detects, and lumpectomy may treat, _____ . [p.436]
13. Hodgkin's disease is a(n) _____ . [p.438]
14. Smoking and smokeless tobacco cause _____ . [p.438]
15. Leukemia is a(n) _____ . [p.439]

Chapter Objectives/Review Questions

This section lists general and detailed chapter objectives that can be used as review questions. You can make maximum use of these items by writing answers on a separate sheet of paper. Fill in answers where blanks are provided. To check for accuracy, compare your answers with information given in the chapter or glossary.

1. If cells overgrow, the result is a defined mass of tissue called a(n) _____ . [p.428]
2. _____ is an abnormal change in the sizes, shapes, and organization of cells in a tissue. [p.428]
3. List the characteristics that distinguish benign and malignant tumors. [p.428]
4. Briefly comment on the meaning of the following characteristics of cancer cells: structural abnormalities, uncontrolled growth, and metastasis. [pp.428–429]
5. Write a definition for the term *proto-oncogene*. [p.430]
6. A(n) _____ is a DNA segment that can induce cancer in a normal cell. [p.430]
7. In addition to an oncogene, the onset of malignant cancer requires the mutation of at least one _____ _____ gene. [p.430]
8. Explain the importance of the p53 gene. [p.430]
9. List various factors that may trigger expression of an oncogene. [p.430]
10. List five routes to carcinogenesis other than oncogenes. [pp.430–431]
11. Prepare brief statements that relate cancer to agricultural and industrial chemicals. [p.432]
12. Define *cancer screening, tumor markers, biopsy,* and *medical imaging.* [p.433]
13. List the recommended cancer-screening tests. [p.433]
14. _____ is the use of drugs to kill dividing cells. [p.434]
15. Why is loss of hair cells, stem cells, lymphocytes, and epithelial cells a side effect of chemotherapy? [p.434]
16. List reasons why immunotherapy holds promise for cancer treatment. [p.434]
17. Identify these types of cancers by their source tissue: sarcomas, carcinomas, adenocarcinomas, lymphomas, and leukemias. [p.435]

Integrating and Applying Key Concepts

1. A large number of chemicals, environmental pollutants, and several types of radiation are recognized as being carcinogenic. Can you think of a common effect these various and different carcinogens might have on a cell to induce cancer?
2. Why do you think that cancer is more common in older people than in younger people?

24

PRINCIPLES OF EVOLUTION

Interactive Exercises

CHAPTER INTRODUCTION [p.443]

24.1. A LITTLE EVOLUTIONARY HISTORY [p.444]

24.2. INDIVIDUAL VARIATIONS — A KEY EVOLUTIONARY IDEA [p.445]

Selected Words: <u>Homo</u> <u>sapiens</u> [p.443], HMS *Beagle* [p.444], "natural selection" [p.444], *morphological* traits [p.445], *physiological* traits [p.445], *behavioral* traits [p.445]

Boldfaced, Page-Referenced Terms

[p.443] evolution _____

_____

[p.443] genus (plural: genera) _____

_____

[p.445] microevolution _____

_____

[p.445] macroevolution _____

_____

[p.445] population _____

_____

[p.445] gene pool _____

_____

Complete the Table

1. Several key figures and events in the life of Charles Darwin led him to his conclusions about natural selection and evolution. Summarize these influences by completing the following table. [p.444]

Key Figures/Events	Importance to Synthesis of Evolutionary Theory
a.	Botanist at Cambridge University who perceived Darwin's real interests; arranged for Darwin to become a ship's naturalist
b.	Institution that granted Darwin a degree in theology
c.	British ship that carried Darwin on a five-year voyage (as a naturalist) around the world
d.	Clergyman and economist who proposed that any population tends to outgrow its resources, so that its members must compete for what is available
e.	Evolutionary process that Darwin proposed after considering his observations and the ideas of other thinkers

Matching

Choose the most appropriate description for each term.

2. _____ gene pool [p.445]

3. _____ behavioral traits [p.445]

4. _____ evolution [p.443]

5. _____ macroevolution [p.445]

6. _____ microevolution [p.445]

7. _____ morphological traits [p.445]

8. _____ physiological traits [p.445]

9. _____ population [p.445]

A. Cumulative genetic changes that may give rise to new species
B. The collective genes with their alleles present in a population
C. Functions of body structures
D. A group of individuals of the same species occupying a given area
E. Responses to certain basic stimuli
F. The large-scale patterns, trends, and rates of change among groups of species
G. General form and function
H. Changes in lines of descent with time

24.3. MICROEVOLUTION: HOW NEW SPECIES ARISE [pp.446–447]

Selected Words: "survival of the fittest" [p.446], *founder effect* [p.446], "reproductively isolated" [p.447], *divergence* [p.447], <u>Homo</u> <u>habilis</u> [p.447], *gradualism* [p.447], *punctuated equilibrium* [p.447]

Boldfaced, Page-Referenced Terms

[p.446] theory of evolution by natural selection _____

_____

[p.446] adaptation _____

_____

[p.446] genetic drift _____

_____

[p.446] gene flow _____

_____

[p.446] species _____

_____

Dichotomous Choice

Circle one of two possible answers given between parentheses in each statement.

1. Change in the relative numbers of different alleles in a population because of chance alone is known as (mutation/genetic drift). [p.446]
2. The variant alleles in species result from heritable changes in DNA called (natural selection/mutations). [p.446]
3. The absence of type B blood in Native Americans can be explained by (natural selection/the founder effect). [p.446]
4. (Gene flow/Natural selection) is the difference in survival and reproduction that has occurred among individuals that differ in one or more traits. [p.446]
5. Allele frequencies can change as individuals leave a population or new individuals enter; this is referred to as (gene flow/genetic drift). [p.446]
6. Charles Darwin presented his ideas in a book called (*On the Origin of Species/The Origin of Man*). [p.446]
7. The major evolutionary process proposed by Charles Darwin was (natural selection/genetic drift). [p.446]
8. The emergence of new species through many small changes in form over long spans of time is called (punctuated equilibrium/gradualism). [p.447]
9. Two species are reproductively isolated when they have enough genetic differences that they cannot produce (offspring/fertile offspring). [p.447]
10. Reproductive isolation develops when (genetic drift/gene flow) between two populations stops. [p.447]
11. The accumulation of differences in allele frequencies among isolated populations is called (convergence/divergence). [p.447]
12. Through evolution over time, organisms come to have characteristics that suit them to conditions in a particular environment; this is (reproductive isolation/adaptation). [p.446]
13. The statement "Most evolutionary changes occur in bursts" refers to (punctuated equilibrium/gradualism). [p.447]
14. Different breeding seasons, different mating rituals, and changes in body structures that prevent mating serve as (hybridization/isolating) mechanisms that reduce the likelihood of successful interbreeding between populations. [p.447]

15. A unit of one or more populations of individuals that can interbreed under natural conditions and produce fertile offspring is a (founder/species). [p.446]
16. Dramatic environmental changes such as an ice age may be the driving force altering the physical environment of adapted populations of organisms; this supports the (gradualism/punctuated equilibrium) model of speciation. [p.447]

Complete the Table

17. Complete the following table to review the major microevolutionary forces.

Microevolution Process	Definition
a.	A heritable change in DNA [p.446]
b.	Chance fluctuation in allele frequencies over time, as in the founder effect [p.446]
c.	Change in allele frequencies as individuals leave or enter a population (immigrate or emigrate) [p.446]
d.	Increased survival and reproduction of some members of a population due to their having traits that are the "fittest" in the population [p.446]

Analyzing a Diagram

18. The following illustration depicts the divergence of one species into two as time passes. Answer the questions below the illustration. [p.447]

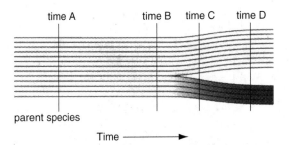

a. In what areas is there distinctly one species? _____ .
b. Between what letters does the divergence begin? _____ .
c. Is the divergence gradual or rapid if the time duration between B and D is hundreds of thousands of years? _____ .
d. What letter represents the time when divergence is probably complete? _____ .

Short Answer

19. Define the term *species* as you now understand it. [pp.446–447] _____

_____

_____

_____

24.4. LOOKING AT FOSSILS AND BIOGEOGRAPHY [pp.448–449]

24.5. COMPARING THE FORM AND DEVELOPMENT OF BODY PARTS [pp.450–451]

24.6. COMPARING BIOCHEMISTRY [p.452]

Selected Words: plate tectonics [p.449], "supercontinent" [p.449], *vestigial* [p.451]

Boldfaced, Page-Referenced Terms

[p.448] fossils _____

_____

[p.448] fossilization _____

_____

[p.449] biogeography _____

_____

[p.450] comparative morphology _____

_____

[p.450] homologous structures _____

_____

[p.450] analogous structures _____

_____

[p.450] morphological convergence _____

_____

[p.452] molecular clock _____

_____

Matching

Choose the most appropriate answer to match with examples of lines of evidence supporting macroevolution.

1. _____ analogous structures [p.450]

2. _____ biogeography [p.449]

3. _____ comparative biochemistry [p.452]

4. _____ comparative embryology [pp.450–451]

5. _____ comparative morphology [p.450]

6. _____ fossilization [p.448]

7. _____ homologous structures [p.450]

8. _____ morphological convergence [p.450]

9. _____ nucleic acid comparisons [p.452]

10. _____ plate tectonics [p.449]

11. _____ stratification [p.448]

12. _____ vestigial structures [p.451]

A. Embryos of all vertebrate lineages go through strikingly similar stages

B. An example of this pattern of change: compare forelimbs of a porpoise (flippers) with those of a penguin (wings)

C. The layering of sedimentary deposits (that may contain fossils)

D. Species of egg-laying mammals are found only in Australia, Tasmania, and New Guinea

E. Involves burial in sediments or volcanic ash, water infiltration, infusion of inorganic compounds, and added sediments with increasing pressure

F. Comparisons of body form with the purpose of reconstructing evolutionary history

G. Gene sequencing and amino acid sequence comparison

H. These structures arise when different lineages evolve in the same or similar environments and different body parts put to similar uses come to resemble one another

I. Organisms ranging from aerobic bacteria to corn plants to humans produce cytochrome c

J. All present continents, including Africa and South America, were parts of supercontinent Pangea

K. Ear-wagging muscles in humans, pelvic girdle in pythons, and the human coccyx

L. Wings of pterosaurs, birds, and bats

24.7. HOW SPECIES COME AND GO [pp.452–453]

24.8. *Focus on Our Environment:* ENDANGERED SPECIES [p.453]

24.9. FIVE TRENDS IN HUMAN EVOLUTION [pp.454–455]

Selected Words: "background extinction" [p.452], Homo habilis [p.453], Homo erectus [p.453], *prehensile* [p.454], *anthropoids* [p.455], *hominoids* [p.455], *hominids* [p.455]

Boldfaced, Page-Referenced Terms

[p.452] mass extinction _____

_____

[p.452] adaptive radiation _____

_____

[p.453] endangered species _____

_____

[p.454] bipedalism _____

[p.455] culture _____

Matching

Choose the most appropriate description for each term.

1. _____ adaptive radiation [p.452]
2. _____ habitat loss [p.453]
3. _____ endangered species [p.453]
4. _____ mass extinction [p.452]
5. _____ background extinction [p.452]

A. Major threat to over 90 percent of endangered species
B. An abrupt, widespread rise in extinction rates above the background level; a global event in which major groups of organisms are simultaneously wiped out
C. Species living in only one region and is extremely vulnerable to extinction
D. New species of a lineage fill a wide range of habitats during bursts of evolutionary activity
E. A relatively stable rate of species' disappearance over time

Dichotomous Choice

Circle one of two possible answers given between parentheses in each statement.

6. The habitual two-legged gait peculiar to the evolved, reorganized human skeleton is called (opposable movement/bipedalism). [p.454]
7. Compared with monkeys and apes, humans have a (longer/shorter), S-shaped, and somewhat flexible backbone. [p.454]
8. The position and shape of the human backbone, knee and ankle joints, and pelvic girdle are the basis of (bipedalism/prehensile movement). [p.454]
9. Two-hand movements developed in ancient (grassland/tree)-dwelling primates. [p.454]
10. Through alterations in hand bones, fingers could be wrapped around objects in (opposable/prehensile) movements. [p.454]
11. Further, the thumb and tip of each finger could touch in (opposable/prehensile) movements. [p.454]
12. Later, when (hominoids/hominids) were evolving, refinements in hand movements led to the precision grip and power grip that permitted unique technologies and cultural development. [p.454]
13. Later primates had (forward-directed eyes/an eye on each side of the head), which allows sampling shapes and movements in three dimensions. [p.454]
14. Through additional modifications, the eyes were able to respond to color variations and light intensity. These visual stimuli are typical of life in the (grasslands/trees). [p.454]
15. Humans have (bow-shaped/rectangular) jaws and (long canine teeth/smaller teeth of about the same length). [p.454]
16. On the road from early primates to humans, there was a shift from (eating insects, then fruit and leaves/a mixed diet) to (eating insects, then fruit and leaves/a mixed diet). [p.454]
17. In many primate lineages, parents started to invest more effort in (fewer/more) offspring. [p.455]
18. As parents formed stronger bonds with their young and maternal care became intense, the learning period grew (shorter/longer). [p.455]
19. The interlocking of brain modifications and behavioral complexity is most evident in the parallel evolution of the human (hand/brain) and culture. [p.455]
20. (Culture/Education) is the sum total of behavior patterns of a social group, passed between generations by learning and symbolic behavior — especially language. [p.455]

21. The capacity for (emotions/language) arose among ancestral humans through changes in the skull bones and expansion of parts of the brain. [p.455]
22. Primates evolved from ancestral (reptiles/mammals) more than 60 million years ago. [p.455]
23. The first primates resembled (dogs/small rodents). [p.455]
24. Between 54 and 38 million years ago, some primates stayed in the (grasslands/trees). [p.455]
25. (Living/Fossil) primates provide evidence that brain size increased, a shorter snout developed, daytime vision was enhanced, and grasping movements were refined. [p.455]
26. By 36 million years ago, tree-dwelling (hominoids/anthropoids) had evolved in tropical forests; they included ancestors of monkeys, apes, and humans. [p.455]
27. From 25 million to 5 million years ago, continents began to assume their current positions, climates became cooler and drier, and an adaptive radiation of apelike forms, the first (hominoids/hominids), occurred. [p.455]
28. By 13 million years ago, (human/hominoid) populations had spread widely. Many became extinct about this time. [p.455]
29. A few fossils, along with genetic studies, indicate that (two/three) lines of divergence occurred between 10 million and 5 million years ago. [p.455]
30. One such line gave rise to early (hominoids/hominids), including the ancestors of humans. [p.455]
31. *Homo sapiens* appeared about (40,000/100,000) years ago and gave rise to hominids, such as the Neandertals, that disappeared. [p.455]
32. Since the emergence of modern humans, evolution has been almost entirely (biological/cultural). [p.455]

24.10. EARTH'S HISTORY AND THE ORIGIN OF LIFE [pp.456–457]

Boldfaced, Page-Referenced Terms

[p.457] chemical evolution _____

_____

True/False

If the statement is true, write a T in the blank. If the statement is false, make it correct by writing the word(s) in the blank that should take the place of the underlined word(s).

_____ 1. The first atmosphere <u>contained</u> gaseous oxygen. [p.456]

_____ 2. Liquid <u>water</u> was essential for the formation of cell membranes. [p.456]

_____ 3. Without an <u>oxygen-rich</u> atmosphere, the organic compounds that led to life would never have formed. [p.456]

_____ 4. The first living cells emerged around <u>2.5 billion</u> years ago. [p.457]

_____ 5. The energy to drive chemical reactions that yielded organic molecules might have come from <u>aerobic respiration</u>. [p.457]

_____ 6. Complex compounds may have formed on <u>clay layers</u>. [p.457]

_____ 7. Complex compounds might have formed at <u>deep sea vents</u>. [p.457]

_____ 8. The first "molecule of life" was possibly not DNA but <u>ATP</u>. [p.457]

_____ 9. Membrane-bound sacs resembling cell membranes <u>have</u> been formed in laboratories. [p.457]

Self-Quiz

_____ 1. The term _____ refers to cumulative genetic changes that give rise to new species. [p.445]
   a. macroevolution
   b. mutation
   c. microevolution
   d. variation

_____ 2. _____ applies to the large-scale patterns, trends, and rates of change among groups of species. [p.445]
   a. Macroevolution
   b. Natural selection
   c. Microevolution
   d. Variation

_____ 3. Mutation, genetic drift, gene flow, and natural selection are the major _____ processes. [p.446]
   a. macroevolutionary
   b. population
   c. microevolutionary
   d. isolating

_____ 4. The idea of natural selection was the principal contribution of _____ . [p.446]
   a. Fox
   b. Oparin
   c. Miller
   d. Darwin

_____ 5. When the relative numbers of different alleles in a gene pool change randomly through the generations because of chance events alone, that change is called _____ . [p.446]
   a. natural selection
   b. genetic drift
   c. gene flow
   d. mutation

_____ 6. The source of new alleles for natural selection to work with is _____ . [p.446]
   a. natural selection
   b. genetic drift
   c. gene flow
   d. mutation

_____ 7. In the model known as _____ , new species emerge through many small changes in form over long expanses of time. [p.447]
   a. reproductive isolation
   b. punctuated equilibrium
   c. gradualism
   d. comparative evolution

_____ 8. Divergence may be the first stage on the road to _____ , the process by which species originate. [p.447]
   a. natural selection
   b. speciation
   c. the founder effect
   d. genetic variation

_____ 9. Related species remain alike in many ways; these shared but diverged characteristics of a common genetic plan are _____ . [p.450]
   a. analogous
   b. vestigial
   c. homologous
   d. adaptive

_____ 10. The physical record for the long history of life comes from _____ . [p.448]
   a. comparative morphology
   b. fossils
   c. comparative embryology
   d. comparative biochemistry

_____ 11. Modern studies of _____ indicate that all present-day continents were once part of the supercontinent called Pangea. [p.449]
   a. adaptive radiation
   b. plate tectonics
   c. meteor impact
   d. mass extinction

_____ 12. Many new mammal species arose and moved into habitats vacated by dinosaurs during a(n) _____ . [pp.452–453]
   a. adaptive radiation
   b. bottleneck
   c. meteor impact
   d. reverse extinction

_____ 13. When gene flow between two populations stops, it leads to _____ .
[p.447]
a. adaptive radiation
b. reproductive isolation
c. punctuated equilibrium
d. natural selection

_____ 14. Which of the following species existed
earliest? [pp.453,455]
a. *Homo sapiens* — Neandertals
b. *Homo erectus*
c. *Homo sapiens* — modern humans
d. *Homo habilis*

_____ 15. What two molecules that were not present on early Earth are essential for the beginning of life? [p.456]
a. H_2 and N_2
b. O_2 and N_2
c. H_2O and CO_2
d. O_2 and H_2O

Chapter Objectives/Review Questions

This section lists general and detailed chapter objectives that can be used as review questions. You can make maximum use of these items by writing answers on a separate sheet of paper. Fill in answers where blanks are provided. To check for accuracy, compare your answers with information given in the chapter or glossary.

1. _____ refers to genetic changes in lines of descent over time. [p.443]
2. Briefly describe Charles Darwin's early life and his contribution to evolutionary theory. [p.444]
3. Contrast the definitions of *microevolution* and *macroevolution*. [p.445]
4. List the three types of traits possessed by members of a population. [p.445]
5. Define the term *gene pool*. [p.445]
6. _____ are the source of new alleles, hence of the heritable variation in traits. [p.446]
7. Define *natural selection*. [p.446]
8. Define each of the following evolutionary forces: genetic drift, gene flow, reproductive isolating mechanisms. [pp.446–447]
9. The occurrence of chance events in bringing about changes in allele frequencies in small populations is known as _____ _____ . [p.446]
10. Name the unit whose individuals can interbreed under natural conditions and produce fertile offspring. [p.446]
11. Why does the explanation of life's history include both the punctuated equilibrium model and the gradualism model? [p.447]
12. Briefly define and cite an example of each of the following lines of evidence supporting macroevolution: the fossil record, biogeography, comparative morphology, homologous and analogous structures, comparative embryology, vestigial structures, comparative biochemistry, and nucleic acid comparisons. [pp.448–452]
13. A(n) _____ _____ is an abrupt, widespread rise in extinction rates above the background extinction level. [p.452]
14. In a(n) _____ _____ , new species of a lineage fill a wide range of habitats during bursts of microevolutionary activity. [p.452]
15. Name and briefly discuss the five trends in human evolution. [pp.454–455]
16. Define *culture*. [p.455]
17. Distinguish among anthropoids, hominoids, and hominids. [p.455]
18. Describe the conditions of early Earth and explain how they contributed to the origin of life. [p.456]

Integrating and Applying Key Concepts

1. Imagine that in the next decade, three more Chernobyl-type disasters happen, the oceans acquire critical levels of carcinogenic pesticides that work their way up the food chains, and the ozone layer shrinks dramatically in the upper atmosphere. Describe the macroevolutionary events that you believe might follow.

2. As Earth becomes increasingly loaded with carbon dioxide and various industrial waste products, how do you think life forms on Earth will evolve to cope with these changes?

25

ECOLOGY AND HUMAN CONCERNS

Interactive Exercises

CHAPTER INTRODUCTION [p.461]

25.1. SOME BASIC PRINCIPLES OF ECOLOGY [pp.462–463]

25.2. FEEDING LEVELS AND FOOD WEBS [pp.464–465]

25.3. HOW ENERGY FLOWS THROUGH ECOSYSTEMS [p.466]

Selected Words: "disturbed" habitats [p.462], *specialist* species [p.462], *generalist* species [p.462], *biotic* [p.462], "climax community" [p.462], *primary succession* [p.462], *secondary succession* [p.462], *autotrophs* [p.464], heterotrophs [p.464], *trophic levels* [p.464], *energy input* [p.474], *nutrient inputs* [p.474], biomass pyramid [p.466], energy pyramid [p.466]

Boldfaced, Page-Referenced Terms

[p.462] biosphere _____

_____

[p.462] ecology _____

_____

[p.462] habitat _____

_____

[p.462] community _____

_____

[p.462] niche _____

_____

[p.462] ecosystem _____

_____

[p.462] succession _____

_____

[p.464] producers _____

_____

[p.464] consumers _____

_____

[p.464] herbivores _____

_____

p.464] carnivores _____

_____

[p.464] omnivores _____

_____

[p.464] decomposers _____

_____

[p.464] food chain _____

_____

[p.464] food webs _____

_____

[p.466] primary productivity _____

_____

[p.466] ecological pyramid _____

_____

Matching

Choose the most appropriate description for each term.

1. _____ biosphere [p.462]

2. _____ ecology [p.462]

3. _____ habitat [p.462]

4. _____ community [p.462]

5. _____ niche [p.462]

6. _____ specialist species [p.462]

7. _____ generalist species [p.462]

8. _____ ecosystem [p.462]

9. _____ biotic [p.462]

10. _____ succession [p.462]

A. In any given habitat, the populations of all species that associate directly or indirectly

B. A community of organisms interacting with one another and with the physical environment

C. Consists of all the physical, chemical, and biological conditions the species requires to live and reproduce in an ecosystem

D. An orderly progression of species replacement in a community until a stable climax community is reached

E. The general type of place where a species normally lives

F. Have broad niches; can live in a range of habitats and eat many types of food

G. Have narrow niches; may be able to use only one or a few types of food or live only in one type of habitat

H. The regions of the earth's crust, waters, and atmosphere in which organisms live

I. Refers to the living portions of an ecosystem

J. The study of the interactions of organisms with one another and with the physical environment

Choice

For questions 11–14, choose from the following:

a. primary succession   b. secondary succession

11. _____ Successional changes begin when a pioneer species colonizes a newly available habitat. [p.462]

12. _____ A community develops toward the climax state after parts of the habitat have been disturbed. [p.462]

13. _____ This pattern occurs in abandoned fields in which wild grasses and other plants quickly take hold when cultivation stops. [p.462]

14. _____ Might occur on a recently deglaciated region. [p.462]

Fill-in-the-Blanks

Nearly every ecosystem runs on energy from the (15) _____ [p.464]. Plants and other photosynthetic organisms are (16) _____ [p.464], or "self-feeders." By securing energy from the physical environment, the self-feeders serve as the (17) _____ [p.464] for the entire system. All other organisms in the system are (18) _____ [p.464] that depend directly or indirectly on energy stored in the tissues of producers. Consumers are "other-feeders," or (19) _____ [p.464]. Within this group are (20) _____ [p.464], such as grazing animals and insects, which eat plants; (21) _____ [p.464], such as lions and snakes, which eat animals; (22) _____ [p.464], such as humans and bears, which feed on a variety of either plant or animal foods; and (23) _____ [p.464], such as fungi, bacteria, and worms, which get energy from the remains or products of organisms.

Choice

For questions 24–31, choose from the following:

a. primary producer   b. herbivore   c. primary carnivore   d. secondary carnivore   e. decomposer

24. _____ the only category lacking heterotrophs [p.464]
25. _____ gain energy directly from sunlight [p.464]
26. _____ a hawk that eats a snake [p.464]
27. _____ fungi and bacteria [p.464]
28. _____ organisms that prey on herbivores [p.464]
29. _____ green plants [p.464]
30. _____ autotrophs [p.464]
31. _____ snails, grasshoppers, and other plant-eaters [p.464]

Dichotomous Choice

Circle one of two possible answers given between parentheses in each statement.

32. Minerals carried by erosion into a lake represent nutrient (input/output). [p.464]
33. Energy from sunlight (can/cannot) be recycled. [p.464]
34. Ecosystems require a continual energy input from the (sun/heterotrophs). [p.464]
35. Nutrients typically (can/cannot) be reused. [p.464]
36. Each species in an ecosystem fits somewhere in a hierarchy of feeding relationships called (niche/trophic) levels. [p.464]
37. A linear sequence of who eats whom in an ecosystem is sometimes called a food (chain/web). [p.464]
38. The amount of energy actually stored in an ecosystem depends on how many (plants/animals) are present and on the balance between energy trapped by photosynthesis and energy used by the plants. [p.466]
39. In a harsh ecosystem environment, productivity would be expected to be (lower/higher) than in a mild ecosystem environment. [p.466]
40. In an ecological pyramid, the primary (producers/consumers) form a base for successive tiers of (producers/consumers) above them. [p.466]
41. A(n) (biomass/energy) pyramid depicts the weight of all of an ecosystem's organisms. [p.466]
42. The amount of available energy (increases/decreases) through successive feeding levels of an ecosystem. [p.466]

25.4. CHEMICAL CYCLES — AN OVERVIEW [p.467]

25.5. THE WATER CYCLE [p.468]

25.6. CYCLING CHEMICALS FROM THE EARTH'S CRUST [p.469]

Selected Words: "blooms" [p.469]

Boldfaced, Page-Referenced Terms

[p.467] biogeochemical cycle _____

_____

[p.468] water cycle _____

_____

[p.468] watershed _____

_____

[p.469] phosphorus cycle _____

_____

[p.469] eutrophication _____

_____

Analyzing Diagrams

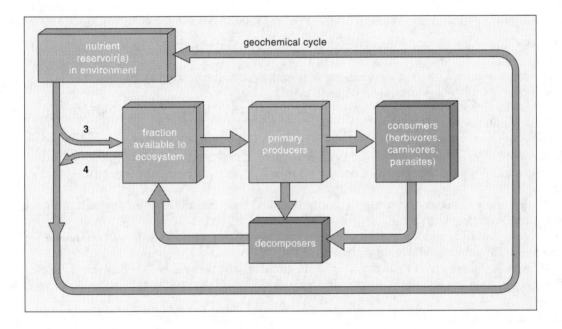

1. _____ The diagram represents the movement of nutrients from the physical environment, through organisms, and back to the environment. What is this known as? [p.467]

2. _____ The amount of nutrient that enters or leaves an ecosystem is far smaller than what cycles through the ecosystem. What component of the diagram is responsible for this internal cycling? [p.467]

3. _____ In addition to rainfall and snowfall, what else is included in the arrow labeled "3"? [p.467]

4. _____ Give an example of what accounts for the arrow labeled "4." [p.467]

5. _____ Which component of the diagram determines the type of biogeochemical cycle? [p.467]

6. _____ In the water cycle, water is the reservoir of what two substances? [p.467]

_____

7. _____ In the atmospheric cycle, the reservoir of carbon and nitrogen is in what form? [p.467]

8. _____ What is the major storehouse of nongaseous nutrients such as phosphorus? [p.467]

Matching

Choose the most appropriate phrase or description to match the following.

9. _____ solar energy [p.468]

10. _____ ocean [p.468]

11. _____ watershed [p.468]

12. _____ forms of precipitation [p.468]

13. _____ important role of plants [p.468]

14. _____ forms of atmospheric water [p.468]

15. _____ phosphorus [p.469]

16. _____ uplift and drainage [p.469]

17. _____ eutrophication [p.469]

A. Mostly rain and snow
B. Nutrient enrichment of a body of water resulting in algal blooms and other negative consequences, accelerated by runoff carrying phosphorus into water bodies
C. Take up minerals dissolved by water and prevent their loss in runoff
D. Water vapor, clouds, and ice crystals
E. Where precipitation of a specified region becomes funneled into a single stream or river
F. Process occurring as crustal plates move that makes minerals in seafloor sediments available
G. Slowly drives water through the atmosphere, on or through land mass surface layers, to oceans, and back again
H. Main reservoir of water
I. Earth's crust is the main storehouse of this mineral, calcium, potassium, and others; key component of many organic compounds

25.7. THE CARBON CYCLE [pp.470–471]

25.8. *Science Comes to Life:* GLOBAL WARMING [pp.472–473]

Selected Words: "greenhouse gases" [p.472]

Boldfaced, Page-Referenced Terms

[p.470] carbon cycle _____

_____

[p.472] greenhouse effect _____

_____

[p.472] global warming _____

_____

Matching

Choose the most appropriate description for each term/phrase.

1. _____ greenhouse gases [p.472]

2. _____ carbon cycle [p.470]

3. _____ way carbon enters the atmosphere [p.470]

4. _____ greenhouse effect [p.472]

5. _____ carbon dioxide (CO_2) [p.470]

6. _____ sediments, gas, petroleum, and coal [p.471]

A. Form of most of the atmospheric carbon
B. Aerobic respiration, fossil fuel burning, and volcanic eruptions
C. CO_2, H_2O, ozone, CFCs, methane, nitrous oxide
D. Buried reservoirs of carbon
E. Carbon moves from reservoirs in the atmosphere and oceans, through organisms, then back
F. Warming of Earth's lower atmosphere due to accumulation of greenhouse gases

Fill-in-the-Blanks

In the (7) _____ _____ [p.472], Earth's surface is warmed by (8) _____ [p.484] energy trapped in the lower atmosphere by "greenhouse gases." Human activity has resulted in high levels of these gases, which is most likely contributing to (9) _____ _____ [p.472]. A warmer planet would mean that (10) _____ [p.472] levels would rise so that waterfronts of major coastal cities would be submerged. (11) _____ [p.472] patterns would be disturbed, reducing crop yields in currently productive regions. If current warming trends continue, the Arctic (12) _____ [p.472] may disappear within a few decades. Scientists have found evidence that atmospheric levels of (13) _____ _____ [p.473] have increased. This buildup may intensify the greenhouse effect over the next century. The global burning of (14) _____ _____ [p.473] is probably the biggest contributor to global warming. (15) _____ [p.473] contributes to the problem because the burning of wood releases carbon, and there is less (16) _____ [p.473] biomass to absorb carbon dioxide in photosynthesis.

25.9. THE NITROGEN CYCLE [p.474]

25.10. *Focus on Our Environment:* THE DANGER OF BIOLOGICAL MAGNIFICATION [p.475]

Selected Words: "denitrification" [p.474]

Boldfaced, Page-Referenced Terms

[p.474] nitrogen cycle _____

_____

[p.474] nitrogen fixation _____

_____

[p.474] nitrification _____

_____

[p.475] biological magnification _____

_____

Choice

For questions 1–8, choose from the following:

    a. nitrogen cycle   b. nitrogen fixation   c. nitrogen   d. nitrification   e. denitrification

1. _____ Ammonia and ammonium in soil are converted to nitrite (NO_2); other bacteria metabolize nitrite and convert nitrite to nitrate (NO_3) [p.474]

2. _____ Bacteria are its key organisms — they convert nitrogen into forms plants can use as well as releasing nitrogen to complete the cycle [p.474]

3. _____ Component of proteins and nucleic acids [p.474]

4. _____ Atmospheric cycle; only certain bacteria can break the bonds of N_2 and convert nitrogen into forms that can enter food webs [p.474]

5. _____ Process in which a few kinds of bacteria convert N_2 to ammonia (NH_3), which dissolves quickly in water to form ammonium (NH_4^+) [p.474]

6. _____ Gaseous form makes up 80 percent of the atmosphere [p.474]

7. _____ Plants called legumes have mutually beneficial associations with bacteria that carry out this process [p.474]

8. _____ Bacteria convert nitrate or nitrite to N_2 (and a bit of nitrous oxide), which escapes into the atmosphere [p.474]

Fill-in-the-Blanks

During World War II, DDT was sprayed in the tropical Pacific to control (9) _____ [p.475] responsible for transmitting the organisms that cause dangerous malaria. In Europe, it was used to control (10) _____ _____ [p.475], which caused typhus. After the war, the use of DDT continued in many applications. DDT spreads through wind and water and accumulates in the fatty (11) _____ [p.475] of organisms. DDT can show (12) _____ _____ [p.475], in which there is an increase in the concentration of a nondegradable substance in organisms as it is passed through (13) _____ _____ [p.475]. As DDT spread through the environment after the war, it had disastrous effects. Where it was used to control Dutch elm disease, (14) _____ [p.475] began dying, as well as (15) _____ [p.475] in the streams flowing through the forests. In croplands where DDT was used to control one kind of pest, new kinds of (16) _____ [p.475] moved in. In addition, DDT was killing off the natural (17) _____ [p.475] that kept pest insect populations in check. Then, through biological magnification, species at the ends of (18) _____ _____ [p.475] were devastated, including bald eagles, peregrine falcons, ospreys, and brown pelicans. One consequence was that birds produced eggs with (19) _____ _____ [p.475], so that chick embryos didn't hatch, placing some species on the brink of (20) _____ [p.475]. DDT is now (21) _____ [p.475] in the United States, but DDT from 20 years ago is still contaminating parts of the (22) _____ [p.475].

25.11. HUMAN POPULATION GROWTH [pp.476–477]
25.12. NATURE'S CONTROLS ON POPULATION GROWTH [p.478]

Selected Words: bubonic plague [p.478], Yersinia pestis [p.478], density-independent controls [p.478]

Boldfaced, Page-Referenced Terms

[p.477] demographics _____

_____

[p.477] population size _____

_____

[p.477] population density _____

_____

[p.477] age structure _____

_____

[p.477] reproductive base _____

_____

[p.478] carrying capacity _____

_____

[p.478] logistic growth _____

_____

[p.478] density-dependent controls _____

_____

Dichotomous Choice

Circle one of two possible answers given between parentheses in each statement.

1. In nations where industrialization is in full swing, population growth (decreases/increases). [p.476]
2. When death rate exceeds birth rate, population size (stabilizes/decreases). [p.476]
3. The world population is expected to reach (six/nine) billion by the year 2050. [p.477]
4. Population (size/density) is the number of individuals per unit of area or volume. [p.477]
5. The general pattern of dispersal of members of a population through the habitat, that is, clumped together or spread out, is its (density/distribution). [p.477]
6. Dividing a population into prereproductive, reproductive, and postreproductive categories characterizes its (age/distribution) structure. [p.477]
7. The reproductive (rate/base) of a population refers to the number of individuals in the prereproductive and reproductive age structure categories. [p.477]
8. Most scientists believe that (overpopulation/industrialization) is the root of many, if not most, of the environmental problems the world now faces. [p.478]
9. The human population has been doubling, or experiencing (exponential/logistic) growth, since the mid-eighteenth century. [p.478]
10. When the course of logistic growth is plotted on a graph, a(n) (J-shaped/S-shaped) curve is obtained. [p.478]
11. Infectious diseases and parasites are a bigger problem in (low-density/high-density) populations. [p.478]

Matching

Select the best description for each term.

12. _____ density-independent controls [p.478]

13. _____ carrying capacity [p.478]

14. _____ limiting factor [p.478]

15. _____ logistic growth [p.478]

16. _____ density-dependent controls [p.478]

17. _____ population size [p.477]

18. _____ demographics [p.477]

19. _____ exponential growth [p.478]

A. A growth pattern in which a population's density increases, reaches overcrowded conditions, then decreases due to density-dependent controls
B. An essential resource such as food and water that, when in short supply, affects population growth
C. Number of individuals in a population's gene pool
D. An example is overcrowding, which results in competition for resources
E. Events such as natural disasters that cause deaths or births regardless of whether crowding exists
F. The number of individuals of a given species that can be sustained indefinitely by the resources in a given area
G. A population's vital statistics, such as its size, age structure, and density
H. Population growth in doubling increments

Graph Construction and Interpretation

Year	Estimated World Population
1650	500,000,000
1850	1,000,000,000
1930	2,000,000,000
1975	4,000,000,000
1987	5,000,000,000
1993	5,500,000,000
1995	5,700,000,000
1997	5,800,000,000
2000	6,000,000,000

20. Construct a graph of the data above, in the space provided on the next page.
    a. Estimate the year that the world contained 3 billion humans. _____ [p.476]
    b. Estimate the year that Earth will house 9 billion humans. _____ [p.476]
    c. Do you expect Earth to house 9 billion humans within your lifetime? [Check one] ☐ Yes ☐ No [p.476]

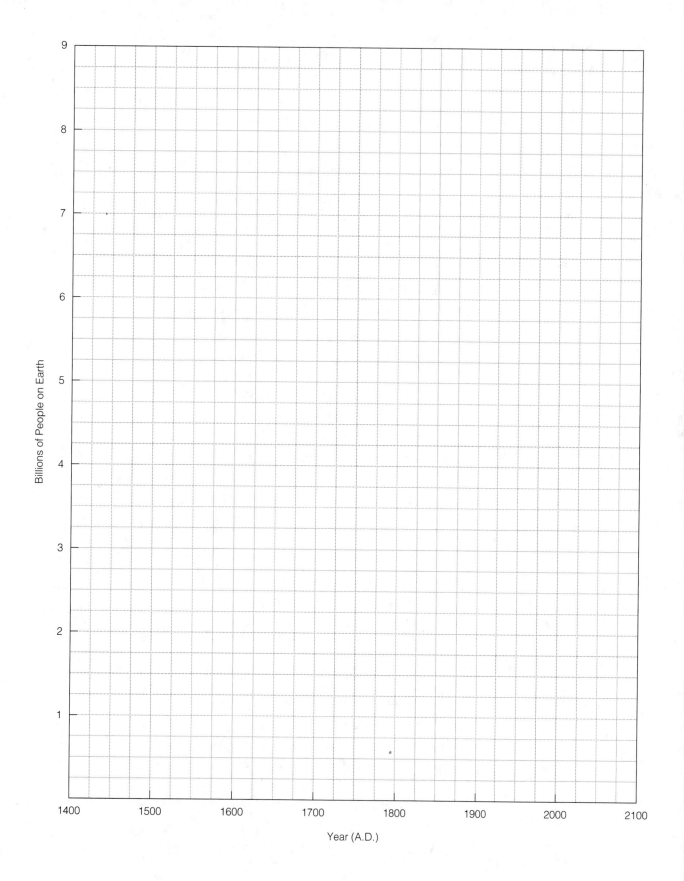

Billions of People on Earth

Year (A.D.)

25.13. ASSAULTS ON OUR AIR [p.479]

25.14. WATER, WASTES, AND OTHER PROBLEMS [pp.480–481]

Selected Words: ozone [p.479], ozone "hole" [p.479]

Boldfaced, Page-Referenced Terms

[p.479] pollutants _____

_____

[p.479] industrial smog _____

_____

[p.479] photochemical smog _____

_____

[p.479] acid rain _____

[p.480] salinization _____

_____

[p.480] green revolution _____

_____

[p.480] desertification _____

_____

[p.481] deforestation _____

_____

Choice

For questions 1–12, choose from the following aspects of atmospheric pollution.

a. chlorofluorocarbons   b. industrial smog   c. photochemical smog   d. acid rain   e. ozone layer

1. _____ Develops as a brown, smelly haze over large cities [p.479]

2. _____ Precipitation containing weak sulfuric and nitric acids [p.479]

3. _____ Contributes to ozone reduction more than any other factor [p.479]

4. _____ Where winters are cold and wet, this develops as a gray haze over many cities that burn coal and other fossil fuels [p.479]

5. _____ Substances from power plants and factories as well as from motor vehicles combine with water, producing a solution with a pH as low as 2.3 [p.479]

6. _____ Each year, from September through mid-October, it thins down by as much as half above Antarctica [p.479]

7. _____ Today most of this forms in cities of China, India, eastern Europe, and other developing countries [p.479]

8. _____ Contains airborne pollutants, including dust, smoke, soot, ashes, asbestos, oil, bits of lead and other heavy metals, and sulfur dioxide [p.479]

9. _____ Its reduction allows more harmful ultraviolet radiation to reach the earth's surface [p.479]

10. _____ Nitric oxide released from vehicles is exposed to sunlight and reacts with hydrocarbons to form harmful oxidants [p.479]

11. _____ Widely used as propellants in aerosol spray cans, refrigerator coolants, air conditioners, industrial solvents, and plastic foams; enter the atmosphere slowly and resist breakdown [p.479]

12. _____ CFCs and methyl bromide from fungicides are contributors to its depletion [p.479]

Fill-in-the-Blanks

There is a tremendous amount of water in the world, but most is too (13) _____ [p.480] for human consumption. Today, we produce about one-third of our food supply on (14) _____ [p.480] land. Irrigation water is often loaded with (15) _____ [p.480] salts. In regions with poorly draining soils, evaporation may cause (16) _____ [p.480]; such soils can stunt growth, decrease yields, and in time kill crop plants. Large-scale (17) _____ [p.480] is also a major factor in groundwater depletion. Farmers take so much water from the Ogallala (18) _____ [p.480] that the overdraft nearly equals the annual flow of the Colorado River. In coastal areas, overuse of groundwater can cause (19) _____ [p.480] intrusion into human water supplies. (20) _____ [p.480] runoff pollutes public water sources with sediments, pesticides, and fertilizers. Power plants and (21) _____ [p.480] pollute water with chemicals, radioactive materials, and excess heat. Many people view the (22) _____ [p.480] as convenient refuse dumps and dump untreated wastes there.

Matching

Choose the most appropriate description for each term.

23. _____ landfills [p.480]

24. _____ green revolution [p.480]

25. _____ deforestation [p.481]

26. _____ recycling [p.480]

27. _____ forested watersheds [p.481]

28. _____ desertification [p.480]

A. More people are participating in this beneficial alternative to landfills

B. Act like giant sponges that absorb, hold, and gradually release water

C. Conversion of large tracts of grasslands, rain-fed cropland, or irrigated cropland to a less productive, more desertlike state

D. Burying wastes in areas that may leak into groundwater supplies

E. Research directed toward improving crop plants for higher yields and exporting modern agricultural practices and equipment to developing countries

F. Removal of all trees from large land tracts; leads to loss of fragile soils and disrupts watersheds

25.15. CONCERNS ABOUT ENERGY [p.482]

25.16. *Focus on Our Environment:* A PLANETARY EMERGENCY: LOSS OF BIODIVERSITY [p.483]

Boldfaced, Page-Referenced Terms

[p.482] fossil fuels _____

_____

[p.482] meltdown _____

_____

Fill-in-the Blanks

(1) _____ [p.482] energy is the energy left over after subtracting the energy used to locate,

extract, transport, store, and deliver energy to consumers. Some sources of energy, such as direct

(2) _____ [p.482] energy, are renewable. Most of the energy used in (3) _____ [p.482]

countries comes from renewable resources. Energy supplies in the form of (4) _____ [p.482] fuels

are nonrenewable and dwindling, and their extraction and use come at high environmental cost.

Dichotomous Choice

Circle one of two possible answers given between parentheses in each statement.

5. Paralleling human population growth is a steep rise in total and per capita energy (conservation/consumption). [p.482]
6. World coal reserves can meet human energy needs for several centuries, but burning it has been the main source of (ozone depletion/air pollution). [p.482]
7. Compared to coal-burning plants, a nuclear plant emits (less/more) radioactivity and pollutants. [p.482]
8. The greatest hazard in the use of nuclear energy as an energy supply during normal operation is with potential (radioactivity escape/meltdown). [p.482]
9. Low-cost (wind harnesses/solar cells) can use sunlight energy to generate electricity and produce hydrogen gas. [p.482]
10. Another energy alternative is (fusion power/wind farms), in which the idea is to bring hydrogen atoms together to form helium atoms, a reaction that releases considerable energy. [p.482]
11. Human actions are leading to the premature extinction of at least six species per (hour/year). [p.483]
12. Globally, (pollution/tropical deforestation) is the greatest killer of species, followed by destruction of coral reefs. [p.483]
13. In the United States, we have cut down 92 percent of old-growth forests and drained (a fourth/half) of the wetlands. [p.483]
14. In (less/more) affluent countries, people use greater than the average amount of resources. [p.483]
15. In (less/more) affluent countries, there is overuse of land and animal resources. [p.483]

Self-Quiz

_____ 1. All the populations of different species that occupy and are adapted to a given habitat are referred to as a(n) _____ . [p.462]
   a. biosphere
   b. community
   c. ecosystem
   d. niche

_____ 2. The _____ of a species consists of all the physical, chemical, and biological conditions that the species needs to live and reproduce in an ecosystem. [p.462]
   a. habitat
   b. niche
   c. carrying capacity
   d. ecosystem

_____ 3. A network of interactions that involve the cycling of materials and the flow of energy between one or more communities and the physical environment is a(n) _____ . [p.462]
   a. population
   b. community
   c. ecosystem
   d. biosphere

_____ 4. _____ get energy from the remains or products of organisms. [p.464]
   a. Herbivores
   b. Parasites
   c. Decomposers
   d. Carnivores

_____ 5. A linear sequence of who eats whom in an ecosystem is sometimes called a(n) _____ . [p.464]
   a. trophic level
   b. food chain
   c. ecological pyramid
   d. food web

_____ 6. In a typical food web, the second trophic level includes primary consumers that are _____ . [p.464]
   a. herbivores
   b. carnivores
   c. scavengers
   d. decomposers

_____ 7. Succession involves the replacement of species by others as each changes the habitat, finally resulting in a more or less stable _____ . [p.462]
   a. climax community
   b. pioneer community
   c. secondary community
   d. pyramid community

_____ 8. A(n) _____ is a way to represent the relative amounts of energy available at each level of an ecosystem. [p.466]
   a. food chain
   b. food web
   c. productivity map
   d. ecological pyramid

_____ 9. In the carbon cycle, carbon enters the atmosphere through _____ . [p.470]
   a. carbon dioxide fixation
   b. aerobic respiration, fossil fuel burning, and volcanic eruptions
   c. oceans and accumulation of plant biomass
   d. evaporation

_____ 10. _____ is the name for the process in which certain bacteria convert N_2 to ammonia, which dissolves giving ammonium, a form that plants can pull into an ecosystem. [p.474]
   a. Nitrification
   b. Ammonification
   c. Denitrification
   d. Nitrogen fixation

_____ 11. _____ refers to an increase in concentration of a nondegradable (or slowly degradable) substance in organisms as it is passed along food chains. [p.475]
   a. Ecosystem modeling
   b. Nutrient input
   c. Biogeochemical cycle
   d. Biological magnification

_____ 12. The average number of individuals of the same species per unit area at a given time is the _____ . [p.477]
   a. population density
   b. population growth
   c. population birth rate
   d. population size

_____ 13. The maximum number of individuals of a population (or species) that can be sustained by a given environment defines _____ . [p.478]
   a. the carrying capacity
   b. exponential growth
   c. logistic growth
   d. density-independent factors

_____ 14. Which of the following is *not* characteristic of logistic growth? [p.478]
   a. S-shaped curve
   b. growth levels off as carrying capacity is reached
   c. unrestricted growth
   d. slow growth of a low-density population followed by rapid growth

_____ 15. Factors that are at work in crowded populations, and tend to reduce population size, are called _____ . [p.478]
   a. density-independent controls
   b. density-dependent controls
   c. limiting factors
   d. logistic factors

_____ 16. _____ result(s) when nitrogen dioxide and hydrocarbons react in the presence of sunlight. [p.479]
   a. Photochemical smog
   b. Industrial smog
   c. A thermal inversion
   d. Eutrophication

_____ 17. Sulfur and nitrogen dioxides dissolve in atmospheric water to form a weak solution of sulfuric acid and nitric acid; this describes the formation of _____ . [p.479]
   a. photochemical smog
   b. industrial smog
   c. ozone and PANs
   d. acid rain

_____ 18. For every million liters of water in the world, only about _____ liters are in a form that can be used for human consumption or agriculture. [p.480]
   a. 6
   b. 60
   c. 600
   d. 6,000

_____ 19. In theory, world reserves of _____ can meet the energy needs of the human population for at least several centuries. [p.480]
   a. carbon monoxide
   b. oil
   c. natural gas
   d. coal

_____ 20. Many biologists believe that extinction is a more serious problem than ozone depletion or global warming because it is happening faster and is _____ . [p.483]
   a. destroying food webs
   b. reducing food supplies for humans
   c. irreversible
   d. causing succession

Chapter Objectives/Review Questions

This section lists general and detailed chapter objectives that can be used as review questions. You can make maximum use of these items by writing answers on a separate sheet of paper. Fill in answers where blanks are provided. To check for accuracy, compare your answers with information given in the chapter or glossary.

1. Define and distinguish between *habitat* and *niche*. [p.462]
2. Distinguish between *primary succession* and *secondary succession*. [p.462]
3. Name and distinguish among the four types of consumers in food webs. [p.464]

4. Biomass pyramids are based on the _____ of all the ecosystem's members at each trophic level; _____ pyramids reflect the energy losses as energy flows through the trophic levels of an ecosystem. [p.466]
5. Give examples of biogeochemical cycles and tell what the reservoir is for each. [pp.467–471,474]
6. In the _____ cycle, these minerals move from land to sediments in the seas and then back to the land. [p.469]
7. Explain the greenhouse effect and the implications this has for life on Earth. [pp.472–473]
8. Define the chemical events that occur during nitrogen fixation, nitrification, ammonification, and denitrification. [p.474]
9. Explain the detrimental effects of biological magnification. [p.475]
10. Define *zero population growth* and explain why it is attained more often in industrialized countries. [p.476]
11. Describe the three stages of a population's age structure. [p.477]
12. Distinguish between exponential growth and logistic growth. [p.478]
13. Define limiting factors and tell how they influence carrying capacity. [p.478]
14. Define *density-dependent controls,* give two examples, and indicate how density-dependent factors act on populations. [p.478]
15. Define *density-independent controls* and give two examples. [p.478]
16. Identify the principal air pollutants, their sources, and their effects. [p.479]
17. Distinguish between industrial smog and photochemical smog. [p.479]
18. Describe the environmental conditions leading to acid rain and list its negative effects. [p.479]
19. Summarize the major problems contributing to contamination of the world's supply of pure water. [p.480]
20. Describe the process of desertification and its effects. [p.480]
21. Explain the repercussions of deforestation. [p.481]
22. List the disadvantages of the use of fossil fuels as energy sources. [p.482]
23. Explain what a meltdown is. [p.482]
24. Describe how our use of fossil fuels, solar energy, and nuclear energy affects ecosystems. [p.482]
25. Cite major reasons for the continuing loss of biodiversity. [p.483]

Integrating and Applying Key Concepts

1. Explain why some biologists believe that the endangered species list now includes all species.
2. Is there a *fundamental niche* that is occupied by humans? If you think so, describe the minimal abiotic and biotic conditions required by populations of humans in order to live and reproduce. (Note that "to thrive" and "to be happy" are not requirements.)

ANSWERS

Chapter 1 Learning about Human Biology

CHAPTER INTRODUCTION [p.1]

1.1. THE CHARACTERISTICS OF LIFE [p.2]
1. b; 2. a; 3. a; 4. c; 5. e; 6. e; 7. d; 8. b; 9. d; 10. f; 11. f; 12. a;
13. c; 14. f.

1.2. HUMANS IN THE LARGER WORLD [p.3]
1. diversity; 2. kingdoms; 3. evolutionary; 4. bacteria;
5. protistans; 6. fungi; 7. plants; 8. animals; 9. animals;
10. primates; 11. mammals; 12. vertebrates; 13. manual;
14. brain; 15. cultures.

1.3 LIFE'S ORGANIZATION [pp.4–5]
1. D; 2. C; 3. F; 4. F; 5. G; 6. I; 7. J; 8. B; 9. H; 10. A; 11. E;
12. I; 13. J; 14. C; 15. A; 16. F; 17. G; 18. H; 19. D; 20. B;
21. sun; 22. photosynthesis; 23. producers; 24. consumers;
25. decomposers; 26. recycling; 27. webs.

1.4. WHAT IS SCIENCE? [pp.6–7]

**1.5. SCIENCE IN ACTION: CANCER, BROCCOLI,
AND MIGHTY MICE** [p.8]
1. a. hypothesis; b. prediction; c. experiment; d. variable;
e. control group; 2. F; 3. A; 4. D; 5. B; 6. C; 7. E;
8. E — question; 9. D — hypothesis; 10. B — prediction;
11. F — test; 12. C — repeat testing; 13. A — conclusion;
14. inductive; 15. deductive.

1.6. SCIENCE IN PERSPECTIVE [p.9]
1.7. CRITICAL THINKING IN SCIENCE AND LIFE
  [p.10]
1.8. *Choices: Biology and Society:* **HOW DO YOU
  DEFINE DEATH?** [p.11]
1. F; 2. F; 3. T; 4. F; 5. T; 6. T; 7. F; 8. correlation; 9. cause;
10. critical; 11. opinion; 12. fact.

Self-Quiz
1. b; 2. d; 3. d; 4. c; 5. a; 6. b; 7. d; 8. e; 9. c; 10. d.

Chapter 2 The Chemistry of Life

CHAPTER INTRODUCTION [pp.14–15]

2.1. THE ATOM [p.16]
2.2. *Science Comes to Life:* **USING RADIOISOTOPES
  TO TRACK CHEMICALS AND SAVE LIVES** [p.17]
1. A; 2. B; 3. K; 4. M; 5. N; 6. I; 7. C; 8. O; 9. D; 10. G; 11. J;
12. E; 13. F; 14. H; 15. L; 16. a. hydrogen, H; b. carbon, C;
c. nitrogen, N; d. oxygen, O; e. sodium, Na; f. chlorine,
Cl; 17. atomic number; 18. mass numbers; 19. protons;
20. electrons.

2.3. WHAT IS A CHEMICAL BOND? [pp.18–19]
**2.4. IMPORTANT BONDS IN BIOLOGICAL
  MOLECULES** [pp.20–21]
2.5. *Focus on Your Health:* **ANTIOXIDANTS —
  FIGHTING FREE RADICALS** [p.22]
1. Both oxygen gas and water are molecules, since they
consist of two or more atoms joined together. Only water
is a compound, because the atoms are different elements.

2.

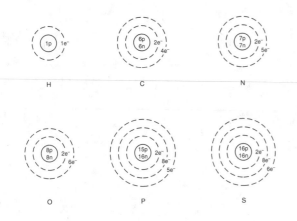

H   C   N

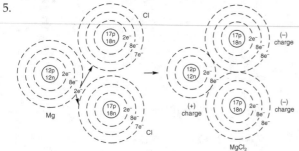

O   P   S

3. Helium will not form chemical bonds because it is inert, with no electron vacancies in its outer shell.
4. oxygen, carbon, hydrogen, and nitrogen; each has one or more electron vacancies in its outer shell
5.

Mg   MgCl₂
Cl

6. In a nonpolar covalent bond, atoms attract shared electrons equally. An example of a nonpolar covalent bond is that found in molecular hydrogen — H_2. In a polar covalent bond, atoms do not share electrons equally, so the bond is slightly positive at one end and slightly negative at the other. An example is water — H_2O.
7.

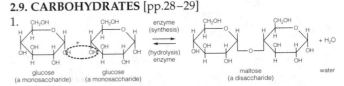

8. electrons; 9. shells; 10. orbitals; 11. energy levels; 12. lowest; 13. higher; 14. two; 15. eight; 16. chemical bond; 17. outer; 18. molecules; 19. compounds; 20. ion; 21. ionic bond; 22. covalent bond; 23. structural; 24. single; 25. two pairs; 26. triple; 27. electronegative; 28. hydrogen; 29. oxygen; 30. nitrogen; 31. hydrogen; 32. dashed; 33. slight; 34. molecules; 35. same; 36. DNA; 37. stabilize; 38. water; 39. Free radicals form by oxidation — the loss of one or more electrons to another atom. These highly unstable molecules have the potential to "steal" electrons from stable molecules, disrupting their structure and function. The effects of such damage to essential molecules in the body can be fatal; 40. An antioxidant is a chemical that

can give up an electron to a free radical before it causes damage. Antioxidants in our diet include ascorbic acid (vitamin C) and vitamin E, as well as carotenoids found in orange and leafy green vegetables.

2.6. LIFE DEPENDS ON WATER [p.23]
2.7. ACIDS, BASES, AND BUFFERS [pp.24–25]
1. two-thirds; 2. hydrogen; 3. hydrophilic; 4. hydrophobic; 5. heat; 6. do not reform; 7. losing; 8. solvent; 9. solutes; 10. dissolve; 11. pH; 12. hydrogen; 13. 7; 14. tenfold; 15. acid; 16. below; 17. base; 18. alkaline; 19. above; 20. buffer; 21. releases; 22. bicarbonate; 23. carbon dioxide; 24. acidosis; 25. alkalosis; 26. salts; 27. It raises the pH as magnesium ions and hydroxide ions are released. The hydroxide ions (OH^-) combine with the excess hydrogen ions (H^+); 28. a. 7.3–7.5, slightly basic; b. 5.0–7.0, slightly acid to neutral; c. 1.0–3.0, acid.

2.8. ORGANIC COMPOUNDS: BUILDING ON CARBON ATOMS [pp.26–27]
1. T; 2. <u>water</u> — nucleic acids; 3. <u>two</u> — four; 4. T; 5. <u>carbohydrate</u> — hydrocarbon; 6. <u>enzymes</u> — functional groups; 7. hydroxyl (sugars); 8. ketone (sugars); 9. amino (amino acids, proteins); 10. phosphate (phosphate compounds); 11. carboxyl (fats, sugars, amino acids); 12. aldehyde (sugars); 13. Enzymes are proteins that speed up metabolic reactions. Each enzyme facilitates a specific reaction required for either building or breaking apart large organic compounds.
14.

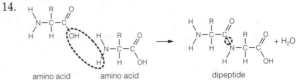

amino acid   amino acid   dipeptide

2.9. CARBOHYDRATES [pp.28–29]
1.

glucose   glucose   maltose   water
(a monosaccharide) (a monosaccharide)   (a disaccharide)

2. a. sucrose; b. glucose; c. cellulose; d. deoxyribose; e. lactose; f. glycogen; g. starch.

2.10. LIPIDS [pp.30–31]
1.

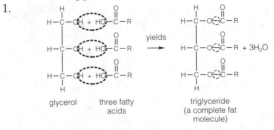

glycerol   three fatty   triglyceride
              acids   (a complete fat
                                             molecule)

2. a. unsaturated; b. saturated; 3. A phospholipid has two hydrophobic fatty acid tails attached to a glycerol backbone. The hydrophilic head consists of a phosphate group and another polar group. Phospholipids are the main structural component of cell membranes; 4. Unlike fats, sterols have no fatty acid tails. Although they differ from one another in number, position, and type of functional groups, all sterols have four carbon rings bonded together; 5. B; 6. E; 7. D; 8. A; 9. C.

2.11. PROTEINS: BIOLOGICAL MOLECULES WITH MANY ROLES [pp.32–33]
2.12. A PROTEIN'S FUNCTION DEPENDS ON ITS SHAPE [pp. 34–35]

1. A; 2. C; 3. B.

4.

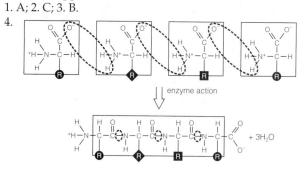

enzyme action

5. C; 6. A; 7. D; 8. B; 9. Lipoproteins; 10. Glycoproteins; 11. cells; 12. Protein denaturation is a change in the three-dimensional structure of a protein occurring as a result of a change in pH or an increase in temperature that breaks weak hydrogen bonds within the molecule.

2.13. NUCLEOTIDES AND NUCLEIC ACIDS [p.36]
2.14. *Focus on Our Environment:* FOOD PRODUCTION AND A CHEMICAL ARMS RACE [p.37]

1. B; 2. A; 3. C; 4. two;

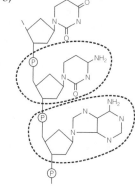

5. B; 6. A; 7. D; 8. C.

Self-Quiz

1. a. lipids; b. proteins; c. proteins; d. nucleic acids; e. carbohydrates; f. lipids; g. proteins; h. lipids; i. nucleic acids; j. lipids; k. proteins; l. carbohydrates; 2. a; 3. d; 4. c; 5. e; 6. e; 7. b; 8. c; 9. c; 10. b; 11. b.

Chapter 3   Cells

CHAPTER INTRODUCTION [p.41]

3.1. CELLS: ORGANIZED FOR LIFE [pp.42–43]
3.2. THE PARTS OF A EUKARYOTIC CELL [pp.44–45]
3.3. THE CYTOSKELETON: SUPPORT AND MOVEMENT [p.46]

1. microscopes; 2. Cell Theory; 3. cells; 4. life; 5. pre-existing; 6. B; 7. A; 8. C; 9. a. DNA; b. cytoplasm; c. plasma membrane; 10. Prokaryotic cells have no distinctly separated internal compartments; the DNA is loose in the cytoplasm. In the eukaryotic cell, there is distinct compartmentalization; the DNA is kept separate from the other cellular components by the nuclear envelope. The cell contains several membrane-bound organelles besides the nucleus; 11. Most cells are small because of the surface-to-volume ratio. If the cell were large, there would not be enough surface area for the necessary high rate of import and export of nutrients and wastes. The cell would die. To compensate, large cells are either long and thin or have folds to increase surface area; 12. The lipid bilayer has the hydrophilic heads of the phospholipids to the outside because the intracellular fluid and extracellular fluid are aqueous and therefore polar, like water. The tails are nonpolar and therefore hydrophobic and so tend to be attracted to each other. They form a "sandwich" of tails inside with the heads outside; 13. organelles; 14. membranes; 15. cytoskeleton; 16. microtubules; 17. microfilaments; 18. tubulins; 19. actin; 20. intermediate filaments; 21. movement; 22. cross-linking proteins; 23. flagella; 24. cilia; 25. centrioles; 26. basal body.

3.4. THE PLASMA MEMBRANE: A LIPID BILAYER [p.47]
3.5. THE CYTOMEMBRANE SYSTEM [pp.48–49]

1. The cell membrane must be "fluid" rather than a solid barrier so that substances can enter and leave the cell. The "mosaic" quality refers to the mixture of proteins and the lipid bilayer; 2. a. recognition proteins; b. lipid bilayer; c. transport proteins; d. adhesion proteins; e. receptor proteins; 3. E; 4. F; 5. B; 6. D; 7. C; 8. A; 9. a. ribosome; b. rough ER; c. rough ER vesicle; d. Golgi body; e. Golgi body vesicle; f. exocytosis into small intestine; 10. phospholipids; 11. proteins; 12. endoplasmic reticulum; 13. cytoplasm.

3.6. MOVING SUBSTANCES ACROSS MEMBRANES BY DIFFUSION AND OSMOSIS [pp.50–51]
3.7. OTHER WAYS SUBSTANCES CROSS CELL MEMBRANES [pp.52–53]
3.8. *Focus on Our Environment:* **REVENGE OF EL TOR** [p.53]

1. E; 2. A; 3. K; 4. B; 5. G; 6. C; 7. D; 8. I; 9. F; 10. L; 11. H; 12. J; 13. M; 14. T; 15. T; 16. greater than— equal to; 17. T; 18. equal— unequal; 19. hypertonic— hypotonic; 20. hypertonic— isotonic; 21. T; 22. hypotonic— hypertonic; 23. fungus— bacterium; 24. phagocytosis— exocytosis; 25. exocytosis— endocytosis; 26. diffusion; 27. lipid bilayer; 28. water; 29. lipid bilayer; 30. passive transport; 31. active transport; 32. transport protein.

3.9. THE NUCLEUS [p.54]
3.10. MITOCHONDRIA: THE CELL'S ENERGY FACTORIES [p.55]

1. G; 2. H; 3. D; 4. B; 5. C; 6. F; 7. A; 8. E; 9. ATP; 10. organic; 11. carbon dioxide; 12. water; 13. oxygen; 14. eukaryotic; 15. Chromatin is a cell's collection of DNA along with associated proteins, existing in an uncondensed state. A chromosome is just one DNA molecule with its associated proteins, condensed or not; 16. Mitochondria resemble bacteria in size. They have their own DNA and some ribosomes. They reproduce on their own.

3.11. METABOLISM: DOING CELLULAR WORK [pp.56–57]

1. C; 2. D; 3. E; 4. G; 5. K; 6. J; 7. F; 8. B; 9. H; 10. A; 11. I; 12. Enzymes speed up reactions by hundreds to millions of times. They are not permanently altered or used up during reactions. Usually they work for both anabolic reactions and catabolic reactions. Each enzyme reacts with a specific substrate; 13. C; 14. A; 15. D; 16. B; 17. anabolic; 18. 7.35–7.4; 19. anabolic; 20. proteins; 21. coenzymes; 22. energy; 23. metabolism; 24. food; 25. ATP; 26. three; 27. covalent; 28. phosphate; 29. phosphorylation; 30. ADP; 31. ATP; 32. ATP/ADP; 33. meta-

bolic pathway; 34. anabolic; 35. catabolic; 36. substrate; 37. reactant or precursor; 38. intermediates; 39. end products; 40. energy carrier; 41. coupled; 42. Enzymes; 43. substrate; 44. active site; 45. Coenzymes; 46. NAD^+; 47. FAD; 48. homeostasis.

3.12. HOW CELLS MAKE ATP [pp.58–59]
3.13. SUMMARY OF AEROBIC RESPIRATION [pp.60–61]
3.14. ALTERNATIVE ENERGY SOURCES IN THE BODY [pp.62–63]

1. aerobic; 2. covalent; 3. electrons; 4. glycolysis; 5. cytoplasm; 6. pyruvate; 7. two; 8. energy-requiring; 9. glucose; 10. PGAL; 11. four; 12. two; 13. intermediates; 14. mitochondria; 15. inner; 16. coenzyme A; 17. oxaloacetate; 18. CO_2; 19. two; 20. coenzymes; 21. electron transport system; 22. electron transport systems; 23. inner; 24. H^+; 25. ATP; 26. water; 27. 36; 28. glycolysis; 29. cytoplasm; 30. Krebs; 31. mitochondrion; 32. ATP; 33. electron transport; 34. inner; 35. Oxygen; 36. electrons; 37. water; 38. 36; 39. More than two-thirds of the blood glucose is used up by the brain. It is the preferred energy source of brain cells; 40. Glycogen is the storage form of sugar in animal cells. It is produced when there is an excess of glucose in the blood stream. It is stored in liver and muscle cells. It makes up only about 1% of the body's energy reserves; 41. Glucagon is a hormone secreted by the pancreas. It promotes the conversion of glycogen stored in the liver to glucose, which is released back into the blood for transport to other body tissues for metabolism; 42. 78%; triglycerides; 43. between meals as an alternative to glucose; 44. The energy in fatty acids is stored in the covalent bonds that bond the hydrogen atoms to the carbon chain. Glucose doesn't have all those bonds; 45. ammonia.

Self-Quiz

1. c; 2. c; 3. a; 4. b; 5. d; 6. c; 7. d; 8. c; 9. a; 10. c; 11. c; 12. d; 13. d; 14. d.

Chapter 4   Introduction to Tissues, Organ Systems, and Homeostasis

CHAPTER INTRODUCTION [p.65]

4.1. EPITHELIUM: THE BODY'S COVERING AND LININGS [pp.66–67]
4.2. CONNECTIVE TISSUE: BINDING, SUPPORT, AND OTHER ROLES [pp.68–69]
4.3. MUSCLE TISSUE: MOVEMENT [p.70]
4.4. NERVOUS TISSUE: COMMUNICATION [p.71]
4.5. *Science Comes to Life:* **THE HUMAN BODY SHOP: HOPE AND CONTROVERSY** [pp.72–73]

1. tissue; 2. organ; 3. organ system; 4. tissue; 5. epithelial; 6. connective; 7. muscle; 8. nervous; 9. organ; 10. organ;

11. internal environment; 12. homeostasis; 13. Epithelial; 14. basement; 15. connective; 16. soft connective; 17. specialized connective; 18. Muscle; 19. skeletal; 20. smooth; 21. cardiac; 22. nervous; 23. neurons; 24. a. stratified columnar; b. simple cuboidal; c. stratified cuboidal; d. pseudostratified columnar; e. simple squamous; f. stratified squamous; g. simple columnar; 25. Exocrine; 26. mucus; 27. saliva; 28. earwax; 29. oil; 30. milk; 31. digestive enzymes; 32. Endocrine; 33. hormones; 34. thyroid; 35. adrenal; 36. pituitary; 37. b; 38. c; 39. a; 40. b; 41. a; 42. c; 43. a; 44. c; 45. a; 46. a; 47. b; 48. c; 49. b; 50. b; 51. c; 52. a; 53. a; 54. b; 55. c; 56. a; 57. a; 58. c;

59. b; 60. a; 61. "Bionic" body parts are made from metal or plastic, while "bioartificial" parts consist of living cells growing inside a matrix of plastic fibers; 62. Cartilage or bone cells are "seeded" onto a plastic model made in the desired shape. The cells multiply and take on the model's shape, after which the plastic model biodegrades; 63. Stem cells from embryos are the most versatile, with the potential to give rise to any type of cell in the body.

4.6. CELL JUNCTIONS: HOLDING TISSUES TOGETHER [p.74]
4.7. MEMBRANES: THIN, SHEETLIKE COVERS [p.75]

1. tight junction (B); 2. adhering junction (A); 3. gap junction (C); 4. a. serous; b. synovial; c. cutaneous; d. mucous; 5. C; 6. D; 7. C; 8. B; 9. D; 10. A.

4.8. ORGAN SYSTEMS [pp.76–77]
4.9. THE SKIN: EXAMPLE OF AN ORGAN SYSTEM [pp.78–79]
4.10. *Focus on Our Environment:* SUN, SKIN, AND THE OZONE LAYER [p.79]

1. E; 2. C; 3. G; 4. I; 5. F; 6. B; 7. H; 8. J; 9. D; 10. K; 11. A; 12. cranial; 13. spinal; 14. thoracic; 15. abdominal; 16. pelvic; 17. midsagittal; 18. frontal; 19. anterior; 20. transverse; 21. inferior; 22. sun; 23. bacteria; 24. moisture; 25. temperature; 26. sensory; 27. D; 28. integument; 29. organ; 30. epidermis; 31. dermis; 32. hypodermis;

33. Fat; 34. stratified squamous epithelium; 35. keratin; 36. stratum corneum; 37. melanocytes; 38. ultraviolet; 39. skin color; 40. Langerhans; 41. Granstein; 42. dermis; 43. follicles; 44. sebaceous; 45. Tanning; 46. elastin; 47. immune; 48. proto-oncogenes; 49. cancer; 50. increasing; 51. 100; 52. sebaceous gland; 53. smooth muscle; 54. melanocyte; 55. hair shaft; 56. adipose (fat) cells; 57. sweat gland; 58. pressure receptor; 59. hair follicle; 60. nerve fiber; 61. hypodermis; 62. dermis; 63. epidermis.

4.11. HOMEOSTASIS: THE BODY IN BALANCE [pp.80–81]

1. Extracellular; 2. interstitial; 3. cells; 4. ions; 5. Homeostasis; 6. stability; 7. receptors; 8. stimulus; 9. integrator; 10. effectors; 11. "set points"; 12. Feedback; 13. negative; 14. positive; 15. stimulus (B); 16. receptor (E); 17. integrator (A); 18. effector (C); 19. response (D); 20. b; 21. a; 22. b; 23. b; 24. a; 25. b; 26. a.

Self-Quiz

1. connective (D) (g); 2. epithelial (G) (b); 3. muscle (I) (e); 4. muscle (J) (l); 5. connective (E) (d); 6. connective (L) (i); 7. connective (B) (k); 8. epithelial (H) (j); 9. connective (K) (f); 10. nervous (M) (c); 11. muscle (C) (a); 12. epithelial (F) (m); 13. connective (A) (h); 14. d; 15. c; 16. c; 17. a; 18. d; 19. c; 20. d; 21. b; 22. c.

Chapter 5   The Skeletal System

CHAPTER INTRODUCTION [p.85]

5.1. BONE — MINERALIZED CONNECTIVE TISSUE [pp.86–87]
5.2. THE SKELETON: THE BODY'S BONY FRAMEWORK [pp.88–89]

1. living; 2. nonliving; 3. compact; 4. spongy; 5. osteoblasts; 6. osteocytes; 7. epiphyseal plates; 8. bone; 9. calcium; 10. phosphate; 11. bone remodeling; 12. osteoblasts; 13. osteoclasts; 14. osteoporosis; 15. bone marrow; 16. blood cells; 17. 206; 18. ligaments; 19. tendons; 20. calcium; 21. phosphorus; 22. nutrient canal; 23. marrow cavity; 24. compact bone; 25. spongy bone; 26. dense connective tissue; 27. osteon; 28. blood vessel; 29. lacuna; 30. cranial bones; 31. clavicle; 32. sternum; 33. scapula; 34. radius; 35. carpals; 36. femur; 37. tibia; 38. tarsals; 39. metatarsals.

5.3. THE AXIAL SKELETON [pp.90–91]
5.4. THE APPENDICULAR SKELETON [pp.92–93]

1. parietal; 2. temporal; 3. occipital; 4. frontal; 5. maxilla; 6. mandible; 7. palatine; 8. zygomatic; 9. foramen magnum; 10. cervical; 11. thoracic; 12. lumbar; 13. sacrum; 14. coccyx; 15. axial; 16. sinuses; 17. Temporal; 18. sphenoid; 19. ethmoid; 20. parietal; 21. occipital; 22. mandible;

23. maxillary; 24. zygomatic; 25. lacrimal; 26. maxillary; 27. palatine; 28. vomer; 29. cervical; 30. thoracic; 31. lumbar; 32. sacrum; 33. coccyx; 34. intervertebral disks; 35. herniate; 36. ribs; 37. sternum; 38. organs; 39. breathing; 40. pectoral; 41. appendicular; 42. carpals; 43. metacarpals; 44. phalanges; 45. coxal; 46. tibia; 47. fibula; 48. tarsals; 49. metatarsals; 50. phalanges.

5.5. JOINTS — CONNECTIONS BETWEEN BONES [pp.94–95]
5.6. SKELETAL DISEASES, DISORDERS, AND INJURIES [pp.96–97]
5.7. *Science Comes to Life:* REPLACING JOINTS [p.97]

1. abduction; 2. adduction; 3. circumduction; 4. rotation; 5. flexion; 6. extension; 7. supination; 8. pronation; 9. Synovial; 10. synovial; 11. hingelike; 12. ball-and-socket; 13. Cartilaginous; 14. fibrous; 15. cartilage; 16. osteoarthritis; 17. rheumatoid arthritis; 18. Tendinitis; 19. sprain; 20. dislocated; 21. simple; 22. complete; 23. compound; 24. prosthesis.

Self-Quiz

1. d; 2. e and g; 3. a; 4. j; 5. i; 6. c; 7. h and k; 8. f; 9. b; 10. i; 11. c; 12. d; 13. b; 14. f; 15. a; 16. c; 17. a; 18. a; 19. b.

Chapter 6   The Muscular System

CHAPTER INTRODUCTION [p.101]

6.1. MUSCLES INTERACT WITH THE SKELETON [pp.102–103]
6.2. HOW MUSCLES CONTRACT [pp.104–105]
1. skeletal; 2. skeleton; 3. joints; 4. 600; 5. origin; 6. insertion; 7. pulls; 8. fibers; 9. connective; 10. tendon; 11. bone; 12. pairs; 13. opposition; 14. contracts; 15. relaxes; 16. contracts; 17. reciprocal innervation; 18. deltoid (C); 19. trapezius (H); 20. gluteus maximus (G); 21. biceps femoris (J); 22. gastrocnemius (A); 23. pectoralis major (F); 24. rectus abdominis (D); 25. sartorius (E); 26. quadriceps femoris (I); 27. tibialis anterior (B); 28. sarcomere; 29. myofibrils; 30. Z; 31. actin; 32. myosin; 33. heads; 34. sliding-filament; 35. cross-bridges; 36. contract; 37. outer sheath (connective tissue); 38. muscle cell bundles; 39. muscle cell; 40. myofibril; 41. sarcomere; 42. Z bands.

6.3. ENERGY FOR MUSCLE CELLS [p.106]
6.4. THE NERVOUS SYSTEM CONTROLS MUSCLE CONTRACTION [pp.106–107]
6.5. PROPERTIES OF WHOLE MUSCLES [pp.108–109]
6.6. *Focus on Your Health:* MUSCLE MATTERS [pp.110–111]
1. Phosphate is transferred from creatine phosphate to ADP in order to quickly provide ATP needed for a muscle contraction; 2. Glycogen stored in the muscle is used for the first 5–10 minutes. Then, glucose and fatty acids from the blood are used for about half an hour.

Beyond that, fatty acids are used; 3. Glycolysis produces a small amount of ATP when there is not enough oxygen for aerobic respiration to continue, such as during hard exercise; 4. Muscle fatigue and heavy breathing result from the oxygen debt that occurs when the muscles produce ATP by glycolysis, not aerobic respiration; 5. motor; 6. T tubules; 7. sarcoplasmic reticulum; 8. calcium; 9. troponin; 10. actin; 11. myosin cross-bridges; 12. calcium; 13. myosin; 14. neuromuscular junction; 15. axons; 16. synapse; 17. neurotransmitter; 18. contract; 19. motor neuron; 20. neuromuscular junction; 21. motor unit; 22. spinal cord; 23. muscle tension; 24. isometrically; 25. isotonically; 26. motor unit; 27. muscle twitch; 28. twitch; 29. temporal summation; 30. tetany; 31. fatigue; 32. all-or-none; 33. muscle tone; 34. "Slow"; 35. myoglobin; 36. capillaries; 37. slowly; 38. long; 39. "fast"; 40. rapidly; 41. short; 42. increase in the number and size of mitochondria, increase in number of capillaries supplying the muscle, and more myoglobin; 43. endurance; 44. Botox causes muscles associated with wrinkles to relax; 45. An anabolic steroid mimics the hormone testosterone. In men, side effects include acne, baldness, shrinking testes, infertility, atherosclerosis, and 'roid rage. In women, side effects include deepening of the voice, obvious facial hair, irregular menstrual periods, shrinking breasts, and enlarged clitoris.

Self-Quiz
1. d; 2. e; 3. a; 4. b; 5. c; 6. b; 7. d; 8. a; 9. e.

Chapter 7   Digestion and Nutrition

CHAPTER INTRODUCTION [p.115]

7.1. THE DIGESTIVE SYSTEM: AN OVERVIEW [pp.116–117]
7.2. CHEWING AND SWALLOWING: FOOD PROCESSING BEGINS [pp.118–119]
7.3. THE STOMACH: FOOD STORAGE, DIGESTION, AND MORE [p.120]
1. mouth (oral) cavity; 2. salivary glands; 3. stomach; 4. small intestine; 5. anus; 6. large intestine; 7. pancreas; 8. gallbladder; 9. liver; 10. esophagus; 11. pharynx; 12. a. mouth cavity; b. salivary glands; c. pharynx; d. esophagus; e. stomach; f. small intestine; g. pancreas; h. liver; i. gallbladder; j. large intestine; k. rectum; l. anus; 13. Mechanical processing and motility (of food along the digestive tract), secretion (of digestive enzymes and

other substances), digestion (or chemical breakdown of food into small molecules), absorption (of nutrients and fluids), and elimination (of undigested matter); 14. The mucosa is the inner lining of the tube. Surrounding the mucosa is the submucosa, a layer of connective tissue containing blood and lymph vessels and networks of nerve cells. Next is the smooth muscle, one circular and one lengthwise layer. The outermost layer is the serosa, a thin, serous membrane; 15. C; 16. E; 17. B; 18. I; 19. H; 20. J; 21. A; 22. D; 23. G; 24. F; 25. pharynx; 26. involuntary; 27. epiglottis; 28. Heimlich; 29. stomach; 30. small intestine; 31. Gastric; 32. pepsins; 33. intrinsic; 34. chyme; 35. Protein; 36. denatures; 37. pepsins; 38. gastrin; 39. "gastric mucosal barrier"; 40. bacterium; 41. peptic ulcer; 42. peristalsis; 43. pyloric; 44. close; 45. small intestine; 46. rugae; 47. alcohol.

1. 20; 2. duodenum; 3. pancreas; 4. liver; 5. gallbladder; 6. peptide; 7. amino acids; 8. bicarbonate; 9. enzymes; 10. bile; 11. gallbladder; 12. emulsification; 13. nutrients; 14. surface area; 15. villi; 16. microvilli; 17. Absorption; 18. segmentation; 19. Transport; 20. Bile; 21. micelles; 22. triglycerides; 23. chylomicrons; 24. exocytosis; 25. blood; 26. lacteals; 27. hepatic portal vein; 28. hepatic vein; 29. glycogen; 30. bile; 31. gallbladder; 32. cholesterol; 33. gallstones; 34. urea; 35. I; 36. L; 37. H; 38. B; 39. N; 40. A; 41. G; 42. J; 43. M; 44. K; 45. F; 46. D; 47. C; 48. E; 49. G; 50. D; 51. E; 52. H; 53. B; 54. I; 55. F; 56. A; 57. C.

1. break apart; 2. fat; 3. glucose; 4. grain products; 5. 2–3 ounces; 6. 3 to 5; 7. teenager; 8. the tip — added fats and simple sugars add calories but have few vitamins and minerals; 9. carbohydrates; 10. fiber; 11. vitamins; 12. membranes; 13. energy; 14. cushion; 15. insulation; 16. vitamins; 17. liver; 18. saturated; 19. 8; 20. Complete; 21. incomplete; 22. young; 23. Vitamins; 24. 13; 25. inorganic; 26. 300,000 — diabetes, heart disease, hypertension; 27. 1950, 3320, 1200; 28. set point; 29. ob; 30. obese; 31. caloric; 32. activity; 33. A; 34. F; 35. D; 36. B; 37. K; 38. I; 39. H; 40. E; 41. C; 42. D; 43. G.

Self-Quiz
1. b; 2. b; 3. a; 4. e; 5. b; 6. c; 7. b; 8. c; 9. d; 10. d; 11. E; 12. D; 13. A; 14. G; 15. B; 16. C; 17. F.

Chapter 8   Blood

CHAPTER INTRODUCTION [p.139]

1. 4–5; 2. plasma; 3. water; 4. albumin; 5. water; 6. Plasma; 7. pH; 8. fluid volume; 9. Red blood; 10. hemoglobin; 11. stem cells; 12. white blood; 13. tissues; 14. bone marrow; 15. granulocytes; 16. neutrophils; 17. eosinophils; 18. basophils; 19. agranulocytes; 20. monocytes; 21. lymphocytes; 22. platelets; 23. clotting; 24. a. 91%–92%; b. Proteins; c. Neutrophils; d. Lymphocytes; e. Phagocytosis; f. Red blood; g. Platelets; 25. a. hemoglobin; b. oxyhemoglobin; c. more O_2, cooler, less acidic; d. lungs; e. less O_2, warmer, more acidic; f. tissues; g. hemoglobin; h. red; i. four; j. iron; k. binds oxygen; l. oxyhemoglobin; 26. stem; 27. red blood (E); 28. platelets (C); 29. neutrophil (A); 30. lymphocytes (D); 31. monocyte; 32. macrophage (B); 33. F; 34. G; 35. A; 36. B; 37. C; 38. D; 39. H; 40. E.

1. I; 2. J; 3. B; 4. A; 5. F; 6. G; 7. H; 8. C; 9. D; 10. E; 11. B; 12. E; 13. C; 14. D; 15. A; 16. F; 17. The series of reactions is triggered outside the blood by substances coming from damaged blood vessels or from surrounding tissues; 18. Both are clots that form in an unbroken blood vessel. A thrombus stays where it forms, while an embolus breaks free and circulates through the bloodstream. An embolus is more dangerous because it may shut down an organ's blood supply; 19. C; 20. D; 21. F; 22. B; 23. E; 24. A.

Self-Quiz
1. d; 2. e; 3. a; 4. e; 5. e; 6. a; 7. b; 8. d; 9. a; 10. e.

Chapter 9   Circulation — The Heart and Blood Vessels

CHAPTER INTRODUCTION [p.149]

9.1. THE CARDIOVASCULAR SYSTEM — MOVING BLOOD THROUGH THE BODY [pp.150–151]
9.2. THE HEART: A DOUBLE PUMP [pp.152–153]
9.3. THE TWO CIRCUITS OF BLOOD FLOW [pp.154–155]
9.4. *Science Comes to Life:* **HEART-SAVING DRUGS** [p.155]
9.5. HOW THE HEART CONTRACTS [p.156]

1. cardiovascular; 2. blood; 3. heart; 4. blood vessels; 5. arteries; 6. arterioles; 7. capillaries; 8. venules; 9. veins; 10. capillary; 11. lymphatic; 12. jugular; 13. superior; 14. pulmonary; 15. hepatic portal; 16. renal; 17. inferior; 18. iliac; 19. femoral; 20. femoral; 21. iliac; 22. abdominal; 23. renal; 24. brachial; 25. coronary; 26. pulmonary; 27. carotid; 28. myocardium; 29. pericardium; 30. endocardium; 31. endothelium; 32. septum; 33. atrium; 34. ventricle; 35. atrioventricular; 36. tricuspid; 37. bicuspid; 38. chordae tendineae; 39. atrium; 40. semilunar; 41. coronary; 42. aorta; 43. superior vena cava; 44. right semilunar valve; 45. right atrioventricular (or tricuspid) valve; 46. chordae tendineae; 47. inferior vena cava; 48. myocardium; 49. left ventricle; 50. arch of aorta; 51. D; 52. I; 53. F; 54. N; 55. C; 56. M; 57. A; 58. K; 59. E; 60. J; 61. G; 62. O; 63. B; 64. L; 65. D; 66. arteries 67. organs; 68. arteries; 69. arterioles; 70. capillaries; 71. venules; 72. veins; 73. superior vena cava; 74. inferior vena cava; 75. capillary beds; 76. sphincters; 77. relaxes; 78. contracts; 79. hepatic portal vein; 80. impurities; 81. hepatic; 82. oxygenated; 83. statins; 84. LDL; 85. HDLs; 86. triglycerides; 87. heart attacks; 88. brain strokes; 89. cardiac conduction; 90. intercalated discs; 91. sinoatrial; 92. atrioventricular; 93. Purkinje fibers; 94. SA.

9.6. BLOOD PRESSURE [p.157]
9.7. THE STRUCTURE AND FUNCTIONS OF BLOOD VESSELS [pp.158–159]

9.8. *Focus on Your Health:* **HEART-HEALTHY EXERCISE** [p.160]
9.9. EXCHANGES AT CAPILLARIES [pp.160–161]
9.10. CARDIOVASCULAR DISORDERS [pp.162–163]
9.11. THE MULTIPURPOSE LYMPHATIC SYSTEM [pp.164–165]

1. blood pressure; 2. systolic; 3. aorta; 4. diastolic; 5. aorta; 6. aorta; 7. hypertension; 8. hypotension; 9. a; 10. c; 11. b; 12. c; 13. a; 14. b; 15. c; 16. e; 17. c; 18. e; 19. a; 20. e; 21. e; 22. d; 23. artery, outer coat; 24. arteriole, endothelium; 25. capillary, basement membrane; 26. vein, smooth muscle with elastic fibers; 27. E; 28. C; 29. A; 30. B; 31. D; 32. capillaries; 33. T; 34. Proteins; 35. T; 36. T; 37. lymphatic; 38. Hypertension; 39. atherosclerosis; 40. heart attacks; 41. strokes; 42. smoking; 43. exercise; 44. obesity; 45. cholesterol; 46. heart failure; 47. hypertension; 48. age; 49. gender; 50. estrogen; 51. Hypertension; 52. elevated; 53. artery; 54. silent killer; 55. arteriosclerosis; 56. angina; 57. electrocardiogram; 58. angiography; 59. coronary bypass; 60. angioplasty; 61. balloon; 62. LDL; 63. atherosclerotic; 64. clot; 65. embolism; 66. homocysteine; 67. B; 68. Arrhythmias; 69. bradycardia; 70. tachycardia; 71. ventricular fibrillation; 72. tonsils; 73. thymus gland; 74. thoracic duct; 75. spleen; 76. lymph vessels; 77. lymph nodes; 78. lymphocytes; 79. C; 80. A; 81. E; 82. F; 83. B; 84. D; 85. Vessels act as *drainage* channels, collecting fluids that have leaked from capillary beds and returning them to the bloodstream. Vessels *deliver* fats absorbed from the small intestine to the bloodstream. The system transports foreign cells and cellular debris from body tissues to lymph nodes for *disposal*.

Self-Quiz
1. c; 2. b; 3. a; 4. e; 5. b; 6. a; 7. b; 8. e; 9. a; 10. e.

Chapter 10   Immunity

CHAPTER INTRODUCTION [p.169]

10.1. THREE LINES OF DEFENSE [p.170]
10.2. COMPLEMENT PROTEINS: PARTNERS IN OTHER DEFENSES [p.171]
10.3. INFLAMMATION — RESPONSES TO TISSUE DAMAGE [pp.172–173]

1. pathogens; 2. surface; 3. nonspecific; 4. tissue damage; 5. immune; 6. complement system; 7. CI; 8. molecules; 9. cascade; 10. membrane attack complexes; 11. lysis; 12. lysozyme; 13. inflammation; 14. white blood cells; 15. phagocyte; 16. stem cells; 17. hours; 18. Neutrophils; 19. ingest; 20. Eosinophils; 21. allergic; 22. Basophils; 23. inflammation; 24. macrophages; 25. inflammation; 26. capillaries; 27. histamine; 28. "leaky"; 29. neutrophils; 30. Macrophages; 31. phagocytes; 32. prostaglandins; 33. clotting.

10.4. AN IMMUNE SYSTEM OVERVIEW [pp.174–175]
10.5. HOW LYMPHOCYTES FORM AND DO BATTLE [pp.176–177]
10.6. ANTIBODIES IN ACTION [pp.178–179]

**10.7. CELL-MEDIATED RESPONSES —
COUNTERING THREATS INSIDE CELLS**
[pp.180–181]
10.8. *Focus on Your Health:* **ORGAN TRANSPLANTS:
BEATING THE (IMMUNE) SYSTEM** [p.181]
1. B and T cells are part of the body's third line of defense.
It is called upon when the first two lines of defense —
physical barriers and inflammation — don't stop an
invader; 2. Self/nonself recognition — B and T cells
attack only "nonself" substances or cells, Specificity —
B and T cells attack only specific invaders, Diversity —
B and T cells can respond to over a billion specific
threats, Memory — Some B and T cells are reserved for
a future battle with a specific invader; 3. D; 4. A; 5. C;
6. D; 7. b; 8. c; 9. a; 10. d; 11. b; 12. e; 13. c; 14. a; 15. c; 16. d;
17. a; 18. antigen; 19. T cells; 20. T; 21. cell-mediated; 22. B
cells; 23. T; 24. antigens; 25. variable regions; 26. shuffled;
27. antibody; 28. project; 29. naive; 30. thymus; 31. TCRs;
32. antigen-MHC; 33. antigen receptor; 34. clonal selection; 35. Immunological memory; 36. Memory cells;
37. terminated; 38. macrophages; 39. lymph; 40. B and T;
41. lymph; 42. lymph nodes; 43. spleen; 44. helper T;
45. immune response; 46. F; 47. C; 48. G; 49. D; 50. E;
51. B; 52. A; 53. E; 54. D; 55. B; 56. A; 57. C; 58. cells;
59. cell-mediated; 60. antigens; 61. general; 62. interferons; 63. MHC; 64. "touch-kill"; 65. perforins; 66. apoptosis; 67. rejection; 68. interleukins; 69. first; 70. foreign;
71. 75; 72. relatives; 73. suppress; 74. antibiotics; 75. pigs;
76. xenotransplantation; 77. eye; 78. testicles.

**10.9. PRACTICAL APPLICATIONS OF
IMMUNOLOGY** [pp.182–183]
10.10. *Science Comes to Life:* **A CAN'T WIN
PROPOSITION?** [p.183]
**10.11. ABNORMAL, HARMFUL, OR DEFICIENT
IMMUNE RESPONSES** [pp.184–185]
1. F; 2. G; 3. A; 4. E; 5. I; 6. H; 7. B; 8. C; 9. D; 10. Strains
of influenza have different surface antigens. In addition,
"gene swapping" increases variation within a strain;
11. allergy; 12. allergen; 13. predisposed; 14. antibodies;
15. mast; 16. histamines; 17. constrict; 18. anaphylactic
shock; 19. epinephrine; 20. Antihistamines; 21. IgG;
22. autoimmune; 23. rheumatoid arthritis; 24. diabetes;
25. systemic lupus erythematosus; 26. lymphocytes;
27. severe combined immune deficiency; 28. human
immunodeficiency virus; 29. acquired immunodeficiency
syndrome; 30. body fluids; 31. helper T; 32. macrophages;
33. cancer.

Self-Quiz

1. d; 2. b; 3. b; 4. a; 5. a; 6. e; 7. e; 8. e; 9. a; 10. a; 11. H;
12. D; 13. E; 14. J; 15. B; 16. G; 17. C; 18. I; 19. F; 20. A.

Chapter 11   The Respiratory System

CHAPTER INTRODUCTION [p.189]

**11.1. THE RESPIRATORY SYSTEM — BUILT FOR
GAS EXCHANGE** [pp.190–191]
1. oral cavity (H); 2. pleural membrane (J); 3. intercostal
muscles (F); 4. diaphragm (B); 5. nasal cavity (E);
6. pharynx (K); 7. epiglottis (D); 8. larynx (A); 9. trachea
(G); 10. lung (L); 11. bronchial tree (C); 12. bronchiole;
13. alveolar duct; 14. alveoli (I); 15. alveolar sac; 16. pulmonary capillaries; 17. F; 18. D; 19. G; 20. E; 21. B; 22. A;
23. C.

11.2. THE "RULES" OF GAS EXCHANGE [p.192]
11.3. *Focus on Our Environment:* **BREATHING AS A
HEALTH HAZARD** [p.193]
11.4. BREATHING — AIR IN, AIR OUT [pp.194–195]
1. Gas exchange; 2. pressure; 3. 21; 4. respiratory; 5. pressure gradient; 6. alveoli; 7. hemoglobin; 8. four; 9. releases; 10. gradient; 11. decreases; 12. hyperventilation;
13. increases; 14. N_2; 15. nitrogen narcosis; 16. hypoxia;
17. c; 18. b; 19. a; 20. b; 21. c; 22. a; 23. c; 24. b; 25. c; 26. b;
27. left; 28. right; 29. intercostals; 30. diaphragm; 31. down;
32. upward and outward; 33. expands; 34. up; 35. inward
and downward; 36. They recoil or relax; 37. thoracic
cavity; 38. pleural sac; 39. enters; 40. expiration; 41. out
of; 42. tidal volume; 43. vital capacity; 44. inspiratory;
45. expiratory; 46. residual; 47. one-half; 48. 350.

**11.5. HOW GASES ARE EXCHANGED AND
TRANSPORTED** [pp.196–197]
11.6. CONTROLS OVER GAS EXCHANGE
[pp.198–199]
11.7. *Focus on Your Health:* **TOBACCO AND OTHER
THREATS TO THE RESPIRATORY SYSTEM**
[pp.200–201]
1. H; 2. J; 3. B; 4. D; 5. K; 6. G; 7. C; 8. A; 9. E; 10. F; 11. L;
12. I; 13. 70; 14. more; 15. decreases; 16. more; 17. bicarbonate; 18. water; 19. falls; 20. lower; 21.oxygen; 22. nervous; 23. rhythm; 24. magnitude; 25. inspiration;
26. expiration; 27. pons; 28. oxygen; 29. carbon dioxide;
30. arteries; 31. diaphragm; 32. medulla; 33. cerebrospinal; 34. respiratory; 35. carotid; 36. aortic; 37. ventilation; 38. secondhand; 39. 80; 40. Each individual will
have his or her own responses.

Self-Quiz
1. c; 2. e; 3. a; 4. b; 5. a; 6. a; 7. a; 8. a; 9. c; 10. d.

Chapter 12   The Urinary System

CHAPTER INTRODUCTION [p.206]

**12.1. THE CHALLENGE: SHIFTS IN
EXTRACELLULAR FLUID** [pp.206–207]
12.2. *Focus on Our Environment:* **IS YOUR DRINKING
WATER POLLUTED?** [p.207]
**12.3. THE URINARY SYSTEM — BUILT FOR
FILTERING AND WASTE DISPOSAL** [pp.208–209]
1. The urinary system maintains stable conditions in the
extracellular fluid by removing substances that enter the
ECF from the intracellular fluid and adding needed
substances that move from the ECF into the intracellular
fluid; 2. a. absorption from liquids and solid food; b.
metabolic reactions; c. excretion in urine; d. evaporation
from lungs and skin; e. sweating; f. elimination in feces;
g. absorption from liquids and solid food; h. secretion
from cells; i. respiration; j. metabolism; k. urinary excretion; l. respiration; m. sweating; 3. brain; 4. urinary excretion; 5. T; 6. respiratory; 7. T; 8. urine; 9. T; 10. protein;
11. urea; 12. T; 13. kidney; 14. ureter; 15. urinary bladder;
16. urethra; 17. kidney cortex; 18. kidney medulla;
19. ureter; 20. kidneys; 21. nephrons; 22. solutes; 23. blood;
24. million; 25. afferent; 26. renal corpuscle; 27. glomerulus; 28. Bowman's; 29. proximal; 30. Henle; 31. distal;
32. efferent; 33. peritubular; 34. reabsorbed; 35. kidney
cortex; 36. kidney medulla; 37. glomerular capillaries;
38. proximal tubule; 39. distal tubule; 40. peritubular
capillaries; 41. collecting duct; 42. Bowman's capsule.

12.4. HOW URINE FORMS [pp.210–211]
12.5. KEEPING WATER AND SODIUM IN BALANCE
[pp.212–213]
12.6. *Focus on Your Health:* **DRINK TO THE HEALTH
OF YOUR URINARY TRACT** [p.213]

**12.7. ADJUSTING REABSORPTION: HORMONES
ARE THE KEY** [pp.214–215]
1. a. glomerulus; b. water, solutes; c. Bowman's, proximal;
d. nephron; e. out; f. capillaries; g. nephron; h. capillaries;
i. urine; 2. urinalysis; 3. urinary bladder; 4. relaxes; 5. external; 6. Kidney; 7. proximal tubule; 8. sodium "pumps";
9. tissue fluid; 10. amino acids; 11. out; 12. peritubular
capillaries; 13. medulla; 14. descending; 15. out;
16. sodium; 17. dilute; 18. aldosterone; 19. sodium;
20. ADH; 21. solutes; 22. blood pressure; 23. water;
24. small; 25. diuretic; 26. suppressing; 27. renin; 28. angio-
tensin; 29. aldosterone; 30. sodium; 31. less; 32. thirst
center.

**12.8. MAINTAINING THE BODY'S ACID–BASE
BALANCE** [p.216]
12.9. KIDNEY DISORDERS [p.217]
**12.10. MAINTAINING THE BODY'S CORE
TEMPERATURE** [pp.218–219]
1. 7.37, 7.43; 2. eliminate; 3. nephron; 4. urine; 5. homeostasis; 6. E; 7. A; 8. B; 9. C; 10. D; 11. 37, 98; 12. metabolic;
13. hypothalamus; 14. contract; 15. peripheral vasoconstriction; 16. pilomotor response; 17. shiver; 18. nonshivering heat production; 19. Hypothermia; 20. dive;
21. peripheral dilation; 22. Evaporative heat loss;
23. sweat; 24. electrolyte; 25. hyperthermia; 26. heat
exhaustion; 27. heat stroke.

Self-Quiz
1. d; 2. c; 3. e; 4. b; 5. d; 6. d; 7. d; 8. b; 9. e; 10. a.

Chapter 13   The Nervous System

CHAPTER INTRODUCTION [p.223]

**13.1. NEURONS — THE COMMUNICATION
SPECIALISTS** [pp.224–225]
13.2. A CLOSER LOOK AT ACTION POTENTIALS
[pp.226–227]
**13.3. CHEMICAL SYNAPSES: COMMUNICATION
JUNCTIONS** [pp.228–229]
1. E; 2. J; 3. B; 4. G; 5. C; 6. D; 7. A; 8. F; 9. H; 10. I; 11. B;
12. C; 13. F; 14. A; 15. E; 16. D; 17. dendrites; 18. cell body;
19. axon hillock; 20. axon; 21. axon ending; 22. negative;
23. resting membrane potential; 24. input zone; 25. action

potential; 26. resting membrane potential; 27. concentrations; 28. proteins; 29. diffuse; 30. gates; 31. closed;
32. electric; 33. active transport; 34. sodium–potassium
pumps; 35. T; 36. graded; 37. trigger; 38. T; 39. into; 40. T;
41. potassium; 42. away from; 43. neurotransmitters;
44. chemical synapse; 45. presynaptic; 46. vesicles;
47. postsynaptic; 48. inhibit; 49. dopamine; 50. synapses;
51. billion; 52. graded; 53. EPSPs; 54. IPSPs; 55. Synaptic
integration; 56. summation; 57. Integration; 58. removal;
59. enzymes; 60. serotonin; 61. B; 62. D; 63. F; 64. A; 65. C;
66. E; 67. action potential; 68. threshold; 69. resting membrane potential; 70. millivolts.

13.4. INFORMATION PATHWAYS [pp.230–231]
13.5. THE NERVOUS SYSTEM: AN OVERVIEW
[pp.232–233]
13.6. *Focus on Our Environment:* **AN ENVIRON-
MENTAL ASSAULT ON THE NERVOUS SYSTEM**
[p.233]
1. circuits; 2. "reverberating"; 3. nerve; 4. myelin sheath;
5. action potentials; 6. Schwann; 7. node; 8. jump; 9. central nervous system; 10. oligodendrocytes; 11. stretch
reflex; 12. reflex arc; 13. contracts; 14. axon; 15. myelin
sheath; 16. nerve fascicle; 17. blood vessels; 18. F; 19. E;
20. G; 21. D; 22. B; 23. C; 24. A; 25. The central nervous
system (CNS) is made up of the brain and spinal cord.
The peripheral nervous system (PNS) consists of nerves
spread through the rest of the body. It is divided into
somatic and autonomic subdivisions; 26. cranial;
27. spinal cord; 28. cervical; 29. thoracic; 30. lumbar;
31. sacral; 32. coccygeal.

**13.7. MAJOR EXPRESSWAYS: PERIPHERAL NERVES
AND THE SPINAL CORD** [pp.234–235]
13.8. THE BRAIN — COMMAND CENTRAL
[pp.236–237]
13.9. A CLOSER LOOK AT THE CEREBRUM
[pp.238–239]
13.10. MEMORY AND STATES OF CONSCIOUSNESS
[pp.240–241]
1. A; 2. S; 3. S; 4. A; 5. A; 6. S; 7. A; 8. Parasympathetic;
9. Sympathetic; 10. Parasympathetic; 11. Sympathetic;
12. norepinephrine; 13. gray matter; 14. gray matter;
15. meninges; 16. do not; 17. autonomic; 18. spinal cord;
19. ganglion; 20. nerve; 21. vertebra; 22. intervertebral

disk; 23. meninges; 24. gray matter; 25. white matter;
26. brain; 27. cranium; 28. meninges; 29. dura; 30. hemispheres; 31. cerebrospinal fluid; 32. brain stem; 33. gray
matter; 34. F, B; 35. F, A; 36. H, G; 37. H, H; 38. M, F; 39. F,
D; 40. H, E; 41. F, C; 42. reticular formation; 43. posture;
44. cerebral cortex; 45. cerebrospinal; 46. ventricles;
47. meninges; 48. blood–brain barrier; 49. lipid; 50. cerebrum; 51. T; 52. right; 53. T; 54. nerve tracts; 55. T;
56. cerebral cortex; 57. T; 58. learned 59. frontal; 60. parietal; 61. T; 62. T; 63. T; 64. self-gratifying; 65. thalamus;
66. hypothalamus; 67. corpus callosum; 68. midbrain;
69. pons; 70. medulla oblongata; 71. cerebellum; 72. short-
term; 73. long-term; 74. skills; 75. amygdala; 76. hippo-
campus; 77. prefrontal; 78. corpus striatum; 79. cerebellum; 80. Amnesia; 81. skills; 82. Parkinson's; 83. learning;
84. Alzheimer's; 85. PET; 86. reticular formation;
87. serotonin.

13.11. NERVOUS SYSTEM DISORDERS [p.242]
13.12. THE BRAIN ON DRUGS [pp.242–243]
1. C; 2. D; 3. H; 4. E; 5. G; 6. B; 7. A; 8. F; 9. drug; 10. Psychoactive; 11. receptors; 12. neurotransmitter; 13. behavior; 14. pleasure; 15. hypothalamus; 16. stimulants;
17. acetylcholine; 18. tolerance; 19. dopamine; 20. immune; 21. Alcohol; 22. analgesic; 23. heroin; 24. tolerance;
25. increases; 26. psychological; 27. addicted; 28. synergistic; 29. antagonistic; 30. potentiating; 31. D; 32. D;
33. D; 34. C; 35. A; 36. B; 37. E; 38. D; 39. A.

Self-Quiz
1. a; 2. d; 3. a; 4. b; 5. a; 6. d; 7. e; 8. b; 9. d; 10. c.

Chapter 14   Sensory Reception

CHAPTER INTRODUCTION [p.247]

**14.1 SENSORY RECEPTORS AND PATHWAYS — AN
OVERVIEW** [pp.248–249]
14.2. SOMATIC "BODY" SENSATIONS [pp.250–251]
1. A sensation is conscious awareness of a stimulus,
while perception is understanding what it means; 2. Instead of involving one stimulus, a compound sensation
integrates different stimuli; 3. The brain's assessment of
the nature of a stimulus depends on (1) which sensory
area receives signals from nerves, (2) the frequency of
signals, and (3) the number of axons that respond; 4. Sensory adaptation occurs with pressure sensations, such
as the pressure from clothing on the skin. It does not
occur with stretch receptors, as when the arm is holding
weight; 5. Somatic sensations involve receptors present at
more than one location in the body, while special senses
depend on receptors restricted to specific sense organs;
6. D; 7. C; 8. A; 9. B; 10. A; 11. E; 12. E; 13. B; 14. D; 15. B;

16. A; 17. C; 18. cerebrum; 19. greatest; 20. slowly;
21. mechanoreceptors; 22. Meissner's corpuscles;
23. Ruffini endings; 24. visceral pain; 25. referred pain.

14.3. TASTE AND SMELL — CHEMICAL SENSES
[pp.252–253]
14.4. *Science Comes to Life:* **A TASTY MORSEL OF
SENSORY SCIENCE** [p.253]
14.5. HEARING: DETECTING SOUND WAVES
[pp.254–255]
**14.6. BALANCE: SENSING THE BODY'S NATURAL
POSITION** [pp.256–257]
14.7. *Focus on Our Environment:* **NOISE POLLUTION:
AN ATTACK ON THE EARS** [p.257]
1. chemoreceptors; 2. thalamus; 3. taste buds; 4. four;
5. salty; 6. bitter; 7. olfactory; 8. olfactory; 9. cerebral
cortex; 10. vomeronasal; 11. pheromones; 12. amplitude;
13. frequency; 14. mechanoreceptors; 15. hairs; 16. action
potential; 17. sound; 18. D; 19. B; 20. J; 21. G; 22. L; 23. K;

24. H; 25. E; 26. A; 27. I; 28. C; 29. F; 30. i; 31. h; 32. d; 33. g; 34. k; 35. f; 36. j; 37. c; 38. b; 39. e; 40. a; 41. tympanic membrane; 42. middle ear bones; 43. oval window; 44. cochlea; 45. auditory nerve; 46. scala vestibuli; 47. cochlear duct; 48. scala tympani; 49. basilar membrane; 50. tectorial membrane; 51. "equilibrium position"; 52. receptors; 53. vestibular; 54. semicircular canals; 55. dynamic equilibrium; 56. deceleration; 57. cupula; 58. static; 59. vestibular; 60. otolith organ; 61. otoliths; 62. medulla oblongata; 63. balance; 64. Motion sickness.

14.8. VISION: AN OVERVIEW [pp.258–259]
14.9. FROM VISUAL SIGNALS TO "SIGHT" [pp.260–261]

14.10. DISORDERS OF THE EYE [pp.262–263]
1. Vision; 2. photoreceptors; 3. brain; 4. eyes; 5. cornea; 6. iris; 7. retina; 8. H; 9. K; 10. A; 11. B; 12. E; 13. G; 14. J; 15. F; 16. D; 17. I; 18. C; 19. choroid; 20. vitreous humor; 21. ciliary muscle; 22. iris; 23. pupil; 24. lens; 25. cornea; 26. aqueous humor; 27. sclera; 28. retina; 29. fovea; 30. optic disk (blind spot); 31. optic nerve; 32. c; 33. e; 34. d; 35. d; 36. cones; 37. T; 38. T; 39. a few; 40. greater; 41. ganglion cells; 42. T; 43. prevents; 44. "visual field"; 45. T; 46. I; 47. D; 48. F; 49. H; 50. G; 51. C; 52. E; 53. A; 54. B; 55. K; 56. J.

Self-Quiz
1. a; 2. c; 3. d; 4. a; 5. e; 6. d; 7. a; 8. c; 9. b; 10. b.

Chapter 15   The Endocrine System

CHAPTER INTRODUCTION [p.267]

15.1. THE ENDOCRINE SYSTEM: HORMONES [pp.268–269]
15.2. HORMONE CATEGORIES AND SIGNALING [pp.270–271]
1. E; 2. D; 3. G; 4. F; 5. A; 6. B; 7. C; 8. a. (9) PTH; b. (4) cortisol, aldosterone; c. (7) melatonin; d. (11) insulin, glucagon, somatostatin; e. (5) estrogens, progesterones; f. (1) produces and secretes releasing and inhibiting hormones and synthesizes ADH and oxytocin; g. (2) ACTH, TSH, FSH, LH, PRL, GH; h. (8) thyroxine, triiodothyronine, calcitonin; i. (10) thymosins; j. (6) androgens (including testosterone); k. (3) stores and secretes ADH and oxytocin, both from the hypothalamus; l. (4) epinephrine, norepinephrine; 9. a; 10. b; 11. a; 12. b; 13. b; 14. a; 15. b.

15.3. THE HYPOTHALAMUS AND PITUITARY GLAND — MAJOR CONTROLLERS [pp.272–273]
15.4. WHEN PITUITARY SIGNALS GO AWRY [p.274]
1. A (G); 2. P (F); 3. A (A); 4. A (H); 5. A (C); 6. P (D); 7. A (B); 8. A (E); 9. hypothalamus; 10. posterior; 11. anterior; 12. releaser; 13. inhibitor; 14. ACTH; 15. FSH; 16. Pituitary dwarfism; 17. gigantism; 18. acromegaly.

15.5. SOURCES AND EFFECTS OF OTHER HORMONES [p.275]

15.6. HORMONES AND FEEDBACK CONTROLS — THE ADRENALS AND THYROID [pp.276–277]
15.7. FAST RESPONSES TO LOCAL CHANGES — PARATHYROIDS AND THE PANCREAS [pp.278–279]
15.8. SOME FINAL EXAMPLES OF INTEGRATION AND CONTROL [pp.280–281]
15.9. *Science Comes to Life:* GROWTH FACTORS [p.281]
1. D, c; 2. I, e; 3. F, h; 4. A, g; 5. H, k; 6. E, d; 7. C, a; 8. K, j; 9. B, i; 10. J, f; 11. G, b; 12. F, l; 13. A, m; 14. B, o; 15. C, n; 16. rises; 17. excessive; 18. energy; 19. insulin; 20. Glucagon; 21. type 1 diabetes; 22. type 1 diabetes; 23. type 2 diabetes; 24. Type 2 diabetes; 25. Type 2 diabetes; 26. E; 27. G; 28. I; 29. C; 30. B; 31. H; 32. A; 33. D; 34. F; 35. thyroid; 36. metabolic; 37. calcitonin; 38. iodine; 39. TSH; 40. goiter; 41. hypothyroidism; 42. overweight; 43. iodized salt; 44. hyperthyroidism; 45. Graves'; 46. parathyroid; 47. parathyroid; 48. kidneys; 49. calcium; 50. rickets; 51. c; 52. f; 53. a; 54. g; 55. b; 56. e; 57. b; 58. d; 59. g; 60. a; 61. e; 62. b; 63. f; 64. a.

Self-Quiz
1. a; 2. e; 3. d; 4. d; 5. c; 6. e; 7. b; 8. a; 9. c; 10. a; 11. H; 12. F; 13. O; 14. N; 15. D; 16. E; 17. A; 18. G; 19. B; 20. L; 21. I; 22. M; 23. Q; 24. C; 25. K; 26. J; 27. P.

Chapter 16   Reproductive Systems

CHAPTER INTRODUCTION [pp.285]

16.1. THE MALE REPRODUCTIVE SYSTEM
[pp.286–287]
16.2. HOW SPERM FORM [pp.288–289]
1. spermatogonia; 2. mitosis; 3. meiosis; 4. primary spermatocytes; 5. meiosis I; 6. secondary spermatocytes; 7. meiosis II; 8. spermatids; 9. sperm; 10. tail; 11. Sertoli; 12. head; 13. acrosome; 14. mitochondria; 15. testes; 16. epididymides; 17. vas deferentia; 18. urethra; 19. seminal; 20. prostate; 21. bulbourethral; 22. Leydig cells; 23. Leydig cells; 24. Testosterone; 25. testosterone; 26. anterior, 27. hypothalamus; 28. decrease; 29. LH; 30. FSH; 31. increase.

16.3. THE FEMALE REPRODUCTIVE SYSTEM
[pp.290–291]
16.4. HOW OOCYTES DEVELOP [pp.292–293]
16.5. VISUAL SUMMARY OF THE MENSTRUAL CYCLE [p.294]
1. ovary; 2. oviduct; 3. uterus; 4. myometrium; 5. endometrium; 6. cervix; 7. vagina; 8. labia majora; 9. labia minora; 10. clitoris; 11. urethra; 12. three; 13. Follicular; 14. Menstruation; 15. Oocyte; 16. Ovulation; 17. Luteal; 18. Corpus luteum; 19. Endometrium; 20. Menarche; 21. Menopause; 22. Endometriosis; 23. ovarian; 24. 300,000; 25. ovary; 26. granulosa; 27. FSH; 28. zona pellucida; 29. estrogens; 30. secondary; 31. polar; 32. follicle; 33. Secondary oocyte; 34. oviduct; 35. Fertilization; 36. ovum; 37. endometrium; 38. progesterone; 39. cervix; 40. corpus luteum; 41. endometrium; 42. follicles; 43. implantation; 44. endometrium; 45. estrogen; 46. grows; 47. LH; 48. decrease; 49. increase; 50. increase; 51. remain stable; 52. fully developed; 53. luteal; 54. menstruation; 55. be maintained; 56. corpus luteum; 57. around the middle of.

16.6. SEXUAL INTERCOURSE, ETC. [p.295]
16.7. CONTROLLING FERTILITY [pp.296–297]
1. F (a); 2. L (d); 3. B (f); 4. D (f); 5. O (c); 6. N (c); 7. P (b); 8. J (c); 9. C (b); 10. M (b); 11. Q (c); 12. K (b); 13. H (b); 14. E (a); 15. I (a); 16. A (a); 17. G (a); 18. Biodegradable implants, contraceptives for males, chemical "sterilizers," reversible sterilization for both males and females.

16.8. COPING WITH INFERTILITY [p.298]
16.9. *Choices: Biology and Society:* **DILEMMAS OF FERTILITY CONTROL** [p.299]
1. c; 2. a; 3. b; 4. a; 5. b; 6. a; 7. c; 8. b; 9. a; 10. c; 11. a; 12. b; 13. a.

Self-Quiz
1. a; 2. b; 3. d; 4. c; 5. a; 6. b; 7. c; 8. e; 9. e; 10. c; 11. d; 12. a; 13. d; 14. a; 15. c; 16. e; 17. b; 18. c; 19. b; 20. a.

Chapter 17   Development and Aging

CHAPTER INTRODUCTION [p.303]

17.1. THE SIX STAGES OF DEVELOPMENT
[pp.304–305]
17.2. THE BEGINNINGS OF YOU — EARLY STEPS IN DEVELOPMENT [pp.306–307]
17.3. VITAL MEMBRANES OUTSIDE THE EMBRYO
[pp.308–309]
1. F; 2. B; 3. C; 4. A; 5. E; 6. D; 7. a. mesoderm; b. ectoderm; c. endoderm; d. mesoderm; e. ectoderm; f. mesoderm; g. endoderm; h. mesoderm; i. mesoderm; 8. organogenesis; 9. specialization; 10. determination; 11. differentiation; 12. morphogenesis; 13. movements; 14. paddle; 15. died; 16. complex; 17. G; 18. D; 19. J; 20. B; 21. F; 22. A; 23. K; 24. H; 25. I; 26. C; 27. E; 28. I; 29. M; 30. C; 31. J; 32. N; 33. E; 34. B; 35. H; 36. K; 37. A; 38. L; 39. F; 40. G; 41. D; 42. O; 43. 31; 44. 30; 45. 29; 46. 40; 47. 33; 48. f; 49. e; 50. e; 51. d; 52. a; 53. b; 54. c; 55. a; 56. b; 57. d; 58. f; 59. f; 60. e; 61. e; 62. f.

17.4. HOW THE EARLY EMBRYO TAKES SHAPE
[pp.310–311]
17.5. THE FIRST EIGHT WEEKS — HUMAN FEATURES EMERGE [pp.312–313]
17.6. DEVELOPMENT OF THE FETUS [pp.314–315]
1. F; 2. J; 3. G; 4. K; 5. A; 6. E; 7. I; 8. C; 9. B; 10. D; 11. H; 12. lanugo; 13. vernix caseosa; 14. second; 15. third; 16. distress; 17. umbilical; 18. umbilical; 19. lungs; 20. liver; 21. collapsed; 22. atrium; 23. foramen ovale; 24. birth; 25. heart; 26. increases; 27. foramen ovale; 28. pulmonary; 29. systemic; 30. venous; 31. yolk sac; 32. embryonic disk; 33. amniotic cavity; 34. chorionic cavity; 35. primitive streak; 36. neural tube; 37. future brain; 38. somites; 39. pharyngeal arches; 40. four; 41. pharyngeal arches; 42. somites; 43. 5–6; 44. forelimb; 45. The embryo is just over 1 inch long and its organ systems are formed. The fetal heartbeat can be detected with a monitor and the genitals are well enough formed to be visible with ultrasound. The embryo has matured into a fetus.

17.7. BIRTH AND BEYOND [pp.316–317]

17.8. *Choices: Biology and Society:* **SHOULD EMBRYOS BE CLONED?** [p.317]

17.9. MOTHER AS PROVIDER, PROTECTOR, AND POTENTIAL THREAT [pp.318–319]

17.10. *Science Comes to Life:* **PRENATAL DIAGNOSIS: DETECTING BIRTH DEFECTS** [p.320]

17.11. THE PATH FROM BIRTH TO ADULTHOOD [p.321]

1. Parturition; 2. labor; 3. cervix; 4. amniotic; 5. birth; 6. two; 7. breech; 8. afterbirth; 9. umbilical cord; 10. carbon dioxide; 11. inhalation; 12. navel; 13. colostrum; 14. lactation; 15. Oxytocin; 16. The developing individual is at the mercy of the mother's diet and health habits. The course of development is greatly influenced by these factors. Infection of the embryo prior to four months can result in malformations. The embryo is highly sensitive to drugs during the first trimester. Fetal alcohol syndrome (FAS) is a constellation of defects that can result from alcohol use by a pregnant woman. FAS is the third most

common cause of mental retardation in the United States. Cigarette smoking harms fetal growth and development. Research shows that women smokers are at greater risk of miscarriage, stillbirth, and premature delivery; 17. CVS; 18. Amniocentesis; 19. preimplantation diagnosis; 20. childhood; 21. newborn or neonate; 22. adult; 23. pubescent; 24. adolescent; 25. senescence; 26. blastocyst.

17.12. WHY DO WE AGE? [p.322]

17.13. AGING SKIN, MUSCLE, BONES, AND TRANSPORT SYSTEMS [p.323]

17.14. AGE-RELATED CHANGES IN SOME OTHER BODY SYSTEMS [pp.324–325]

1. G; 2. D; 3. F; 4. E; 5. H; 6. A; 7. C; 8. I; 9. B.

Self-Quiz

1. d; 2. b; 3. b; 4. a; 5. d; 6. e; 7. F; 8. B; 9. J; 10. I; 11. H; 12. L; 13. D; 14. M; 15. A; 16. N; 17. E; 18. C; 19. G; 20. K.

Chapter 18   Life at Risk: Infectious Disease

CHAPTER INTRODUCTION [p.329]

18.1. VIRUSES AND INFECTIOUS PROTEINS [pp.330–331]

18.2. BACTERIA — THE UNSEEN MULTITUDES [pp.332–333]

1. e; 2. b; 3. c; 4. a; 5. a; 6. d; 7. a; 8. c; 9. a; 10. b; 11. d; 12. a; 13. Many more humans are living on the planet and interacting more with each other and the environment; Easier travel carries disease anywhere that travelers go; Misuse and overuse of antibiotics leads to resistance; 14. The "bad news" component is a nucleic acid, either DNA or RNA; 15. is not; 16. reproduce; 17. T; 18. capsid; 19. T; 20. host cells; 21. E; 22. B; 23. F; 24. C; 25. A; 26. D; 27. receptors; 28. kills its host quickly; 29. reactivated; 30. Epstein-Barr; 31. retroviruses; 32. prions; 33. protein; 34. prokaryotic; 35. peptidoglycan; 36. coccus; 37. bacillus; 38. spirillum; 39. flagellum; 40. Pili; 41. prokaryotic fission; 42. one; 43. 20; 44. plasmids; 45. fertility.

18.3. INFECTIOUS PROTOZOA AND WORMS [p.334]

18.4. *Science Comes to Life:* **MALARIA — HERE TODAY . . . AND TOMORROW AND TOMORROW** [p.335]

18.5. THE SPREAD OF DISEASE AND PATTERNS OF OCCURRENCE [pp.336–337]

18.6. *Focus on Your Health:* **TB, CONTINUED** [p.337]

1. B; 2. D; 3. C; 4. A; 5. E; 6. 3 million; 7. *Plasmodium;* 8. *Anopheles;* 9. fever; 10. anemia; 11. Africa; 12. resistant; 13. mosquitoes; 14. C; 15. G; 16. A; 17. B; 18. J; 19. F; 20. D;

21. E; 22. H; 23. I; 24. About a third of the world's people are infected with *Mycobacterium tuberculosis*. For antibiotics to be effective, they must be taken for at least a year.

18.7. THE HUMAN IMMUNODEFICIENCY VIRUS AND AIDS [pp.338–339]

18.8. TREATING AND PREVENTING HIV INFECTION AND AIDS [p.340]

1. a constellation of more than one; 2. immune; 3. T; 4. T; 5. can; 6. AIDS; 7. RNA; 8. E; 9. B; 10. A; 11. D; 12. C; 13. F; 14. T; 15. T; 16. DNA; 17. protease inhibitor; 18. cell; 19. rapidly; 20. T.

18.9. *Focus on Your Health:* **PROTECTING YOURSELF — AND OTHERS — FROM SEXUAL DISEASE** [p.341]

18.10. A TRIO OF COMMON STDS [pp.342–343]

18.11. A ROGUE'S GALLERY OF VIRAL STDS AND OTHERS [pp.344–345]

1. b; 2. a; 3. b; 4. b; 5. a; 6. c; 7. b; 8. a; 9. d; 10. c; 11. a; 12. b; 13. a; 14. c; 15. e; 16. a; 17. d; 18. b; 19. b; 20. b; 21. e; 22. b.

Self-Quiz

1. g; 2. a; 3. c; 4. e; 5. b; 6. f; 7. d; 8. a; 9. b; 10. c; 11. f; 12. c; 13. h; 14. f.

Chapter 19   Cell Reproduction

CHAPTER INTRODUCTION [p.349]

19.1. DIVIDING CELLS: THE BRIDGE BETWEEN GENERATIONS [pp.350–351]
1. J; 2. L; 3. A; 4. F; 5. G; 6. B; 7. D; 8. I; 9. H; 10. K; 11. E; 12. C.

19.2. THE CELL CYCLE [p.352]
19.3. *Science Comes to Life:* **IMMORTAL CELLS — THE GIFT OF THE HENRIETTA LACKS** [p.353]
1. interphase; 2. mitosis; 3. G_1; 4. S; 5. G_2; 6. prophase; 7. metaphase; 8. anaphase; 9. telophase; 10. cytokinesis; 11. 5; 12. 2; 13. 4; 14. 3; 15. 1; 16. 10; 17. 1; 18. HeLa cells originated in the 1950s when cancer cells taken from Henrietta Lacks were cultured. These cells are still dividing today. They have been invaluable in cancer research.

19.4. A TOUR OF THE STAGES OF MITOSIS [pp.354–355]
19.5. HOW THE CYTOPLASM DIVIDES [p.356]
19.6. *Focus on Our Environment:* **CONCERNS AND CONTROVERSIES OVER IRRADIATION** [p.357]
19.7. A CLOSER LOOK AT THE CELL CYCLE [pp.358–359]
1. interphase — daughter cells (F); 2. anaphase (A); 3. late prophase (G); 4. metaphase (D); 5. interphase — parent cell (E); 6. early prophase (C); 7. prometaphase (B); 8. telophase (H); 9. Prophase; 10. interphase; 11. sister chromatids; 12. centromere; 13. chromatids; 14. chromosomes; 15. microtubules; 16. pole; 17. centromere; 18. centrioles; 19. prophase; 20. Prometaphase; 21. chromosomes; 22. poles; 23. chromatids; 24. metaphase; 25. anaphase; 26. chromatid; 27. Telophase; 28. nuclear envelope; 29. number; 30. telophase; 31. mitosis; 32. B; 33. D; 34. A; 35. C; 36. Natural sources include cosmic rays from outer space and radioactive radon gas in rocks and soil; 37. Ionizing radiation damages cells by breaking apart chromosomes and altering genes. If the damage occurs in germ cells, the resulting gametes may give rise to infants with genetic defects. If somatic (body) cells are damaged, the damage may include burns, miscarriages, eye cataracts, and cancers of the bone, thyroid, breast, skin, and lungs; 38. In medicine, X-rays, MRI, and PET scanning are valuable uses of irradiation. Irradiation therapy is useful in treating cancer. Food is irradiated to kill harmful microorganisms and to prolong shelf life by preventing vegetables like potatoes from sprouting; 39. A; 40. C; 41. D; 42. B.

19.8. MEIOSIS — THE BEGINNINGS OF EGGS AND SPERM [pp.360–361]
19.9. A VISUAL TOUR OF THE STAGES OF MEIOSIS [pp.362–363]
1. T; 2. gametes; 3. T; 4. Diploid; 5. T; 6. E ($2n$); 7. D ($2n$); 8. B ($2n$); 9. A (n); 10. C (n); 11. anaphase II (H); 12. metaphase II (F); 13. metaphase I (A); 14. prophase II (B); 15. telophase II (C); 16. telophase I (G); 17. prophase I (E); 18. anaphase I (D); 19. 2 ($2n$); 20. 5 (n); 21. 4 (n); 22. 1 ($2n$); 23. 3 (n); 24. Spermatogenesis: gamete formation in males. A single diploid germ cell — the primary spermatocyte — undergoes meiosis, producing four haploid spermatids of equal proportions. Oogenesis: gamete formation in females. A primary oocyte contains many more cytoplasmic components than a primary spermatocyte. The primary oocyte undergoes meiosis to produce four cells of different sizes and functions. Only one will develop into an ovum.

19.10. THE SECOND STAGE OF MEIOSIS — NEW COMBINATIONS OF PARENTS' TRAITS [pp.364–365]
19.11. MEIOSIS AND MITOSIS COMPARED [pp.366–367]
1. C; 2. E; 3. A; 4. F; 5. G; 6. B; 7. H; 8. D; 9. a. mitosis; b. mitosis; c. meiosis; d. meiosis; e. meiosis; f. meiosis; g. mitosis; h. meiosis; i. meiosis; 10. C; 11. F; 12. D; 13. A; 14. B; 15. E; 16. four; 17. eight; 18. four; 19. eight; 20. two.

Self-Quiz
1. a; 2. d; 3. a; 4. d; 5. c; 6. e; 7. a; 8. a; 9. b; 10. c; 11. b; 12. c; 13. b; 14. c.

Chapter 20   Observable Patterns of Inheritance

1. F; 2. A; 3. C; 4. G; 5. E; 6. I; 7. D; 8. H; 9. J; 10. B; 11. monohybrid; 12. one; 13. gene; 14. segregation; 15. Punnett; 16. probability; 17. likelihood; 18. a. *C* and *c*; b. *c*; c. *C* and *c* go across the top of the Punnett square; *c*'s go down the side of the square; d. 1 *Cc*: 1 *cc*; e. 1 chin fissure: 1 smooth chin; f. ½; the predicted ratio remains constant for each fertilization event; 19. a. ½ × 1 = ½; b. ½ × 1 = ½; c. ½ chin fissure and ½ smooth chin.

1. The taster parent is heterozygous, *Aa*. The nontaster child must be homozygous and had to have received one nontasting gene from each of his parents; 2. The man is *cc* and the woman is *Cc*. The probability that the child will have either a smooth chin or a chin fissure is ½ for each; 3. Albino = *aa*, normal pigmentation = *AA* or *Aa*. The woman of normal pigmentation with an albino mother is genotype *Aa*; the woman received her recessive gene (*a*) from her mother and her dominant gene (*A*) from her father. The woman's husband is *aa* and will give each of their offspring an *a* allele. It is likely that half of the couple's children will be albinos (*aa*) and half will have normal pigmentation but be heterozygous (*Aa*); 4. a. 9/16; b. 3/16; c. 3/16; d. 1/16 (note the following Punnett square).

	BR	Br	bR	br
BR	BBRR	BBRr	BbRR	BbRr
Br	BBRr	BBrr	BbRr	Bbrr
bR	BbRR	BbRr	bbRR	bbRr
br	BbRr	Bbrr	bbRr	bbrr

1. a. pleiotropy; b. codominance; c. polygenic; d. codominance; 2. f; 3. d; 4. f; 5. g; 6. b; 7. a; 8. c; 9. g; 10. e; 11. a.

Self-Quiz

1. d; 2. b; 3. a; 4. d; 5. d; 6. c; 7. a; 8. a; 9. c; 10. d.

Integrating and Applying Key Concepts (genetics problem answer)

The first mating could only produce one genotype (*HhPp*) and one phenotype, with all individuals having normal hair and normal growth. Matings between genotypes like those of their children (*HhPp* × *HhPp*) would produce offspring with the following probability ratios and phenotypes: 9 normal hair, normal height; 3 normal hair, dwarf; 3 hypotrichotic, normal height; 1 hypotrichotic, dwarf.

Chapter 21   Chromosomes and Human Genetics

1. genes; 2. homologous; 3. Alleles; 4. crossing over; 5. genetic recombination; 6. sex; 7. X; 8. Y; 9. autosomes; 10. karyotype; 11. metaphase; 12. sex chromosomes.

1. Two blocks of the Punnett square should be XX and two blocks should be XY; 2. sons; 3. mothers; 4. daughters; 5. nonsexual; 6. D; 7. I; 8. C; 9. E; 10. F; 11. A; 12. H; 13. B; 14. J; 15. G; 16. d; 17. c.

21.9. SEX-INFLUENCED INHERITANCE [p.396]

1. b; 2. a; 3. c; 4. a; 5. b; 6. b; 7. a; 8. e; 9. c; 10. c; 11. a; 12. e; 13. c; 14. d; 15. d; 16. The woman's father is homozygous recessive, *aa*; the woman is heterozygous normal, *Aa*. The albino man, *aa*, has two heterozygous normal parents, *Aa*. The two normal children are heterozygous normal, *Aa*; the albino child is *aa*; 17. Assuming the father is heterozygous with Huntington disorder and the mother is normal, the chances are ½ (50%) that the son will develop the disease; 18. If only male offspring are considered, the probability is ½ (50%) that the couple will have a color-blind son; 19. The probability is that half of the sons will have hemophilia; the probability is 0 that a daughter will express hemophilia; the probability is that half of the daughters will be carriers; 20. If the woman marries a normal male, the chance that her son would be color blind is ½ (50%); if she marries a color-blind male, the chance that her son would also be color blind is also ½ (50%); 21. The nonbald man is homozygous *b+b+*, and his wife must be heterozygous *b+b* because her mother was bald. The first child can be bald only if it is a male (probability = ½) and also heterozygous for this gene (probability = ½). The chance of having a bald son is (½ × ½) = ¼; 22. a. autosomal dominant; b. X-linked recessive; c. sex-influenced; d. autosomal dominant; e. X-linked recessive; f. X-linked dominant; g. X-linked recessive; h. autosomal recessive; i. autosomal recessive; j. autosomal recessive; k. autosomal dominant.

21.10. HOW A CHROMOSOME'S STRUCTURE CAN CHANGE [pp.396–397]
21.11. CHANGES IN CHROMOSOME NUMBER [pp.398–399]

1. duplication (B); 2. deletion (A); 3. translocation (C); 4. A mutation is a change in one or more of the nucleotides that make up a particular gene; a mutation may have no phenotypic effect, may produce a positive change in phenotype, or may cause a genetic disorder; 5. a. aneuploidy; b. polyploidy; c. nondisjunction; d. trisomy; e. monosomy; 6. Most changes in the number of chromosomes (aneuploidy) arise through nondisjunction during gamete formation; 7. c; 8. b; 9. c; 10. d; 11. a; 12. b; 13. d; 14. c; 15. a.

Self-Quiz

1. d; 2. d; 3. b; 4. c; 5. b; 6. a; 7. c; 8. b; 9. b; 10. c.

Chapter 22   DNA, Genes, and Biotechnology

CHAPTER INTRODUCTION [p.403]

22.1. DNA: A DOUBLE HELIX [pp.404–405]
22.2. PASSING ON GENETIC INSTRUCTIONS [pp.406–407]

1. A five-carbon sugar (deoxyribose), a phosphate group, and one of the four nitrogen-containing bases found in DNA; 2. a. adenine; b. guanine; c. thymine; d. cytosine; e. double-ring; f. double-ring; g. single-ring; h. single-ring; i. thymine; j. cytosine; k. adenine; l. guanine; 3. four; 4. T; 5. five-carbon sugar, nitrogen-containing base; 6. T; 7. T; 8. double helix; 9. T; 10. hydrogen; 11. deoxyribose; 12. phosphate group; 13. double; 14. single; 15. double; 16. single; 17. nucleotide; 18. covalent; 19. hydrogen;

20.

T - A	T - A
G - C	G - C
A - T	A - T
C - G	C - G
C - G	C - G
C - G	C - G

21. unit of heredity; more specifically, a sequence of nucleotides in a DNA molecule that codes for a specific polypeptide chain; 22. DNA replication; 23. bases; 24. nucleotide; 25. semiconservative; 26. DNA polymerases; 27. repair; 28. two million; 29. mutation; 30. ultraviolet radiation; 31. thymine dimer; 32. gene mutations; 33. base pair substitution; 34. inserted; 35. deleted; 36. expansion; 37. transposable elements; 38. proteins; 39. gonads; 40. beneficial.

22.3. DNA INTO RNA — STEP ONE IN MAKING PROTEINS [pp.408–409]
22.4. READING THE GENETIC CODE [pp.410–411]
22.5. TRANSLATING THE GENETIC CODE INTO PROTEIN [pp.412–413]

1. a. deoxyribose; b. adenine, thymine, guanine, cytosine; c. ribose; d. adenine, uracil, guanine, cytosine; 2. transcription; 3. translation; 4. transcription; 5. translation; 6. polypeptide; 7. protein; 8. a. rRNA: nucleic acid chain that combines with certain proteins to form ribosomes, which are involved in assembly of polypeptide chains; b. mRNA: linear sequence of nucleic acids that deliver protein-building instructions to ribosomes for translation into polypeptide chains; c. tRNA: nucleic acid chain that picks up a specific amino acid and delivers it to the ribosome where it will pair with a specific mRNA code for that particular amino acid; 9. In transcription, only the gene segment serves as the template — not the whole DNA strand (as in DNA replication). Enzymes called *RNA polymerases* are involved instead of DNA polymerases. Transcription results in only a single-stranded molecule — not one with two strands (as in DNA replication); 10. D; 11. B; 12. E; 13. A; 14. C; 15. AUG UUC UAU UGU AAU AAA GGA UGG CAG UAG; 16. DNA (E); 17. intron (B); 18. cap (F); 19. exon (A); 20. tail (D); 21. mature RNA transcript (C); 22. F; 23. G; 24. H; 25. A; 26. E; 27. D; 28. C; 29. B; 30. a. initiation; b. elongation; c. termination; 31. The tRNA anticodon sequence is UAC AAG AUA ACA UUA UUU CCU ACC GUC AUC;

32. amino acids: met phe tyr cys asn lys gly try gln stop;
33. transcription (E); 34. mRNA (A); 35. ribosome subunits (D); 36. tRNA (G); 37. translation (B); 38. ribosome (F); 39. polypeptide (C).

22.6. TOOLS FOR "ENGINEERING" GENES [pp.414–415]
22.7. "SEQUENCING" DNA [p.416]

1. The bacterial chromosome, a circular DNA molecule, contains all the genes necessary for normal growth and development. Plasmids are small, circular molecules of "extra" DNA that carry only a few genes and are self-replicating; 2. mutation; 3. recombinant DNA; 4. species; 5. replicate; 6. genetic engineering; 7. plasmids; 8. restriction enzyme; 9. "sticky"; 10. clone; 11. amplify; 12. d; 13. g; 14. h; 15. e; 16. f; 17. b; 18. a; 19. c; 20. PCR: polymerase chain reaction. This method is used to amplify DNA fragments in a test tube, using primers instead of using bacteria as cloning vectors; 21. These short nucleotide chains base-pair with any complementary DNA sequences. DNA polymerases recognize primers as "start" tags; 22. DNA sequencing; 23. probe; 24. gene library.

22.8. THE HUMAN GENOME PROJECT [pp.416–417]
22.9. SOME APPLICATIONS OF BIOTECHNOLOGY [pp.418–419]

22.10. *Choices: Biology and Society:* ISSUES FOR A BIOTECHNOLOGICAL SOCIETY [p.430]
22.11. ENGINEERING BACTERIA, ANIMALS, AND PLANTS [p.421]
22.12. *Science Comes to Life:* MR. JEFFERSON'S GENES [p.422]

1. 25,000 to 45,000; 2. fifteen; 3. T; 4. T; 5. 1.5, 6. T; 7. T; 8. has; 9. gene therapy; 10. viruses; 11. transfection; 12. virus; 13. vectors; 14. cancer; 15. interleukins; 16. attack; 17. Lipoplexes; 18. DNA fragments; 19. tandem repeats; 20. gel electrophoresis; 21. restriction enzymes; 22. Transgenic bacteria or viruses could mutate and become pathogenic; Bioengineered plants might escape as "super weeds"; Pest-resistant engineered crop plants could select for more resistant pests; Transgenic fish could replace native species and disrupt ecosystems; Genetic screening could allow insurance companies to discriminate against individuals carrying problem alleles; 23. "Designer plants" will improve agriculture and enhance crop yields; Genetically enhanced "oil-eating" bacteria help with oil spills; 24. transgenic; 25. Plasmids; 26. proteins; 27. insulin; 28. micro-injected; 29. therapeutic drugs; 30. resistant.

Self-Quiz

1. d; 2. d; 3. d; 4. b; 5. c; 6. b; 7. c; 8. a; 9. c; 10. a; 11. d; 12. b; 13. a; 14. b. 15. b; 16. c; 17. c; 18. a; 19. d; 20. b.

Chapter 23   Genes and Disease: Cancer

CHAPTER INTRODUCTION [p.427]

23.1. CANCER: CELL CONTROLS GO AWRY [pp.428–429]
23.2. THE GENETIC TRIGGERS FOR CANCER [pp.430–431]
23.3. *Focus on Our Environment:* ASSESSING THE CANCER RISK FROM ENVIRONMENTAL CHEMICALS [p.432]

1. F; 2. G; 3. H; 4. B; 5. E; 6. C; 7. A; 8. D; 9. carcinogenesis; 10. Proto-oncogenes; 11. oncogene; 12. does not; 13. oncogene; 14. tumor suppressor gene; 15. one; 16. prevent; 17. promote; 18. activated; 19. triggering; 20. normal cell; 21. proto-oncogenes; 22. tumor suppressor genes; 23. abnormal cells; 24. breakdown; 25. proliferates; 26. tumor; 27. metastasis; 28. activation; 29. destroyed; 30. a. Mutations in germ cells may remove controls over a proto-oncogene; the defect may be passed on to offspring; patterns of familial breast, colon, and other types of cancer suggest several genes may be involved; b. A viral infection may alter a proto-oncogene when viral DNA becomes inserted at a certain position in the host cell DNA; viruses may carry oncogenes in their DNA and insert them into a host's DNA to disrupt

controls of cell division; c. Asbestos, coal tar, vinyl chloride, benzene and many other related compounds, dyes, pesticides, cigarette hydrocarbons, chimney soot, aflatoxin; some chemicals may be precarcinogens, others may act as promoters; d. UV from sunlight and sunlamps, medical and dental X rays, background radiation from cosmic rays, radon gas in soil and water, gamma rays from nuclear reactors or radioactive wastes; e. Proteins on cell surfaces become altered and are recognized as "nonself"; these cells are destroyed by cytotoxic T cells and natural killer cells; some tumor antigens may be chemically disguised or masked — they escape destruction and are free to divide uncontrolled.

23.4. DIAGNOSING CANCER [p.433]
23.5. TREATING AND PREVENTING CANCER [p.434]

1. bowel, bladder; 2. sore; 3. bleeding, bloody; 4. lump; 5. Indigestion; 6. wart, mole; 7. cough, hoarseness; 8. J; 9. F; 10. I; 11. D; 12. G; 13. A; 14. H; 15. K; 16. C; 17. B; 18. E; 19. L.

23.6. SOME MAJOR TYPES OF CANCER [p.435]
23.7. CANCERS OF THE BREAST AND REPRODUCTIVE SYSTEM [pp.436–437]

23.8. A SURVEY OF OTHER COMMON CANCERS
[pp.438–439]
1. sarcomas; 2. carcinomas; 3. adenocarcinomas; 4. Lymphomas; 5. leukemias; 6. h; 7. g; 8. e; 9. d; 10. a; 11. d;
12. g; 13. e; 14. g; 15. g; 16. h; 17. e; 18. g; 19. d; 20. c; 21. a;
22. b; 23. c; 24. d; 25. d; 26. f; 27. h; 28. e; 29. a; 30. g; 31. a;
32. b; 33. g; 34. d; 35. g; 36. g.

Self-Quiz
1. b; 2. d; 3. a; 4. c; 5. b; 6. a; 7. e; 8. d; 9. b; 10. e; 11. b;
12. e; 13. d; 14. a; 15. d.

Chapter 24   Principles of Evolution

CHAPTER INTRODUCTION [p.443]

24.1. A LITTLE EVOLUTIONARY HISTORY [p.444]
**24.2. INDIVIDUAL VARIATIONS — A KEY
EVOLUTIONARY IDEA** [p.445]
1. a. John Henslow; b. Cambridge University; c. HMS
Beagle; d. Thomas Malthus; e. natural selection; 2. B;
3. E; 4. H; 5. F; 6. A; 7. G; 8. C; 9. D.

**24.3. MICROEVOLUTION: HOW NEW SPECIES
ARISE** [pp.446–447]
1. genetic drift; 2. mutation; 3. the founder effect;
4. Natural selection; 5. gene flow; 6. *On the Origin of
Species;* 7. natural selection; 8. gradualism; 9. fertile
offspring; 10. gene flow; 11. divergence; 12. adaptation;
13. punctuated equilibrium; 14. isolating; 15. species;
16. punctuated equilibrium; 17. a. mutation; b. genetic
drift; c. gene flow; d. natural selection; 18. a. A and B;
b. B and C; c. gradual; d. D; 19. A unit of one or more
populations of individuals that can interbreed under
natural conditions and produce fertile offspring.

24.4. LOOKING AT FOSSILS AND BIOGEOGRAPHY
[pp.448–449]
**24.5. COMPARING THE FORM AND
DEVELOPMENT OF BODY PARTS** [pp.450–451]
24.6. COMPARING BIOCHEMISTRY [p.452]
1. H; 2. D; 3. I; 4. A; 5. F; 6. E; 7. L; 8. B; 9. G; 10. J; 11. C;
12. K.

24.7. HOW SPECIES COME AND GO [pp.452–453]
24.8. *Focus on Our Environment:* **ENDANGERED
SPECIES** [p.453]
24.9. FIVE TRENDS IN HUMAN EVOLUTION
[pp.454–455]
1. D; 2. A; 3. C; 4. B; 5. E; 6. bipedalism; 7. shorter;
8. bipedalism; 9. tree; 10. prehensile; 11. opposable;
12. hominids; 13. forward-directed eyes; 14. trees;
15. bow-shaped, smaller teeth of about the same length;
16. eating insects, then fruit and leaves, a mixed diet;
17. fewer; 18. longer; 19. brain; 20. Culture; 21. language;
22. mammals; 23. small rodents; 24. trees; 25. Fossil;
26. anthropoids; 27. hominoids; 28. hominoid; 29. three;
30. hominids; 31. 100,000; 32. cultural.

**24.10. EARTH'S HISTORY AND THE ORIGIN OF
LIFE** [pp.456–457]
1. did not contain; 2. T; 3. oxygen-free; 4. 3.8 billion;
5. sunlight, lightning, or heat escaping from the crust;
6. T; 7. T; 8. RNA; 9. T.

Self-Quiz
1. c; 2. a; 3. c; 4. d; 5. b; 6. d; 7. c; 8. b; 9. c; 10. b; 11. b;
12. a; 13. a; 14. d; 15. d.

Chapter 25   Ecology and Human Concerns

1. H; 2. J; 3. E; 4. A; 5. C; 6. G; 7. F; 8. B; 9. I; 10. D; 11. a;
12. b; 13. b; 14. a; 15. sun; 16. autotrophs; 17. producers;
18. consumers; 19. heterotrophs; 20. herbivores; 21. carnivores; 22. omnivores; 23. decomposers; 24. a; 25. a; 26. d;
27. e; 28. c; 29. a; 30. a; 31. b; 32. input; 33. cannot; 34. sun;
35. can; 36. trophic; 37. chain; 38. plants; 39. lower;
40. producers, consumers; 41. biomass; 42. decreases.

1. geochemical cycle; 2. decomposers; 3. weathering of
rocks; 4. runoff; 5. nutrient reservoir in environment;
6. oxygen and hydrogen; 7. gas; 8. Earth's crust; 9. G;
10. H; 11. E; 12. A; 13. C; 14. D; 15. I; 16. F; 17. B.

1. C; 2. E; 3. B; 4. F; 5. A; 6. D; 7. greenhouse effect; 8. heat;
9. global warming; 10. sea; 11. Rainfall; 12. icecap;
13. carbon dioxide; 14. fossil fuels; 15. Deforestation;
16. plant.

1. d; 2. a; 3. c; 4. a; 5. b; 6. c; 7. b; 8. e; 9. mosquitoes;
10. body lice; 11. tissues; 12. biological magnification;
13. food chains; 14. songbirds; 15. salmon; 16. pests;
17. predators; 18. food chains; 19. brittle shells; 20. extinction; 21. banned; 22. ecosystem.

1. decreases; 2. decreases; 3. nine; 4. density; 5. distribution; 6. age; 7. base; 8. overpopulation; 9. exponential;
10. S-shaped; 11. high-density; 12. E; 13. F; 14. B; 15. A;
16. D; 17. C; 18. G; 19. H; 20. a. 1962–1963; b. 2025 or
sooner; c. Depends on age and optimism of the reader.

1. c; 2. d; 3. a; 4. b; 5. d; 6. e; 7. b; 8. b; 9. e; 10. c; 11. a;
12. e; 13. salty; 14. irrigated; 15. mineral; 16. salinization;
17. irrigation; 18. aquifer; 19. saltwater; 20. Agricultural;
21. factories; 22. oceans; 23. D; 24. E; 25. F; 26. A; 27. B;
28. C.

1. Net; 2. solar; 3. developing; 4. fossil; 5. consumption;
6. air pollution; 7. more; 8. meltdown; 9. solar cells;
10. fusion power; 11. hour; 12. tropical deforestation;
13. half; 14. more; 15. less.

Self-Quiz
1. b; 2. b; 3. c; 4. c; 5. b; 6. a; 7. a; 8. d; 9. b; 10. d; 11. d;
12. a; 13. a; 14. c; 15. b; 16. a; 17. d; 18. a; 19. d; 20. c.